NIELS BOHR

The Man, His Science,

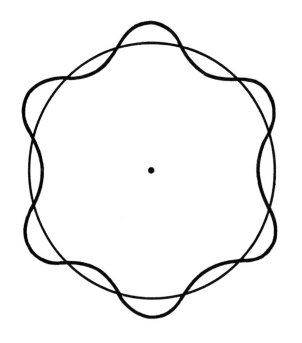

NIELS BOHR:

& the World They Changed

RUTH MOORE

With Drawings by Sue Richert Allen

The MIT Press
Cambridge, Massachusetts
London, England

Second printing, 1985
First MIT Press paperback edition, 1985

Originally published by Alfred A. Knopf, Inc.

Printed and bound in the United States of America

Library of Congress Cataloging in Publication Data

Moore, Ruth E.
  Niels Bohr, the man, his science & the world they changed.

  Originally published: New York: Knopf. c1966.
  Includes bibliographical references and index.
  1. Bohr, Niels Henrik David, 1885–1962.
  2. Physicists—Denmark—Biography.   I. Title.
  QC16.B63M6   1985   530′.092′4 [B]   84-23411
  ISBN 0-262-63101-6 (paperback)

Preface

Niels Bohr is a compelling figure for a biographer. If he had been a scientist alone his life might well have been most suitably recorded in the annals of science. Bohr's influence, however, went far beyond the laboratory or the scientific world. The revolution in science that Bohr did so much to create made this an era of science. Bohr also stepped outside the formerly confined realm of science to set a direction toward peace and survival and away from what could be annihilation.

Added to this was his unique system of work. Few other schools have so united and influenced a generation as did his institute. Here was a method that stretched beyond the mobilizing of knowledge to show how national and other barriers might be surmounted for the benefit of the world at large.

As a human being too Bohr held an interest as wide as his associations with the world's leaders. He struggled against insuperable odds, and yet he never gave up. He did what could be done. And with all the recognition and acclaim that came to him, he remained unpretentious and gentle.

A biography of Niels Bohr forces a writer to deal not only with the man's life but also with one of the most difficult of all subjects, quantum physics, for as Einstein said, what a man of his or Bohr's kind thinks is more important than what he suffers. I have attempted to discuss quantum physics only in terms of its principles. Few non-physicists, certainly including myself, can follow into its far reaches. In justification of this course I will again quote Einstein, who said: "Most of the fundamental ideas of science are essentially simple and may be expressed in a language comprehensive to everyone," and again "Books on physics are full of complicated mathematical formulae. But thought and ideas, not formulae, are the beginning of every physical theory."

Language—in this case the tongues we speak—proved not to be a major difficulty. Virtually everything that Bohr wrote was translated into English. Nearly all of the articles written about him are also available in English. The major, and probably the most significant part of his correspondence, particularly his twenty-five year correspondence with Lord Rutherford, is also in this language. Mrs. Niels Bohr, members of the Bohr family, and the staff of the institute, as well as most Danes, speak excellent English.

I want to express my deepest appreciation to Mrs. Bohr, Professor Aage Bohr, and the staff of the institute for their assistance and their hospitality. I also want to add a special word of appreciation for the assistance of Professor John A. Wheeler of Princeton University.

I am indebted to many others for materials, advice, and assistance of many kinds. A warm *thank you* is extended to them all.

I cannot end a note of this kind without a word of gratitude to my editor, Harold Strauss, of Alfred A. Knopf, Inc., for his unfailing and invaluable aid.

RUTH MOORE

Chicago, Illinois
March 1966

Note on Books and Materials

Relatively little of the full and remarkable life of Neils Bohr is recorded in his own books. In his entire life Bohr did not write a single full-length book. The four slender volumes that bear his name are collections of his speeches and articles, with, in some cases, an added short introduction by Bohr.

Nor was Bohr a prodigious producer of scientific papers, though the complete listing of all of his publications carries about 150 notations. A number of the listings represent duplicate publications and translations into various languages.

Bohr's cautious approach to scientific writing carried over into his correspondence: it was steady rather than prolific. Although his personal correspondence fills many filing drawers, it is not voluminous for a man of his interests and world-wide associations.

Bohr's genius rather was expressed in his personal relations, his seminars, his friendships, his collaborations. Those who knew him were deeply influenced by him. They never forgot him, their impressions are detailed and cherished, and many have been set down on paper. Thus the portrait of Bohr has been written primarily in the words of those around him rather than in his own.

Bohr's own books are: *On the Constitution of Atoms and Molecules* (Copenhagen: Munksgaard Ltd.; 1963), *Atomic Theory and the Description of Nature* (Cambridge: The University Press; first printed 1934 and reprinted 1961), *Essays 1958/1962 on Atomic Physics and Human Knowledge* (New York: Interscience Publishers; 1963), and *On the Line Theory of Spectra* (Copenhagen: Andr, Fred Host & Son; 1918).

Bohr made major contributions to a number of books published by others. His "Discussion with Einstein on Epistomological Problems in Atomic Physics" forms part of the two-volume *Albert Einstein Philosopher and Scientist*, edited by Paul Arthur Schilpp (New York: Harper & Brothers Publishers; 1949). Four articles by Bohr appear in *Rutherford at Manchester* (New York: W. A. Benjamin Inc.; 1963), the volume that grew out of the Rutherford Jubilee International Conference of 1961. Bohr made two original contributions: "Reminiscences of the Founder of Nuclear Science and of Some Developments Based on His Work," and the relatively brief "The General Significance of the Discovery of the Atomic Nucleus." Two papers which Bohr began at Manchester also are reprinted in the volume—"On the Constitution of Atoms and Molecules" (1913), and "On the Quantum Theory of Radiation and the Structure of the Atom" (1915).

The complete list of Bohr's papers is maintained by the institute, now renamed the Niels Bohr Institute for Theoretical Physics, Blegdamsvej 17, Copenhagen. I shall not attempt to reproduce the list here. Some of the most notable papers and speeches are reproduced in the Bohr books. Virtually all are available in English.

Other material comes from the Bohr correspondence. I completely reviewed the Bohr-Rutherford correspondence over its twenty-five years. Mrs. Bohr was so kind as to translate for me some of Bohr's personal letters to her.

Essential for any study of Bohr are the articles contributed by his closest friends and associates to a memorial volume: *Niels Bohr His Life and Work* (New York: John Wiley & Son; 1966). I am deeply appreciative that Professor Aage Bohr made copies available to me prior to publication.

Niels Bohr and the Development of Physics (Oxford: Pergamon Press; 1955, 1962), a book edited by Wolfgang Pauli, brings together ten papers written in honor of Bohr on his seventieth birthday. As the editor says, they "evoke various aspects of the general problems of physics with which Bohr was most deeply concerned." Papers written to celebrate the fiftieth anniversary of Bohr's first papers on the atomic constitution were presented posthumously at the institute on July 8–13, 1963, and mimeographed copies are available. An article "Niels Bohr: An Essay Dedicated to Him on the Occasion of His 60th Birthday" by L. Rosenfeld has been published by Nordita, Publication No. 57.

Many other articles about Bohr have appeared in the world's publications. It would be impossible to list all of them, but among the outstanding ones are "Niels David Bohr" by J. D. Cockcroft (Biographical Memoirs of Fellows of the Royal Society, London: Royal Society; Vol. 9, 1963); "Niels Bohr—The Man Who Explained the Atom" by George Gamow (*Science Digest*, May, 1963); Gamow's books *Biography of Physics*, and *Thirty Years that Shook Physics* (New York: Doubleday & Company, Inc.; 1966); three lectures on "Niels Bohr and His Times" by Robert Oppenheimer (mimeographed); "The philosophy of Niels Bohr" by Aage Petersen (*Bulletin of the Atomic Scientists*, September, 1963), and Petersen's address on Bohr on the Danish radio, 1963 (mimeographed); Rosenfeld's "Foundations of Quantum Theory and Complementarity" (Nordita Publications, No. 91).

Others are: "Niels Bohr at Work" by Stefan Rozental (Nordita Publications 100/101, 1963); Hommage Prononcé le 23 Novembre 1962 par le professeur Victor F. Weisskopf, directeur général, devant le personnel de l'organization Européen pour la Recherche Nucléaire (CERN)" and "Niels Bohr—a memorial tribute" (*Physics Today*, October, 1963) also by Weisskopf; and "Niels Bohr and Nuclear Physics" (*Physics Today*, October, 1963); "No Fugitive and Cloistered Virtue" (*Physics Today*, January, 1963); and talk at Columbus, Ohio (tape); the three last by John A. Wheeler.

Other phases of Bohr's life are reported in articles about him that have appeared in newspapers and magazines.

Two books are indispensable to an understanding of Bohr's work on avoiding a nuclear arms race. They are *Britain and Atomic Energy, 1939–45* by Margaret Gowing (London: St. Martin's Press;

1964) and *The New World* by Richard G. Hewlett and Oscar E. Anderson, Jr. (University Park, Pa.: The Pennsylvania State University Press; 1962). Mrs. Gowing is archivist to the United Kingdom Atomic Energy Authority and had full access to the official records, and also was able to review her chapter with Bohr. Anderson is on the staff of the American Atomic Energy Commission. The two books for the first time make public the documentation for the negotiations in which Bohr was involved.

Other discussion of the events of these years and Bohr's part in them appear in the papers and books of Winston Churchill, Franklin D. Roosevelt, Harry Truman, Sir John Anderson, Henry L. Stimson, David E. Lilienthal.

Other books in which Bohr figures or which throw light upon various parts of his career are *J. J. Thomson* by George Thomson (New York: Doubleday & Company, Inc.; 1965); *Rutherford and the Nature of the Atom* by E. N. daC. Andrade (New York: Doubleday & Company, Inc; 1964); *Atoms in the Family* by Laura Fermi (Chicago: University of Chicago Press; 1954); *Brighter Than a Thousand Suns* by Robert Jungk (New York: Harcourt, Brace & World, Inc.; 1958); *Atomic Quest* by Arthur Holly Compton (New York: Oxford University Press; 1956); and *Now It Can Be Told* by Leslie R. Groves (New York: Harper & Brothers; 1962).

The writings of other scientists about the scientific developments in which they and Bohr participated are numerous. To list the full scientific books and reports would be beyond the scope of these notes. But among the books useful to the layman are: *Physics and Philosophy* by Werner Heisenberg (New York: Harper & Brothers; 1958); *Knowledge and Wonder* by Weisskopf (New York: Doubleday & Company, Inc.; 1963); *The Atom and Its Nucleus* by Gamow (New York: Prentice-Hall, Inc.; 1963); *The Evolution of Physics* by Albert Einstein and Leopold Infeld (New York: Simon and Schuster; 1961); *Explaining the Atom* by Selig Hecht (New York: Viking Press, Inc.; 1947).

Among important articles are: "The Physicist's Picture of Nature" by P. A. M. Dirac (*Scientific American*, May, 1963); "Looking Back" by Lise Meitner (*Bulletin of the Atomic Scientists*,

November, 1964); "The Discovery of Fission," by Otto Hahn (*Scientific American*, February, 1958); Columbia University news release, January 26, 1964, on the 25th anniversary of the splitting of the uranium atom at the university.

For the political and world events that swirled around Bohr and affected his life, I made use of general histories of the period, but most particularly Winston Churchill's *The Second World War* (New York: Houghton Mifflin Company; 1948); William R. Shirer's *The Rise and Fall of the Third Reich* (New York: Simon and Schuster; 1960); and Barbara Tuchman's *The Guns of August* (New York: The Macmillan Company; 1962). For Danish History I used Palle Lauring's *A History of the Kingdom of Denmark* (Copenhagen: Host & Son; 1963), and *Denmark 1940–45* (Copenhagen: The Museum of the Danish Resistance Movement; 1960).

Much material for a book of this kind must be gathered in interviews. I am particularly grateful for the lengthy interviews that were afforded me by Mrs. Niels Bohr, Professor Aage Bohr, and Drs. Rosenfeld, Rozental, and Rudinger at the institute, and Dr. John A. Wheeler of Princeton University.

Most of the descriptions of the scene were drawn from my own visits to the institute, the House of Honor, Tisvilde and Copenhagen generally, to Stockholm, and to Princeton, N.J.

Contents

Illustrations

NIELS BOHR

1 /

Niels Bohr

IT IS NOT GIVEN TO MANY MEN to change the course of the world.

Niels Bohr, though, once altered the course of history. The age of the atom came into being largely through his scientific work and influence, and few men have had a more directive effect on the lives of their fellow men and the earth.

But for a second time Bohr nearly changed the course of history. He came within a touch of changing the postwar march that has embroiled the world in an atomic arms race, leading no one yet knows where. Whether the different direction Bohr advocated could have averted the cold war and the multiplication of nuclear powers, no one can say. It can only be said that when his counsel of fruitful international cooperation was rejected, primarily through the agency of one man—Winston Churchill—exactly the dangers Bohr had foreseen did come to pass.

It is an incredible record for one man and particularly for a modest man born in the small and pleasant, though politically not powerful, country of Denmark. Bohr had no influence other than the massive influence of his mind and gentle, tenacious disposition.

With only this he won the allegiance and guided some of the most powerful figures and the most salient scientific minds of his day. This made him the disciple and collaborator of Lord Rutherford; the respected, indispensable, lifelong challenger of Einstein; the opponent of Churchill; an advisor of Franklin Delano Roosevelt, Henry L. Stimson, Sir John Anderson, the doughty Scot who headed the British atomic energy project, and Lord Halifax, the former British ambassador to the United States; the mentor and counselor of such preeminent scientists as Heisenberg, Pauli, Dirac, and Oppenheimer; and the rallier, the resource, and friend of nearly all physicists in nearly all countries.

Bohr was both one of the great discoverers (of the structure of the atom, the structure of the nucleus, and the principle of complementarity) and one of the most influential of men. To find another who had a comparable effect on his time, it is necessary to turn back to Aristotle. It is not incidental that both headed schools devoted to the search for the deepest truths of Nature. Both Bohr's Institute for Theoretical Physics on the edge of a Copenhagen park and Aristotle's Lyceum on the outskirts of Athens in a grove sacred to Apollo were centers to which the most able came and from which learning radiated. There were few schools like either in the twenty-three centuries that separated them.

Bohr was a Dane of the Danes. Heavy-boned faces structured somewhat like his can be seen now and again among the Danish people. Denmark always was his base, although Bohr lived in England and the United States for significant periods of his life; and he played almost as much a part in the life of both countries as he did in his own. Bohr actually was an internationalist of a new stamp. His work and his life belonged not only to the three countries, but to the world. That Bohr lived through two wars, both worldwide, and a global revolution in physics certainly contributed to making him a world, rather than a national, figure.

The tumultuous years of his lifetime also drew this man of intellect into strange circumstances. Secret messages were delivered to him; he made one escape in a special bomber; he was involved in underground plots when for several years he had to use a "cover name." Few spy melodramas outmatch the experiences of the Danish professor.

There was also tragedy, sorrow, and defeat. Einstein once observed: "The essential in the being of a man of my type lies precisely in what he thinks and how, and not in what he does or suffers." This was true of Bohr, and yet what he did and suffered had a bearing on what he thought and how. Both emotionally and professionally Bohr's life was a very well-rounded one.

That so remarkable a man should appear on the Danish and world scenes certainly could not have been predicted. Nevertheless the Bohr family provided a promising genetic base from which a rare human being might spring.

2 /

High to the Ceiling

T HE DANES HAVE AN OLD SAYING to describe a family of a rarified intellectual standing combined with goodness. They are, it is said, "high to the ceiling." The expression was nearly always applied to the Bohrs. Niels Bohr was the third Professor Bohr, and though he carried the family to a new illustriousness, the rise to the ceiling had long been building up.

H. G. C. Bohr, Niels Bohr's grandfather, was the head of a grammar school on the island of Bornholm, and the most respected man of his community.

Christian Bohr, the father of Niels Bohr, was professor of physiology at the University of Copenhagen. He was known for his scientific achievements and was one of the men around whom the intellectual and cultural life of Copenhagen revolved. In addition, when he died, a special story was carried in the sports section of the newspaper. Professor Bohr, an admirer of all things English, had founded the Akademisk Boldklub, the university soccer football club, and thus had helped to make soccer the national sport of Denmark.

As a boy Christian Bohr, like many future scientists, was an ardent collector of animal skeletons and insects. At the university he entered the medical school, but it was experimental physiology, not the practice of medicine, that interested him. As a physiologist on the university faculty, Christian Bohr contributed many articles to the university publications, and became intensely involved in the great philosophical dispute of the day: is life a unique special thing (vitalism) or is it only a complex working of the laws of physics and chemistry (mechanism). Christian Bohr emphasized the physical and chemical base of all living things, and yet as a biologist he also had to appreciate the indispensable part played by biological function. Not everything could be explained by physics and chemistry, though he favored the scientific side.

In the anti-materialism prevailing at that time, as this indicated, he was a liberal in his views. He supported a liberal candidate in the parliamentary elections of 1884 rather than the conservative, his maternal uncle Christian Rimestad, and he strongly advocated equality for women. Christian Bohr did more than advocate; he introduced classes for "adult female students." One of these students was a young girl whose perfect oval face was emphasized by dark hair parted in the center and drawn simply back into a knot at the nape of the neck. She was as gentle as she was beautiful, and her name was Ellen Adler. She did not complete the courses laid out for graduation; instead she and the professor were married in 1881.[1]

Ellen Adler was the daughter of D. B. Adler, a banker and financier who had founded the Commercial Bank of Copenhagen and been an initiator of the Jutland Provincial Credit Association. He was interested in politics and had been elected to parliament as a member of the National Liberal Party.

The young couple lived at first in the Adler town house, 14 Ved Stranden, an imposing stone-fronted house in a curving row of large town houses facing the Christianborg Palace, the seat of the Danish government. Just across the wide cobbled street stood the statue of Bishop Absalon, who had founded Copenhagen on that very site in 1166. All around stretched the Old Town with its clustered old houses and their peaked red tiled roofs. There were tiny shops hung with gleaming copper pots and pans and quiet bookstores. Nearby ran the Christianborg Slot where the fishing boats came in each morning and the white bonneted, ample fishwives, well wrapped up against a day in the open, cleaned and

---

[1] See *Niels Bohr His Life and Work* (New York: John Wiley & Son; 1966), a commemorative volume written by family and friends.

sold the fish. Up the canal on the other side came boats bringing in masses of flowers each summer morning to supply the flower loving city.

Here in the heart of Copenhagen the Bohr's first child, a daughter Jenny was born, and on October 7, 1885, their first son Niels Henrik David, and two years later their second son Harald. From the moment of birth these were cherished children, cherished by parents and grandparents. As soon as they could toddle, Christian Bohr took his children to see the palace with the gold crown at the top of the spire, the boats, and the fishwives at their scaly work. Both boys had sailor suits, just like those they saw at the palace or on the boats. Ellen Bohr dressed them in these favored white suits when a photographer came to make a picture of them—healthy, alert children, spotless for the occasion, and with high buttoned shoes shined to a fine polish.

At the age of seven, the age at which all Danish children entered school, Niels went to the Gammelholm School; but it was hard for him to leave Harald at home, for even then the two of them were always together. In a woodworking class Niels at once began to make a puppet theater for Harald and was heartbroken when he found that he could not take it home to finish it. The fond and considerate father soon remedied this problem, building a workbench for the boys and equipping it with tools that he taught them to use. Another puppet theater was soon constructed. As the boys grew older the workbench was furnished with a lathe and Niels, very adept at working with his hands, acquired a skill in working with metal that stayed with him, most usefully, all of his life.

Niels's class was the last to go through school under the "old act" (it was replaced by the Education Act of 1903) and school was disciplined and formal. But whatever the educational severity, school presented no problems for this bright, even-tempered child. Besides, the lessons of the school were richly supplemented by the home.

When Christian Bohr became professor of physiology, he took up residence, as did all Danish professors, in a house adjoining his laboratory and classrooms. The Bohrs thus moved into the old Surgical Academy, from which Professor Bohr continued to take the children on walks and expeditions. He wanted to make certain that they understood and appreciated the beauties of nature. On one walk he called the children's attention to a tree—see how beautifully the trunk divides into branches and the branches into twigs and how the leaves bud at the end of the twigs. Niels listened intently and answered: "Yes, but if it weren't like that it wouldn't be a tree."

Professor Bohr, in addition to his lifelong absorption in physiology and natural science, was a disciple of Goethe and could recite whole sections of Faust from memory. Niels, as he walked at his father's side or sat at his feet on long winter evenings, learned Goethe almost by absorption. The majestic lines stayed with him all his life and he frequently quoted from Goethe and the other great German poets to whom his father had introduced him at a very early age. To only a slightly lesser degree, Professor Bohr admired Shakespeare and Dickens and frequently gathered the children around him to hear these authors. He wanted his children to know both the German and English cultures, and he felt that Danes were in a special position to appreciate the best of both.

The children also had an early and remarkable exposure to the best scientific and philosophic thought of Denmark, which meant the whole European tradition. Professor Bohr was a member of the Royal Danish Academy of Sciences and letters. Following its meetings, he, Harald Höffding, professor of philosophy at the university, and C. Christiansen, professor of physics, formed the habit of stopping at a cafe to continue the discussions of the meeting or to discuss their own deep interest in the relation of science to life.

The trio though soon tired of meeting in cafes and decided to shift their meetings to their homes for dinner every other Friday night. The famous philologist Vilhelm Thomsen joined them.

When the four professors met at the Bohr house, Niels and Harald, were permitted, to sit in the dining room and listen. They drank in the spirit of the discussion—the serious consideration of serious subjects, discussion as a means to clarification, unity in the search for knowledge. Nor did they miss the laughter that often rang out, as one or another of the professors told a story or made some comment that amused his colleagues. Niels, sitting quiet and awe-struck, took it in almost like the air he breathed. It became a part of him and he was getting his grounding in the methods that he would follow throughout his life.

Occasionally another colleague or a foreign visitor to Denmark joined the group. The boys met these visitors, too, and learned a little of the habits, appearance, and strange tongues of those who lived outside the orderly, small kingdom of Denmark. This lesson also left its mark. Many others came to the Bohr house, particularly artists, writers, and musicians, for the Bohr household was a very hospitable one.

But life in the lively Bohr household was not all high thought. There was fun and sport too. Professor Bohr had laid out the football fields at Tagensvej, and whenever he went out for a game the brothers, as everyone called the inseparable Niels and Harald, went along. They played too, and thus learned the game when they were very young, as nearly all expert athletes must. They also were taught to ski when they were young.

Sometimes on Sundays the professor would rent a boat and row them over to the canals at Christianshavn. They always stopped to climb the wonderful outside stairs that spiraled around the tower of Our Savior's Church. From the top one could look out over Copenhagen, over the massed roofs to the string of lakes that had replaced the ancient walls or out to the harbor and the sea where ships passed on their way to and from many other lands. Niels liked almost as much to see the works of the big tower clock, and would study the wheels and gears as long as his father could be persuaded to remain.

Niels was already fixing the family clocks and anything else that needed repair. When the hub sprocket of one of their bicycles broke, Niels volunteered to fix it. Despite some protests that it should be left to a mechanic, Niels took it all apart, and discussion arose as to whether it would go together again. All the members of the family soon were taking part, but the professor insisted: "Let the boy alone. He knows what he is doing." After a thorough study of each of the parts, Niels, with the participation of most of the family, successfully put the sprocket back together again and the bicycle worked perfectly. A cousin who had taken part in the consultations said that they all experienced a sense of triumph. Niels was learning how to interest a large number of people in working with him.

Most of the summers were spent at Naerumgaard, his grandparents' large country home just north of Copenhagen. Following the death of D. B. Adler, the household revolved around the strong warm personality of Jenny Adler, the children's grandmother. She sat at one end of the long table in the spacious dining room with the children clustered around her. The parents had their domain at the opposite end of the table. Without resorting to "correction" Grandmother Adler kept all running smoothly and correctly. When Niels at one meal covered his fruit jelly with a mountainous heap of sugar, the professor remarked warningly from the other end of the table "Really, Niels." Grandmother settled the matter quickly and firmly: "Maybe he needs it."

Aunt Hanna Adler also was often at Naerumgaard during the summers. The older sister of Ellen Bohr, she was one of Copenhagen's best known teachers. In the late 1890s she went to the United States and upon her return founded the "Hanna Adler's Coeducational School." Later Niels's own children were to attend it. Hanna had always been devoted to her younger sister and this affection extended to her niece and nephews. Niels and Harald were "her children" and she put all of her great skills as a teacher as well as her love into their education. For a book published in 1959 to commemorate the centenary of her birth, Niels Bohr wrote the introduction:

> From my earliest childhood I have vivid memories of her active and loving participation in everything concerning her brothers and sisters and their children. Although my brother Harald and I were not among her pupils at school we shared with them "Aunt Hanna's" educational influence. When she could spare time from her school work, she took us on Sundays around

Copenhagen's natural history and ethnological exhibitions and art museums and in the summer holidays at Naerumgaard where she accompanied us often on foot or on a bicycle in the woods and fields of the district we learned both about nature and human life, while she jokingly or seriously talked to us about everything that could catch our imagination.

At the Gammelholm School Niels did very well in all of his studies except Danish composition. An essay, according to the requirements of the school, had to have a formal introduction and conclusion, and this proved not at all to the taste and style of young Niels Bohr. This was not the way his mind worked then or later: Bohr papers open with a review and march along until they stop; there is no formal introduction or conclusion. Niels wrote one essay entitled "A Walk Around the Harbor." It began and ended: "My brother and I went for a walk around the harbor. There we saw ships loading and landing." Niels had arrived at the alliteration after long thought and trial, and could not understand the teacher's failing to note it. Another essay on metals, dropped fine hints that there might be more, but ended inconclusively: "In conclusion I would like to mention aluminum." On still another day Niels was assigned an essay on "The Uses of the Forces of Nature in the Home." This was too much for the future scientist; he threatened to conclude: "In our home we do not use the forces of nature." Harald persuaded him to refrain.

The repressed conclusion sounded much more like Harald than Niels. Coming from Niels it undoubtedly was a straight scientific observation. Originating from Harald it would have been a witticism. Harald had a quick, sharp wit, and was generally considered the more brilliant of the brothers. Niels though, the professor once confided to a close friend, was "the special one."

Harald liked to tease and most often teased Niels. One day when one of the adults intervened, Niels at once came to Harald's defense: "Harald is the finest of brothers. I like it." It was as impossible for Niels to "tease back" as it was for him to follow a rigid rule for composition. Harald once proposed that they play "teasing one another." Harald had the first turn and went at it until

Niels begged "Stop, please no more." It was then Niels' turn; he stood there silently pondering; minutes went by. He managed to come forth with: "You've got a small spot on your coat." It was so naïve and unteasing that Harald bent double with laughter. Niels was never to be capable of the quick edged retort. He could laugh, he could enjoy Harald's wit and the banter of others, but nothing more completely conveyed his own dearth of such a sense than his own statement that he "lacked the gift of impertinence." If malice is the soul of wit, Niels had no malice in him.

Niels remained at the Gammelholm School from the first grade until he was ready for the university. During his last six years there he shared a double desk with Ole Chievitz, whose father was professor of anatomy at the university and thus a colleague of Professor Bohr's. The impulsive Ole and the bemused Niels became close friends and the friendship lasted throughout their lives. Chievitz, who had become a professor and surgeon was interviewed by the newspapers on Bohr's sixtieth birthday. When asked "What characteristic of Niels Bohr do you rate the highest?" the old schoolmate did not hesitate: "His goodness . . . Let us not give examples. Bohr would not care for that. You must be satisfied with my word when I tell you that he is as good in big things as in small. I am not exaggerating just because it is his birthday when I say that I consider him the best human being in the world."

Despite the difficulties with composition, Niels did so well in history, languages, Latin, and poetry, which he recited in a low, dreamy, almost chanting tone, that he stood first in his class. Alb V. Jorgensen, another classmate, said that no one minded Niels's excellence. He did not struggle to be first or seem "ambitious." Neither was he glued to his books. If he was first and set the tone of the class, it came about seemingly without special effort on his part. He was not what Jorgensen called a "swot."

As Niels advanced into the upper grades and began to study mathematics and physics his remarkable ability was unmistakable. High school physics was taught very sketchily in the closing years of the last century and the opening of the twentieth. But Niels began to read well beyond the textbooks. Soon he began spotting statements in the textbooks that had been outmoded or made incorrect by the newer physics he was reading in the journals. One of his classmates asked what he would do if they were questioned in their examination about one of the "incorrect" sections. Niels looked surprised: "Tell them, of course, how things really are."

In 1903 Niels entered the University of Copenhagen. The university was in the center of the old city and traffic flowed busily around the buildings. Only the streets and a few cobbled courtyards set the buildings aside from the shops, bookstores, and residences pressing close around them. There was no campus in the American or British sense of the word.

As a sophomore Bohr enrolled for a course in mathematics with Professor T. N. Thiele, the astronomer. Another student was Helga Lund, who had taught for several years before entering the university and who was later to become a headmistress. She and the young man with the heavy bowed head, who was always carrying a briefcase, were assigned to work together on some mathematical problems. In between the Thiele lecture and the next class they often studied together at the Student Union.

Helga Lund had no doubt about the capacity of the young man with whom she was working. In a letter written to a Norwegian cousin in December, 1904, she told in a burst of enthusiasm about knowing a genius, and working with him every day. She said that his name was Niels Bohr, and that his exceptional qualities were more and more evident. But she also mentioned his kindness and extreme modesty. She told her cousin that Bohr's brother was also at the university and was just as bright; he was working in mathematics. The two Bohrs were always together; they were only 17 and 19 years old, but she rarely spent time with any other students because she liked the Bohrs so much.

Though Harald was two years younger than Niels, he had made such rapid progress through school that he entered the university only one year behind his brother.

Niels enrolled for classes in philosophy with Professor Höffding, his father's close friend and to whose tempered words he had listened with such awe in the dining room at home. He took courses in the history of philosophy and logic. Höffding led his students through the main systems of philosophy from the sixteenth century to the eighteenth; however, he did not try to persuade them of any one system. He put his emphasis on the proposition of problems rather than on their solution. "The solutions," he explained, "may die, but the problems are still living; otherwise philosophy would not have had so long a life as she has had."

Bohr was attracted to Spinoza's psychophysical parallelism. He also read Kierkegaard, though he was more impressed with the Danish philosopher's style than with his ideas.

What struck a fundamental chord in Bohr was a deceptively simple little book by Poul Martin Moller, *Tale of a Danish Student*. In the soul-searching student trying to sort out the dualities that confounded him, and to bring his conflicts into some kind of comprehensible order, Bohr undoubtedly recognized himself. In one scene the student explains his dilemma to his untroubled, down-to-earth cousin Fritz:

> "Certainly I have seen before thoughts put on paper; but since I have come distinctly to perceive the contradiction implied in such an action, I feel completely incapable of forming a single written sentence. And although experience has shown innumerable times that it can be done, I torture myself to solve the unaccountable puzzle, how one can think, talk, or write. You see, my friend, a movement presupposes a direction. The mind cannot proceed without moving along a certain line; but before following this line, it must already have thought it. Therefore one has already thought every thought before one thinks it. Thus every thought, which seems the work of a minute, presupposes an eternity. This could almost drive me to madness.
>
> "How could then any thought arise, since it must have existed before it is produced? When you write a sentence, you must have it in your head before you write it; but before you have it in your head, you must have thought, otherwise how could you know that a sentence can be produced? And before you think it, you must have had an idea of it, otherwise how could it have occurred to you to think it? And so it goes on to infinity, and this infinity is enclosed in an instant."
>
> "Bless me," said Fritz, "while you are proving that thoughts cannot move, yours are proceeding briskly forth!"

"That is just the knot," replied the student. "This increases the hopeless mix-up, which no mortal can ever sort out. The insight into the impossibility of thinking contains itself an impossibility, the recognition of which again implies an inexplicable contradiction."

Bohr was having exactly the same trouble with sentences. It led him into reflections on the use of language for the objective communication of experience. Some of the difficulty came because we "use the same word in different contexts to denote aspects of human experiences which are not only different, but mutually exclusive." "I think" thus was applied both to the state of consciousness and to the concomitant working of the brain. And this was a factor in leading Bohr into the dialectical problem of cognition which long had been obscured, almost since the time of Newton, by the conflict between rationalism and mysticism.

He was groping his way. Perhaps the two planes of thought could be traced to man's position in the universe and to the necessity this imposed of learning by making himself the object of his observations. Again *Tale of a Danish Student* drew him deeper into these subtleties. The student explained: "Thus on many occasions man divides himself into two persons, one of whom tries to fool the other, while a third one, who in fact is the same as the other two, is filled with wonder at this confusion. In short, thinking becomes dramatic and quietly acts the most complicated plots with itself, and the spectator again and again becomes actor." Bohr would use this final sentence repeatedly. The actor-spectator comparison came closest to simplifying for others the subtle, puzzling interplay.

The problem so engaged Bohr that he considered for a while turning to epistomology, and he debated the possibility of one day writing a book on the theory of thought. His studies set him to dreaming about the great dualities and about achieving harmony between form and content, and about the "great interrelationships" in all fields of knowledge. It was the tribulations of Moller's *Danish Student*, rather than physics or formal courses, that first aroused his interest in philosophical problems.

Niels and Harald eagerly accepted an invitation to join some of the other students in Höffding's courses in a group which would discuss some of the philosophical and scientific questions raised in class. They chose the name "Ekliptika" and limited their number to twelve. Meetings were held several times a month at the "a Porta" or some of the other cafes where the talk could go on, over a cup of coffee or a glass of beer, until early in the morning. Sometimes the two Bohrs would carry the discussion. Vilhelm Slomann, later an art historian, has described how it went:

> When the discussions were beginning to tail off, it often happened that Niels said a few generous words about the lecture and continued in a low voice and at a furious pace and intensity. He was often interrupted by his brother. Their way of thinking seemed to be coordinated. One improved on the other's or his own expression, or defended in a heated and yet at the same time good humored manner his choice of words. Ideas changed their tone and became more polished. There was no defense of preconceived opinion, but the whole argument was spontaneous. This way of thinking à deux was so deeply ingrained in the brothers that nobody else could join it. The chairman used to put his pencil down quietly and let them carry on. But when everyone moved in closer to them, he might say, ineffectually, "Louder, Niels."

Even over a cafe table Niels's soft voice was hard to hear. It was seldom raised even out on the soccer field. Harald, playing halfback, was becoming one of the best players in Denmark and making all the championship teams. Niels was not quite so good. He played goalie and Harald said that he did not charge out fast enough. Niels was only a substitute on the first team. Nevertheless they both came out religiously for practice and worked hard.

When Niels was not philosophizing or playing football, he was often in the laboratory. In a course in inorganic chemistry he was establishing the year's record for the breaking of glass. One day when the laboratory was rocked by a series of small explosions, Niels Bjerrum, the instructor, without investigating the source, said "It's Bohr." He was right. Despite skill in the manipulation of equipment, Niels had pushed his curiosity a little too far. The transition from philosopher to experimentalist had its perils, but Niels saw nothing strange in it. When asked about it, he said that he was not only a dreamer; "I was willing to work hard."

There was no question but that Niels's main interest, despite the allure of philosophy and epistomology and football, was physics. He could not know then that the new physics he would eventually help to create would prove to be a "philosophical treasure chamber, containing in a new form, the very thoughts and relationships of which he had dreamed."

Professor Christiansen, the physicist of his father's circle, taught most of the physics. In his courses he emphasized the contributions of both the German and British schools. Denmark with its proximity to Germany and its ties to England was in an advantageous position to appreciate the work in both countries. Niels thus received a foundation in German theory and German emphasis on the wave problem, and in the British experiment and work with the atom. The combination proved invaluable for him.

Niels easily decided that he would take his master's and doctor's degrees in physics. However, relatively few advanced courses were offered at Copenhagen, and there was little compulsory laboratory work. He largely was free to proceed on his own.

In 1905 the Academy of Sciences and Letters offered a prize in physics for the best paper on the surface tension of liquids. Lord Rayleigh had shown that it was theoretically possible to determine the surface tension of a liquid by measurement of the length of waves "formed on the surface of a jet [of liquid] of known speed and cross section."

Niels quickly decided to enter the competition, but he would have to work out a method of producing the jet—like a stream of water from the nozzle of a hose—and a way of making the measurements. His father gave him permission to do the work in his laboratory.

To obtain a long stable jet of water, Niels saw that he would need unusually long glass tubes with orifices specially treated to produce the waves on the surface of the jet—the openings had to be elliptical to make the stream of water twist. This took difficult manipulating of the glass and explained some of Niels's glass breakage in the chemical laboratory.

"It was necessary during the heating and drawing of the tubes," Niels explained in a report on his techniques, "to have both ends of the tube fastened on slides which could be displaced along a metal prism. When the glass tubes were drawn out and cut off they were examined under a microscope and only those with orifices of uniform elliptic section were used. . . ."

Niels's method required keeping the twisting jet of water stable for several hours—each measurement took at least this much time. To obtain such complete stability in the daytime when there was other movement in the laboratory proved almost impossible, thus, Niels did most of his work at night when the laboratory was quiet and deserted.

As he worked he found that to obtain a quantitative measurement of the surface tension, Rayleigh's theory had to be extended to take in the viscosity of the liquid, the final amplitudes of the vibrations, and the effect of the surrounding atmosphere. The time for the submission of the paper was approaching. Niels had studied only water, and there was always one more measurement or another variation, or an adjustment that he had to try. Unless he broke off the work, the paper would never be written. Professor Bohr insisted that he end the experiments at a suitable point, and sent his son out to Naerumgaard to write his paper. The little stratagem worked, and Niels completed his paper.

Only two papers were submitted. After careful appraisal the contest committee decided to award both the academy gold medal. P. O. Pedersen, the other competitor, had used a more simple method to determine the surface tension of a number of liquids. The paper was excellent and clearly deserved a medal. Bohr's paper also was excellent. The academy commented:

A single determination according to the method of the author requires constant work over many hours. For that reason the jet must be maintained over a long period under stable conditions. The length of time limits the application of the method for liquids which change their nature on contact with the air. . . .

Although the work does not solve the problem as completely as the first [Pedersen's paper] as it deals only with a single liquid, namely water, its author on the other hand deserves considerable merit for having furthered the solution on other points, so that we feel justified in suggesting that this paper too should be rewarded the Society's gold medallion.

Niels at the age of 21 actually had added to the basic theory of one of the most famous physicists of the age, Lord Rayleigh. He had shown quite unexpectedly that additional factors had to be taken into consideration to determine surface tensions.

The Danish society had chosen a subject that few physicists studied. No one could have foreseen or imagined that knowledge of the surface tension of water would supply a clue, some thirty-five years later, to the structure of the atom and thus to the development of the atom bomb and nuclear energy. Niels Bohr left few threads dangling; nearly everything he touched worked in again, through the most devious and unforeseeable of developments.

Each of the advanced students of physics had to deliver a lecture to his fellow students on a subject of his own choosing, and Niels made a student presentation on radioactivity. In 1895 the German physicist Wilhelm K. Roentgen had discovered that unknown rays coming from a Geissler tube could cause a zinc sulphide screen to fluoresce. He named them X rays. A year later the French physicist Henri Becquerel, attempting to determine if fluorescence was produced by the rays of the sun, on a rainy day wrapped a piece of uranium ore in a bit of black paper and laid it in a drawer until the sun should be shining again. When a few days later he opened the drawer and developed the plate, Becquerel was astounded to see that it was fogged. Even in the dark the mineral had given off rays that passed through the paper and affected the photographic plate. Pierre and Marie Curie, trying to track down the element that gave off the extraordinary radiation, discovered radium and received the Nobel Prize in 1903, the year Bohr entered the university.

In England Ernest Rutherford, a young graduate student from

before Harald, found it difficult to conclude the work for his thesis. As in his work on the competition there was always another investigation to be made or another formula to be tested. Again the only way was to get him away from the laboratories and into a place where he could write without any other distractions.

Naerumgaard was no longer available as the retreat. In 1908 the Adler family had presented the big house and grounds to the Municipality of Copenhagen for a children's home. In accordance with the will of D. B. Adler and his wife Jenny the first two generations of descendants were to serve on the board of directors. Many years later Niels was called to add this pleasant family duty to the many others he carried.

Another "working place" had to be found for Niels. Professor Bohr's young assistant Holgar Mollgaard suggested his father's vicarage at Vissenbjerg on the island of Funen. Niels accepted the invitation, and on the tranquil island where Hans Christian Andersen found the setting for many of his fairy tales, made steady progress on his thesis.

"Here in the vicarage . . ." he wrote to Harald

Everything is fine for me in every way. I eat and sleep a tremendous amount to the satisfaction of mother (sorry to talk nonsense, I meant to my own satisfaction); but I also get quite a lot done. I have now finished . . . dynamics and read most of what Abraham has to say about vector calculation (very interesting, and I have begun Christiansen's manuscript [the manuscript of a textbook on physics published by Professor Christiansen several years later]. I am enjoying it and the dynamic introduction contains many interesting things, but it makes not the slightest attempt to comply with the requirements which you (and I as well) would expect from a properly grounded theory of motion. . . .

Bohr, even as a student looking at the major issues, could not overlook inconsistencies and inadequacies. He had a very keen critical sense, though it took the form of trying to discover the correct answer, rather than to criticize for the sake of criticism. Almost an intuition led him to faults or errors or omissions.

Though Niels was only then completing his thesis, Harald was

New Zealand, had set about measuring the penetrating power of radiation given off by uranium. He discovered, as he reported, t "the uranium radiation is complex, and that there are two distir types of radiation—one that is very readily absorbed will be term for convenience the alpha radiation, and the other of a mo penetrating character, which will be termed the beta radiation."

The beta rays, Rutherford showed, penetrated exactly as di Roentgen's X rays, whereas few of the alpha particles in the earl experiments could penetrate a sheet of aluminum a thousandth of an inch in diameter. "The cause and origin of the radiations continuously emitted by uranium is a mystery," said Rutherford— though he was to solve it not many years later.

Very little of these new findings had made its way into the textbooks, but Bohr reported and reviewed this work that was radically changing the physics they were studying. It was bringing new terms and new ideas to a science that had previously been thought to offer a fairly satisfactory explanation of the universe. In the midst of these new developments Niels took his degree in 1907.

The year also brought other excitements. The AB (Akedemisk Boldklub) was winning victory after victory. Harald was becoming the idol of Danish soccer fans. When Denmark decided to enter a team in the 1908 Olympics in England, there was no question about Harald's choice to play halfback. Niels did not make the Olympic team.

In London at the Olympics the competition was fierce but Denmark went through to the semi-finals and won the silver medal. The British sports writers were high in their praise of Harald the "shock-headed Dane" who did so much to carry his team to victory. Tremendous acclaim greeted Harald on his return to Copenhagen. Niels, as a good player too, shared in some of the fame. The two brothers were known throughout the land. Little boys watched as they played, and the newspapers reported all of their sports exploits. The Bohrs were Denmark's hero athletes.

Soon after the Olympics Harald began work for his master's degree. Niels also was preparing for his, for both had moved on to graduate work without interruption. Niels, though he had started taking his examinations in March, 1909. As soon as Harold had successfully completed his examination he went to Goettingen to study with the famous German mathematician Edmond Landau. He was also celebrating his twenty-second birthday. On April 20, Niels wrote to him:

Heartiest congratulations. This time it is not an ordinary birthday, but the beginning of something entirely new. I shall be so pleased for your sake if you do really well at Goettingen both in respect to your development of your mathematical personality and also with regard to your personal self.

I am sending you herewith (besides what Mother is so good as to send you in my name) Kierkegaard's "Stages on Life's Journey." It is the only thing I have to send, but I do not believe that it would be very easy to find anything better. In any case I have had very much pleasure in reading it; I even think that it is one of the most delightful things I have ever read. I am now looking forward to hearing your opinion of it sometime.

I am doing very well here in Vissenbjerg. It is lovely now, the spring is really here and the first wood anemones have already come out.

The spring was bringing another development as welcome as the white anemones and more exciting. Niels's gold medal paper, "Determination of the Surface Tension of Water by the Method of Jet Vibration," was, he learned, to be printed in the *Philosophical Transactions* of the Royal Society of London. It was a heady success for the then 24-year-old Danish student to have a paper of his appear in the distinguished English journal, and it was besides his first appearance in print. The arrival of the proofs at Vissenbjerg was a memorable moment.

"The treatise was beautifully printed," Niels continued in his letter to Harald, "and so carefully checked over (there was not a single figure printed incorrectly) that it was easy for me to finish it."

In addition a friend had sent him a copy of the abstract of his paper that had appeared in the Society's *Proceedings*. The abstract proved simply to be his conclusions, and Niels was greatly gratified by this and that there had been no mistakes or changes in his meaning.

Despite Niels's enthusiasm about Kierkegaard's book, he by no means agreed entirely, either then or later, with Kierkegaard's philosophy. He hastened to explain this in another letter to Harald:

> When you have finally read the *Stages*, and there really is no reason why you should hurry with it, you shall hear a little about it from me. I have written a few notes about it (not in agreement with K.) but I do not intend to be so banal as to attempt, with my poor nonsense, to interfere with your impression of so nice a book. I am getting on fine and I am looking forward more than I can say to the lovely time I shall spend in Copenhagen when I have finished my examination, before I go abroad. . . .

It was accepted in the education-minded Bohr family that both sons should have the best possible education for the academic careers for which both were then destined. This meant completion of a doctor's degree and study abroad. This was as necessary as grammar school.

In due time Harald reported from Goettingen that he had read a bit "here and there" in the birthday Kierkegaard.

> Whether it is because I, at the moment or perhaps generally do not have the right attitude to K., I must say that although naturally I have to admire his great art and superior gifts, yet I do not feel really drawn to it. . . . No doubt, too, all the muddle I am in and all the bustling around I have had to do recently as a result, has made me need to read something different from K.
>
> It would be really delightful if we, when I come home, could read something really good together, could for example sit with Mother in the sitting room about the gilded "piece" with the three legs and if one or the other of us read aloud to the others. . . .

While Niels was studying physics, changes and developments in the science were coming rapidly. A few years earlier, just before the close of the nineteenth century J. J. Thomson at Cambridge University had found that a narrow beam of cathode rays could be deflected in both an electric and a magnetic field. It looked as though the beam must be made up of particles of some hitherto unknown kind, though many strongly disagreed with this heretical idea. Soon afterward Thomson proved that he was indeed dealing with particles, particles with a mass that he first thought was 1/770th of that of the hydrogen atom but later proved to be 1/1840th of the mass of the smallest of all atoms. Thus there were particles of a mass much less than that of the lightest atom! And for the first time it was learned that the "little hard round ball" that the atom had always been thought to be, contained a number of smaller bodies. Thomson at first referred to the tiny history-making particles as "corpuscles." Though the discovery was regarded as only of scientific interest—no one could foresee that the whole electronics industry and such manifestations of it as television would spring from the little *corpuscles*—studies began all around the world.

In an evacuated glass tube the *corpuscles* would stream out of a negative metal plate called the cathode and travel across the space in between to a positive metal plate called the anode. But the question was, where were the particles when an electric current was not flowing through the vacuum tube? Soon it became evident that they must be near the surface of all matter—they could even be rubbed off by walking swiftly across a thick carpet, and if something metallic were touched they would jump from the finger to the metal surface.

Ordinarily, it was clear, most matter must be electrically neutral. However, if the *corpuscles* were pulled away from any bit of matter it became positively charged and the object acquiring the *corpuscles*, negatively charged. Within a few years it began to look as though atoms in their normal neutral state might be made up of something positive combined with the negative *corpuscles*. In 1909 H. A. Lorentz of the Netherlands suggested that the *corpuscles* might properly be called electrons. The old picture of the atom as a solid "billiard ball" vanished. The atom had become some kind of electrically neutral structure with easily detachable electrons on its surface.

At the annual dinner of the Cavendish Laboratory at Cambridge they drank a toast, "To the electron. May it never be of any use to anybody." Niels Bohr was not concerned about its usefulness, but as he plunged deeper into physics he was interested in the electrons coming from the interior of the atom.

Bohr pored over Thomson's articles and his book, as well as over the publications of Lorentz and Drude in Germany. For his master's thesis he decided to work on the various physical properties of metals, their electrical and thermal conductivity, their magnetic and thermo-electric phenomena, in the light of the electron theory. It was a large undertaking. But Bohr assumed that the characteristic properties of metals were associated with the free movement and collision of the electrons and stationary molecules of the metals. He would try to test the assumption by comparing its consequences with the observed properties of metals.

Characteristically Bohr first examined the fundamental assumptions, and asked if they were reflected in what was known. Though he was still a student, Bohr with a sure instinct or intuition, went to the questions that lay at the heart of the matter. In commenting on one of Lorentz's basic assumptions, he wrote:

> There seem to be certain at least formal flaws in the ideas put forth by Lorentz which derive from the fact that the law whereby metal molecules and electrons are assumed to interact does not lead to thermal equilibrium. Thus one might ask whence the electrons receive their velocities which vary according to the temperature at the different places in the metal. Indeed they only collide with metal molecules and these can, according to the law on which the calculation is founded neither give energy to, nor take it from the electrons.

As he worked at the vicarage Niels wrote to Harald "I am wildly enthusiastic about H. Lorentz's (Leyden) electron theory."

By the beginning of July, Niels finished the written work on his master's thesis. In the quiet of the vicarage it had taken him just six weeks. He wrote to Harald:

> Many thanks for all your cards. Now I have luckily finished with all the writing. That is very nice indeed, although I cannot say, as a certain Magister [Harald] did, that I am fully satisfied

with the result. The problem was so broad and my pen so easily runs away with me that I had to be content with treating just a few parts of it. But I hope the examiners will let it go through, and I think I have put in a few minor details which are not dealt with anywhere else.

These details are mostly of a negative kind (you know I have the bad habit of thinking I can find mistakes in others). On the more positive side I think I have given some indication of the reason why, and this is perhaps less known to you, alloys do not conduct electricity as well as the pure metals of which they are composed.

I am now very excited about hearing what Christiansen will say to it all. I shall go up and have a talk with him tomorrow and will let you know how it all went off. . . .

It went off very well and Niels had a few days free for pleasure and friends before starting on his doctorate. He went one weekend to visit Harald's close friend Niels Erik Norlund, also a member of Ekliptika. There he met Norlund's sister Margrethe, a beautiful young student with curly blond hair and a manner as gentle as that of Niels Bohr's mother. From that moment on no other girl counted with Niels. And Margrethe for her part when she saw "those wonderful eyes and heard him talk and saw how modest and kind he was," was equally taken with Niels. It was the beginning of an almost idyllic love that never wavered for the remainder of their lives.

But Niels had to get back to his studies. The electron theory of metals had proved so rich in possibilities that he decided to continue working with it for his doctor's degree. He and Harald, who was then back in Copenhagen to complete the work for his Ph.D. degree often studied side by side.

Harald was ready first, to defend, as the Danes say, his doctor's dissertation, "Contribution to the Theory of Dirichlet's Series," in January, 1910. Under the Danish system the candidate was examined on all his proposals by various members of the faculty, and had to defend successfully all the propositions he had set forth. Held in a large room the examination was attended by friends, relatives, and classmates. Harald's audience certainly was one of the most unusual to attend a doctoral examination in mathematics. In it were most of the members of the Olympic football team, who cheerfully admitted that they did not understand either the subject or a word that was said, but they were there to back up Harald just as they would on the football field. The "shock-headed" Dane—with his hair carefully combed this time—went through his examination as successfully and easily as he did a soccer game. His "silver medal" this time was a doctor's degree.

Niels was making such rapid progress that he was ready to start writing his dissertation in the spring of 1910. He went back to the vicarage on Funen to get the work done, but even in that quiet setting he found himself working with such intensity that he wrote an imploring letter home begging his family to understand that he could not take the time to answer their letters. They understood; they knew that Niels did not easily dash off letters even to them but generally made a rough draft before the final copy. Harald had once seen a seemingly finished letter lying on Niels's desk and asked if it should not be mailed. Niels glanced up "Oh no, that is just one of the first drafts for a rough copy."

Niels began his work on the doctorate with a paradox. Electricity flows easily through metals, but its passage is not followed by any sensible transport of chemical matter. As Bohr dug into the prevailing theories he found that some of the calculations were not "perfectly rigorous" and that in some cases the agreement between theory and experiment on "many essential points" was "very unsatisfactory." "It therefore seems of interest," Bohr said, "to treat the electron theory of metals on more general assumptions."

Niels undertook to develop the principles of electron movement. He went into the conditions that would affect their movement and calculated that the spaces between atoms would be very large in comparison to the areas in which the electron and the atom would affect each other and where most of the collisions would occur. When he made these calculations he found that they agreed with some of the observed properties of metals. At other points fundamental difficulties seemed to arise. The stability of metal atoms could not be explained by the prevailing laws of motion and radiation, and Niels said straight out that no explanation of magnetism seemed possible with the existing electron theory.

Niels Bohr had begun with a paradox and he was ending with another—a paradox that would not be wholly resolved for fifty years. Bohr also had adopted a lifelong principle—the answer had to be sought in a study of the general principles.

The day of Bohr's examination was set early in 1911. One of the Copenhagen newspapers carried a revealing report of an examination that was as unusual in its way as Harald's had been in its special audience. "Yesterday," said the news account,

the late Professor Bohr's other son defended his doctor's thesis "Studies in the Electron Theory of Metals" for the doctorate of philosophy.

This was the 26-year-old master of science Niels Bohr, who after only one and a half hours was able to leave the University as a Ph.D. Professor Heegaard was his first opponent. He dealt with the linguistic side of the thesis and had nothing but praise for the erudite treatise. Professor Christiansen continued with a more specialized opposition, but it can be called such only in the figurative sense of the word.

Professor Christiansen spoke in his usual pleasant way, told little anecdotes and went so far in his respect for Niels Bohr's word as to regret that the treatise had not appeared in a foreign language. Here in Denmark there was hardly anybody well informed enough about the electron theory of metals to be able to judge a thesis on this subject.

Dr. Bohr, a pale and modest young man, did not take much part in the proceedings, the short duration of which is a record. The little Auditorium III was overflowing and people were standing right out in the corridor of the University.

The words Bohr had written and the questions he had raised were literally so new and unusual that no one was equipped to question them.

A wonderful summer followed with Margrethe and Harald. They went sailing, for walks in the country, and laughed and played at Tivoli. All the while young Dr. Bohr was planning for the fall when he would go abroad to continue his studies. Well informed of the work going forward in both Germany and England, Bohr did not hesitate. There was no question in his mind but that he wanted to go to Cambridge to study with Thomson, the discoverer of the electron. In that direction lay the future—the future of science and of Niels Bohr.

3 /

Cambridge and Manchester

$\mathbf{A}$ s the London train came to a stop and Bohr stepped to the platform his eyes turned to the simply lettered sign "Cambridge." Cambridge—the center of science and learning, the Cambridge of Thomson, of Darwin and Newton, of tomorrow and a thousand years. He, Niels Bohr, in Cambridge! Spenser's words came to his mind:

> My mother Cambridge whom as with a crowne
> He doth adorne and is adorned of it
> With many a gentle muse and many a learned wit . . .

That night Bohr wrote to Margrethe: "I must tell you a little how excited I was when I saw the name Cambridge." He continued to gaze wonderingly at the magic name whenever he saw it on a sign or shop window.

With only the minimum delay to find rooms and unpack his belongings Bohr went to call on the man whose brilliant work had brought him to Cambridge, Joseph John Thomson. In that autumn of 1911, J.J., as he was known to his colleagues and to his students when he was not within hearing, was only 55, but he had been the head of the Cavendish laboratory for twenty-seven years. He had been selected, at the phenomenal age of 28, to succeed the great Lord Rayleigh and Rayleigh's predecessor the first Cavendish professor, the illustrious Clerk Maxwell.

J.J. lifted a thin, intent face to greet the young Dane. The professor's hair was longer than the fashion of the day permitted, his toothbrush mustache was shaggy. His tweed coat and wing collar were correct for the head of the Cavendish, though they were carelessly worn. But as Rutherford, as a new arrival, had observed: "Thomson was pleasant in conversation and . . . not fossilized at all." Thomson's cordial greeting relieved the strain and trepidation Bohr felt at meeting the man who stood at the top of his world.

Bohr carried with him a copy of his dissertation and a copy of one of Thomson's papers. The young Dane knew very little English and thought that the best way he could return the cordial greetings of the Cavendish head and get around language difficulties would be to point to some errors he had discovered in Thomson's work. He was sure Thomson would be eager to know about them. Opening Thomson's paper Bohr put a finger on what he thought was wrong.

Bohr also expressed the hope that the professor would be interested in his own work and that if he should find it worthy, that it might be published in England. (A non-physicist friend of Bohr's had made the translation and, for example, had turned Bohr's "charged particles" into "loaded particles.") Thomson, whatever his feelings, accepted Bohr's paper and put it on top of a stack of papers on his desk.

That night Bohr wrote to his fiancée. He told her how excited he was about the meeting with Thomson and how kind Thomson had been. But Bohr had an inkling that his beginning might not have been of the best. "I wonder," he wrote, "what he will say to my disagreement with his ideas." In after years when he did not have to wonder but knew, Bohr often told the story as an illustration of how not to begin an interview with a distinguished foreigner. Bohr would laugh heartily at his own naïveté.

Bohr began work on studies and experiments proposed by Thomson. As he did, he waited to hear what Thomson would say about his dissertation. When he heard nothing he decided to seek another interview. Thomson had not had an opportunity to read the paper, he explained, but once again he was cordial. He spoke of Bohr's father and invited the young scientist to dinner.

Bohr's work in the laboratory was not going very well, and he began to doubt as the autumn went along that it would be productive. Nevertheless he remained hopeful; perhaps when Thomson read his paper everything could be adjusted.

Again Niels wrote to Margrethe about his meeting with Thomson: "He was extremely kind. I believe he thought there was some sense in what I said. He asked me to dine at Trinity on Sunday. You can imagine how happy I am." The new uplift in Bohr's spirits was relatively short-lived. A week later he was writing to Margrethe "I'm longing to hear what Thomson will say. He's a great man. I hope he will not get angry with my silly talk."

Though his work and his efforts to get his ideas before the great J.J. were trying, Bohr was reveling in Cambridge itself. By October the weather had turned crisp. The shrubs in the meadows were red with ripe berries and the great oaks on the rolling meadows were taking on the reds and yellows of the autumn. "I went for the most beautiful walk," Niels wrote Margrethe. "The meadows were lovely, but most wonderful of all was the autumn sky, with its great driving masses of clouds."

Even a Dane who knew the bite of the winds that could sweep across the North Sea found the cold penetrating. But glowing from a fast walk in the fresh air, Bohr would come in with a heightened sense of well-being and a keen appreciation of shelter and a crackling fire. "With a fine fire and your picture smiling at me," he said in another of his frequent letters to his fiancée, "I felt I was all right."

On some evenings Bohr settled down before his fire to read *David Copperfield*. Dickens was his preferred teacher of English. Bohr had studied English at school and spoke it with a fair degree of proficiency, but he was constantly aware of how far he was from getting all of the words in the right order, or finding the exact expression that he needed. He remembered his father's reading Dickens to him and his brother as they were growing up, and with pleasure both in the language and the story he returned to Dickens to improve his English. Whenever he encountered a word he did not know he stopped and looked it up in the dictionary.

His rapid improvement in English served him well when he took up the required social rounds in Cambridge, for in 1911 calling was obligatory. Although the social duty fell most heavily on the wives of the faculty, even a young bachelor scientist was not excused from it. Bohr went to tea at some of the Cambridge houses and found that despite his apprehensions he could get along very well. "It is strange that I who am so stupid in these things could converse," he wrote with obvious wonder to Margrethe. "But it was not my doing. The English ladies are geniuses in drawing one out."

Getting along in the laboratories was more difficult. The Cavendish then had been accepting foreign students for some sixteen years, but there were no procedures for helping them with the special problems they encountered. And Dickens often did not yield the words needed in the laboratories of 1911.

Equipment was another problem; that available was simple and scarce. It was expected that everyone would devise and make most of his own apparatus. The entire expenditure for galvanometers, ammeters, pumps, and similar apparatus did not exceed £250 a year; and all of this money came from laboratory fees earned for teaching and examining. No one would have thought of seeking private support; in fact, J.J. had a decided aversion to what he called "begging."

When Bohr needed some special glass tubes for his experiments he had to blow them himself. He had developed considerable skill working in his father's laboratories, and in making the glass tubes he used for his experiments on the surface tension of water; but the glassware necessary here was another matter. Bohr struggled to blow the tubes he needed, but could not quite succeed and had to take some lessons in glass blowing.

Once the tubes were ready the experiment still did not work out. Should he go to Thomson about it? He also continued to worry about his paper. If he sought Thomson out, about all that he could ask was, "Have you read my paper?" Nevertheless Bohr forced himself to go. When he walked into Thomson's office he saw his paper sitting at the bottom of a pile on Thomson's desk. Bohr was disheartened; he did not know that Thomson was notorious for being careless about papers and letters. J.J., who undoubtedly was suffering qualms about not having read Bohr's paper, said that he was just reading it, and even talked about the possibility of having it published. Once again Bohr's spirits rose and he wrote Margrethe "I am so happy."

Thomson did submit Bohr's paper to the Philosophical Society, whose publications had carried scientific material since the time when science was called natural philosophy. The editor, however, held that the dissertation was too long for their pages. He suggested that Bohr cut it in half. Bohr believed that the points he was making could not be substantiated in half the space, and unhappily refused the emasculation. Thus the paper was not published in English then or later. Several years later other scientists repeated much of the work Bohr had done, having no way of knowing that Bohr had already covered the ground.

A snowy Christmas came to Cambridge. Bohr, alone and far from the warmth and gaiety of the Danish holidays, let his memories travel back to his childhood. In this mood of reminiscence he wrote to Margrethe: "I see a little boy in the snow covered street on his way to church. It was the only day his father went to church. Why? So the little boy would not feel different from other little boys. He never said a word to the little boy about belief or doubt, and the little boy believed with all of his heart. . . ." Bohr wanted to share his childhood with Margrethe.

Soon after Christmas Rutherford came to Cambridge to speak at the annual Cavendish dinner. The songs rang out. To the tune of "Clementine," the Cambridgians lustily sang:

> In the dusty lab'ratory
> Mid the coils and wax and twine,
> There the atoms in their glory,
> Ionize and recombine.

> Chorus: Oh my darlings! Oh my darlings!
> Oh my darling ions mine!
> You are lost and gone forever
> When just once you recombine!

Bohr sat far back on the benches flanking the long oak tables, singing as heartily as the rest and watching Rutherford. Rutherford, at the high table, was broad-shouldered, ruddy, a dominant figure even against the carved oak paneling and the portraits of the great of England and Cambridge that covered the walls.

Only a short time before, in May, 1911, Rutherford had discovered that the atom had a nucleus. It was comparable to discovering an unknown continent, though the scientific world as a whole still had not fully realized the immensity of the finding.

As Rutherford was being introduced, there were a few words about the discovery, of which Rutherford would speak, and the introducer added that Rutherford held another distinction. Of all the young physicists who had studied at the Cavendish none could match him in swearing at the apparatus. Rutherford's laugh boomed out across the hall to fill the room and echo from the high beamed ceiling. Bohr immediately liked the big hearty man whose work he already regarded with reverence.

For several years previously, Rutherford and his assistant Hans Geiger had been shooting beams of alpha particles at various targets. If there was nothing in the way, the particles flew as straight as bullets to the target, a screen of zinc sulphide. As each atom struck the screen a tiny point of light, a scintillation, could be seen and counted. However when a thin sheet of metal was placed in the way of the beam, some of the particles did not strike the screen as they would have if moving in the usual straight-line path. Some of them were deflected and hit somewhat to the side.

Rutherford suggested one day that young Ernest Marsden, then 20, try to find if any alpha particles could be scattered through a large angle. Two or three days later Marsden rushed in to say that he had discovered some of the alpha particles coming backward! Rutherford soon confirmed this occurrence. "It was quite the most incredible event that ever happened to me in my life," said Rutherford. "It was almost as incredible as if you fired a fifteen-inch shell at a piece of tissue paper and it came back and hit you."

What threw the alpha particles back? What could stop a particle rushing through space at a speed of some 10,000 miles a second? What was there in the atom to halt such a projectile? Rutherford experimented carefully and thought deeply. He did not rush to conclusions, but one day early in 1911, he walked into Geiger's office humming "Onward Christian Soldiers" as he did in moments of singular well-being or triumph. "I know what the atom looks like," he told the astounded Geiger.

The atom Rutherford could "see" was made up of a tiny center, with a swarm of electrons wheeling around it at great distances. In one of the great feats of scientific imagination Rutherford visualized the atom as a miniature solar system. It was an imposing insight. No one before had suggested that the smallest unit of matter might repeat the architecture of the greatest.

If this were the pattern of the atom, the atom they had so long thought of as solid and unalloyed, a speeding particle encountering the small, massive, charged center might well be hurled back, exactly as the alpha particles were in the laboratory. But out of the 34,000,000,000 alpha particles emitted by one gram of radium in one second, only a few would collide with the little center, and only those that hit would be deflected. The others would shoot unimpeded through the vast openness of atomic spaces. The chances of a nuclear hit, Rutherford calculated, would be very small.

"Since the alpha and beta particles traverse the atom, it should be possible from a close study of the nature of the deflection to form some idea of the constitution of the atom," Rutherford argued.

On the same day Geiger began an experiment to test Rutherford's startling but logical idea. Was there a relation between the number of scattered particles and the angle of scattering? Confirmation came rapidly. In February of 1911 Rutherford gave a brief report of his epoch-making theory, and in June his complete theory was presented in the *Philosophical Magazine*. A few months later when Rutherford spoke at Cambridge about the work, Bohr decided as quickly as he would about a move in a soccer game that he wanted to work with this brilliant man whose ability to penetrate to the heart of scientific truth was almost unerring.

Bohr, however, was never rash or precipitate in his actions. A few weeks later he told a few friends in Cambridge that he was going to Manchester to call on a friend of his father's. It was not accident that this colleague of the late professor's also knew Rutherford and quickly arranged an introduction for the young Dane.

Rutherford greeted Bohr warmly. He had just returned from the first Solvay Conference at Brussels, that conference of top physicists sponsored by Ernest Solvay—and there for the first time he had met Einstein and Planck. Bohr listened entranced to Rutherford's account of the meeting. Rutherford was more convinced than ever that the prospects for physics were nearly unlimited. Rutherford was a man of a natural, swift-flowing optimism, but he was also a realist. He cited the developments of the last few years—the discovery of radioactivity, of X rays, of the electron, of the center of the atom which he was beginning to call the nucleus. Bohr drank in every word, and Rutherford, whose judgment of men was almost as accurate as his scientific insight, quickly consented to Bohr's joining his laboratory group in the early spring. By that time, Bohr said, he could finish the work he had started at Cambridge.

Bohr arrived in the industrial city at the beginning of April. The contrast with the mellowness and stateliness of Cambridge was marked. Factories crowded the streets and their smoking chimneys often turned white to black in a winter's time. But the population was about 600,000 and the city was a bustling one. Market Street was said to be the most crowded in Europe. Heavily loaded drays filled the roadways and sparks flew as the shoes of the horses struck against the cobblestones.

"This is a pretty active place," Rutherford explained. "Except for the climate it has a number of advantages—a good set of colleagues, a hospitable and kind people, and no side anywhere." Even more, it had one of the best physics laboratories, a laboratory established by Arthur Schuster, who, in addition to being a physicist, was a rich man and willing to use some of his own funds to equip the laboratory. He had also reached out to Montreal to choose Rutherford as his successor. Rutherford had accepted a post in Canada after completing his graduate work at Cambridge, but was induced to return to Europe, the center of physics. Rutherford was pleased with Manchester and Manchester with Rutherford.

Bohr was equally happy there. Rutherford suggested that he enroll in a course on experimental methods of radioactive research which was being given by Geiger, Marsden, and Walter Makower. The class, whatever its announced subject matter, did not long stay away from the new atomic model. The new structure of the atom dominated all of the talk and a large part of the experiments.

Late each afternoon the whole laboratory staff got together for tea. The tea, the cakes, and the thin slices of bread and butter were spread out on one of the laboratory tables and everyone gathered around. Generally Rutherford perched on a stool, and generally he led the talk. Most of the time the discussion concentrated on the atom and radioactivity, but it could range off in any direction, to the state of the world, or the latest production at the Gaiety, Manchester's repertory theater, to Monsieur Colbert's excellent dinners at the Midlands Hotel, or to Rutherford's new car, a 1911 Wolseley-Siddeley which did fourteen to sixteen miles an hour. Nearly always the talk returned to physics.

The newest student was free to speak up. Rutherford was willing to listen to anything that "made sense." The only thing he could not abide was what he called "pompous talk." Bohr was often in the middle of the discussion, thinking out, as he talked, the implications of the newly discovered structure of the atom. Now that the structure was known, he argued, it should be possible for the first time to arrive at some explanation of the properties of the elements. The answer to their differentness should lie in the way the atom is made. Perhaps, he said, it might be possible to determine why some elements are metals, why others are gases, why some easily combine and others do not, and thus why all the variety of the earth ultimately has its distinctive form. As Lucretius had said, Nature "resolves everything into its component atoms," but there the trail had ended. No one could say why the atoms differed.

Bohr suggested that Rutherford's alpha and beta radiation came from the nucleus, and that the arrangement of the electrons wheeling around the nucleus like planets might determine the ordinary physical and chemical characteristics of all the elements. Bohr's tea often grew cold while he developed these radical ideas.

At the moment George von Hevesy, a young Hungarian who had worked with Rutherford at Montreal and had come to Manchester with him, was trying to separate some of Rutherford's new radioactive materials. Rutherford had given him the assignment and with a clap on the back had told him that if he were worth his salt, he would get them apart.

Despite the jibe Hevesy found the job impossible. His failure, though, gave him another idea. The radioactive material, in combination with its nonradioactive counterpart, might supply a means of tracing the two through any body in which they might appear. The radiation emitted by the radioactive member of the pair would provide a clue to the position of both.

The inseparables also gave Bohr an idea. He connected Hevesy's problem with the new structure Rutherford had discovered in the atom. Perhaps the inseparables differed only in the nucleus. Perhaps they had exactly the same number of electrons circling the nucleus. If the chemical properties of an element were determined by its electrons, any two with the same number of electrons would be chemically alike. They would be inseparable.

Hevesy was half convinced and Bohr decided to discuss his daring ideas with Rutherford. Rutherford as usual was quite willing to listen. As Bohr said, he was always interested in "any promising simplicity." But Rutherford, too, was unconvinced. As an experimentalist he had an innate distrust for theorizing, however logical it might seem. He cautioned Bohr against building up too much theory from the comparatively meager experimental evidence.

Soon, however, the "meager" evidence ceased to be meager. Frederick Soddy, Kasimir Fajans, and A. S. Russell all independently pointed out that when an element emits a beta particle it is transformed into an element with the chemical properties of the next element above it in the atomic table. Suddenly everyone saw that the loss of one unit of negative charge would be the equivalent of a gain of one unit of positive charge. Suddenly it was also obvious that the loss of an alpha particle with two units of positive charge would lessen the atomic number by two. Thus an element could be transformed into another element one place above it or two places below it; it could, according to the circumstances, move up one or down two. It was further apparent that if part of the nuclear weight were lost, but that there was no change in charge, the element would change in weight, but not in appearance and properties. It would look the same.

Russell presented a paper on the subject to the Chemical Society in the late autumn of 1912. A few months later Soddy announced the working out of what was called the "radioactive displacement" law. The word "isotope" was invented to describe the twin elements, and some years later Soddy received the Nobel Prize for chemistry for his discovery.

Bohr always remembered that he had on his own arrived at a general idea of radioactive displacement and isotopes, but he did not complain. His admiration for Rutherford was not even slightly diminished because Rutherford had discouraged him from going ahead with his "theory."

At the time Bohr was absorbed in another problem, or in a different phase of the same problem. He had another "little idea," as he wrote to Harald on June 12:

> Things are not too bad at the moment. A few days ago I had a little idea about the absorption of alpha rays (it so happened that a young mathematician here, C. G. Darwin—the grandson of the right Darwin—has just published a theory on this question. It seemed to me that it was not only not quite right mathematically (this was however rather trifling) but quite unsatisfactory in its basic conception.) I have worked out a theory about it, which however modest, may perhaps throw some light upon a few things concerning the structure of atoms. . . .
>
> I am considering publishing a little paper about it. You can imagine it is fine to be here, where there are so many people to talk with . . . and this with those who know most about these things; and Professor Rutherford takes such a lively and effective interest in all that it seems to him there is something in. In late years he has worked out a theory of the structure of atoms, which seems to be quite a bit more firmly founded than anything one

has had hitherto. And not because mine is anything of the same significance or the same kind, yet my result does not agree so badly with his (you understand that I only mean that the foundation of my little calculation can be brought into agreement with his ideas). . . . I have so many things that I should like to try . . . but they must wait. . . ."

Fajans and Soddy had explained the stepping up and stepping down of atoms as beta and alpha particles were thrown off, but they did not see that this change had anything to do with the structure of the atom as Rutherford had pictured it. Fajans even insisted that his finding argued against the alpha and beta rays coming from the nucleus. Bohr thought, on the contrary, that the change in the elements after radiation constituted clear proof that the particles came from the nucleus. The nucleus, he maintained with growing conviction, might be the seat of radioactivity.

Bohr again went farther. The electrons orbiting around the nucleus would determine all the chemical properties of an element, they would make it what it is, a material with the rigidity of iron, or a soft malleable solid, or a gas. Here would lie the explanation of the characteristics of each of the elements.

But could there be such a structure, could electrons orbit around the nucleus? This question had to be answered first. According to Newtonian mechanics, an electron whirling around the nucleus would lose energy as it whirled. As its energy was dissipated in radiation the electron would begin to spiral downward until in the end it collapsed into the nucleus.

If this happened there would have been a complete collapse of matter; Newton's "Nature of things" would have been wholly changed. Obviously it did not happen. The world and the universe are remarkably stable, elementally unchanging. Iron remains iron, and the other ninety-two known elements are themselves, even when their atoms are broken. Somehow the elements always seemed to reconstitute themselves. Bohr argued with his friends in the laboratory that scientists simply could not reconcile reality and the classical principles of mechanics and electronics, for according to classic theory collapse was inevitable.

Bohr with his European training and awareness was not especially surprised. Twelve years before, in 1900, Max Planck had demonstrated another case in which the classical theories did not work. Six years later Albert Einstein had confirmed and extended it. Planck and Einstein, through their calculations, proved that thermal radiation and light are not continuous, but are made up of individual packets of energy. The units were named quanta, and the scientists showed that each kind of radiation might have its own special size and shape of unit, just as each country may have its own set of coins. The new theory basically shook scientific belief and the long honored assumptions of science. So novel and strange was the quantum theory that it was accepted very slowly.

As a young student of physics free to read as he chose, Bohr had studied the Planck and Einstein quantum theories with close attention. In 1912 Bohr had not in the least forgotten their lesson. He saw that the usual rules also might not apply to the structure of the atom. The atom was another scale, another world in which the proved rules might not hold.

Looking back over these days, Bohr later wrote: "In the spring of 1912 I became convinced that the electronic constitution of the Rutherford atom was governed throughout by the quantum of action."

Bohr began to make the calculations that would show his theory of atomic structure right or wrong. He worked on them day and night. His English friends trying to drag him away for dinner made no headway and accused him of never emerging from the laboratory.

Niels stopped only to write to Margrethe. "It doesn't look so hopeless with those little atoms, though the outcome of the calculations has its ups and downs," he wrote on July 5.

But there was always one more question or point to face. Bohr drove ahead. By the 17th of July he was writing his theory and figures down on long, ruled sheets. To keep the sheets from becoming mixed as he finished them, he pasted them together in a long roll.

He wrote to Harald: "Things are going rather well, for I

believe I have found out a few things. To be sure I have not been so quick to work them out as I stupidly thought."

Niels and Margrethe had set the date of their wedding for August 1, and Bohr was going to leave for Copenhagen on July 24. He was determined to put the paper in shape to show to Rutherford before he had to leave. As he said to Harald in a letter, he was not "suffering from any lack of something to do."

By July 22 Niels felt that he had his paper sufficiently ready for Rutherford. He sought out the scientist in his office, and though Rutherford was deep in work on his own soon-to-be-famous book *Radioactive Substances and Their Radiations* he willingly stopped to hear what Bohr had to present. Rutherford had told some of his associates "This young Dane is the most intelligent chap I've ever met," and besides Rutherford liked him. Bohr's modesty, his total lack of "side," and his friendliness were all to Rutherford's taste. Bohr began with a bow to his professor, though as both of them knew it was not a simple compliment, but a statement of fact.

With a little misspelling of English, Bohr said: "According to the atom-model proposed by Professor Rutherford to explain the 'big scattering' of alpha particles, the atoms consist of a positive charge concentrated in a point . . . which total charge is equal to that of the positive kern; the kern is also assumed to be the seat of the mass of the atom."

In such an atom there could be no equilibrium unless the electrons surrounding the *kern*—their early name for the nucleus— were in motion. Rutherford nodded in agreement. The question then was what kind of motion? This, Bohr explained in his own brand of English, was what he had set out to "determinate." Bohr reviewed the classical law which postulated collapse. Of course this did not happen; matter retained its form. Rutherford still was in agreement. "The question of stability must therefore be treated from a different point of view," Bohr continued.

Bohr had to prove or demonstrate that electrons circling a nucleus could continue in their orbit without losing energy. He showed the professor his calculations indicating that from one to seven electrons could circle in a stable ring. If there were more than seven there would be an imbalance, he thought. "It is therefore a very likely assumption that an atom consisting of a single ring cannot contain more than sewen electrons," said Bohr, spelling by ear. Every time an extra electron was added, Bohr argued, a new element would be formed, and when one ring had acquired its maximum of seven elements another ring would be created and this might mark the emergence of a new group of elements.

"This," said Bohr reading from his pasted-up sheets, "seems to offer . . . a possible explanation of the periodic law of the chemical properties of the elements by help of the atom-model in question."

In these few guarded and tentative words Bohr was proposing an answer before an answer was actually possible, to one of the oldest and most baffling of problems—the differences in the elements and why each is distinctive. Bohr was also suggesting a new explanation for the stability of the atom and thus of all that is made out of the atom. Such basic ideas have seldom been broached in so short and modest a way.

Bohr went a little farther—here might lie a clue to the chemical reaction between two atoms, say, how water is formed from hydrogen and oxygen. He sketched for Rutherford what might happen when an oxygen atom with its rings of electrons combined with two hydrogen atoms and their rings. Two such systems could fit together, he ventured to say, though he had no proof.

Bohr was forming clear if preliminary concepts of the atom and its orbiting electrons and all that these planetary systems implied to the nature of matter. He still had no clue to the nature of the radiative process. Nor did he use the word quantum, though his calculations and his own later accounts show that he was thinking in Planck's terms.

Rutherford listened carefully as Bohr outlined his ideas. The professor as always distrusted theory. He cautioned Bohr against putting too much weight on his model of the atom. He again warned him against extrapolating from the comparatively meager experimental evidence. But having thus done his duty to his student, he eagerly discussed Bohr's ideas and urged him to prepare a paper for publication. Bohr was elated. As he rushed to collect all his materials at the laboratory and to pack he dashed off a note to Margrethe saying that Rutherford had liked his first presentation, and that in another day he would be on his way to Copenhagen. He sailed as planned on July 24.

Bohr had spent only four months in Manchester. In that brief time he had formulated ideas that would soon direct the revolution in physics. He had also found the model for his life and work, though he could not sense it any more than he did the effect of his "little ideas."

Bohr thereafter thought of the laboratory as a place where the ablest young students from all around the world gathered around a scientist who encouraged them to reach the greatest heights of which they were capable. A laboratory also was a place where the truth was sought in the freest of research and discussion and with the most rigorous adherence to the facts as they can be developed. And however undeviating the purposes, there could be laughter, jokes, and pleasure with friends and colleagues. Nor was the laboratory an ivory tower—politics was followed there with almost as much interest as an alpha particle. As he left Manchester, Bohr knew that he knew what a laboratory should be.

Bohr arrived in Copenhagen in a happy flurry, and on August 1 he and Margrethe were married. Bohr had good reason to believe that Margrethe of all brides was one of the loveliest. By any and all definitions she was beautiful, slender, with nearly classic features, and ashen blond hair that curled softly around her face. But Margrethe's was no surface beauty alone. She was the sister of a mathematician and she had a keen intelligence that would make it easy for her to take part in her husband's work. And added to all of this was an aliveness and warmth that drew even strangers to her.

Immediately after the wedding the Bohrs went to Norway to spend a few days. There Bohr finished the work on his first paper. He dictated it to his bride who copied it in a clear, legible hand, and smoothed out the English as she went along. The system worked so well that a pattern was set. Margrethe became her husband's secretary, and all of his early papers and correspondence are in her writing. She was not a physicist and always insisted that she did not attempt to follow the intricacies of the science, but in fact she had a very excellent understanding of the principles and listened intelligently as her husband talked. She knew the scientific language; she also knew all of the people with whom Niels worked and was his constant consultant. In addition she became the most gracious of hostesses.

The honeymooners went on to England to take the completed manuscript to Rutherford. "Both Rutherford and his wife received us with a cordiality which laid the foundation of the intimate friendship that through the years connected the families," Bohr later explained.

The formal sentence did not begin to describe how delighted the Rutherfords were with Margrethe. Rutherford was so taken with the bride that he almost neglected talking physics with the young husband. It was clear to the Rutherfords that Bohr had found exactly the right wife. Rutherford also had praise for the completed paper.

Filled with happiness, Niels and Margrethe went on to Scotland to finish their wedding trip. A prospect almost too good for two modest people to contemplate seemed to open before them. At the same time Bohr knew that the work ahead would be overwhelmingly difficult. Innumerable points still had to be explained about the atom, but he knew that he would try.

4 /

The Trilogy

THE LEAVES WERE FALLING from the Copenhagen beeches when Niels and Margrethe returned from their wedding trip. But their welcome from families and friends was as warm as the fall winds were chill. They soon found a small flat and settled there as quickly as possible, for Niels was eager to get on with his work with the least possible delay.

There was no question about what Bohr would do. His scholarship, his whole training and background had prepared him for the university. He was at once appointed to an assistant professorship in the University of Copenhagen, and it was agreed that he would give a course of lectures on "The Mechanical Foundation of Thermo-Dynamics." The lectures would begin on October 16 and run through December 18.

Preparing them took time, and Bohr struggled to find additional time to develop his ideas about the atom. Before he could put his theories into an additional formal paper, innumerable points had to be tested and reconsidered from every possible angle.

The work proved difficult. On November 4, 1912, Bohr wrote to Rutherford: "I am very sorry that I have not been able to finish my paper on the atoms and send it to you." In his rough draft of the letter Bohr said, "I have made some small progress with regard to the question of dispersion," but before he gave it to Margrethe to copy and put into final form for mailing he crossed out the "small."

He had in fact made considerable progress, and he added "I hope to finish the paper in a few weeks."

Rutherford was a prompt correspondent. He liked to act on the moment, and at the time this was relatively easy for him; he had a secretary. In his reply exactly one week later on November 11, he suggested that Bohr did not have to feel rushed: "I do not think that you need to feel pressed to publish in a hurry your second paper on the constitution of the atom, for I do not think that anyone is likely to be working on that subject. . . . I hope you will be successful in overcoming your difficulties."

Bohr thought exactly the opposite about the likelihood of others working on the same subject. Papers were beginning to appear on closely related themes and he believed it inevitable that others would ask the most profound, and the most essential question of them all: what is the form of the atom and thus of the universe? The question in fact was so momentous that Bohr did not see how others could avoid it.

A few days before Christmas proof came that others were coming near or at least were working on phases of the atom's structure. Bohr received copies of a series of papers by J. W. Nicholson. The English scientist was studying the sun, but in the last analysis the sun and atoms were one. Nicholson argued that it is possible to account for certain spectral lines in the solar corona and the stellar nebulae by assuming the presence of atoms made up of a small nucleus and a surrounding ring of electrons. He also suggested that the solar atoms might emit "pulses" of energy, or packets equivalent to Planck's quanta.

At first reading Bohr's heart sank. Nicholson seemed to deny all that he was postulating. "Am I altogether wrong?" Bohr agonizingly asked himself.

If the solar atoms continued to give out bursts or pulses of energy, as Nicholson held, ultimately there would be collapse. No other ending seemed possible, whereas all of his own work indicated that the atom did not deflate into nothingness, but rather maintained its form and structure.

Or was there another implication? Bohr snatched up Nicholson's papers and began to read them again. He jotted some calculations on his blackboard. Perhaps there was a way out. If the systems Nicholson was considering existed only in places where atoms were being continuously broken up and formed again, say in a star or in a excited vacuum tube, there might be no conflict. His own theory applied only to atoms in their permanent state, not to atoms in the extreme conditions of death and birth. And if this were correct, the Nicholson papers would not be a denial of his theory, but on the contrary a confirmation of it. At this insight Bohr was jubilant.

It was only two days before Christmas and Bohr's mood matched the carols and the playing of the Copenhagen bells. He must tell Harald what had happened and send him Christmas greetings too. Bohr picked up a plain gray card and began: "Dear Harald, a Very, very merry Christmas from Margrethe and Niels." "P.S." he added in a small script to match the limited space: "Although it does not belong on a Christmas card, one of us would like to say that he thinks Nicholson's theory is not incompatible with his own."

Niels explained that Nicholson dealt with radiation in pulses, but that his own calculations would be valid for the final chemical state of atoms. And then he added another "P.S.": "again a very merry Christmas." It was one of the most unusual of Christmas cards, as Harald well realized. He saved it.

With the hardest of work, Bohr was able by the end of January to send Rutherford a final copy of his paper on alpha rays and to report that he was making progress on his paper on the atoms.

Early in February a letter arrived from Hevesy enclosing two of his papers on the chemical properties of the radioactive elements. They confirmed what Bohr called his "point of vieuw." Hevesy had asked too how Bohr was getting along.

In a reply on February 7 in which he expressed his appreciation of Hevesy's "beautiful results" Bohr reported on his own work: ". . . In answer to your question, I shall try to characterize the ideas I have used as the foundation of my calculations.

"As you mention in your letter the main difficulty is the question of the stability, or as it can also be expressed from a somewhat different point of view, the question of the dimensions of the system of electrons surrounding the nucleus."

The stability question, Bohr said with new firmness, simply could not be solved by the old rules. He was also ready to say that the problem could be solved in a new way.

"It can now be shown," Bohr continued, "that taking Planck's theory of radiation into account, we can in a simple way get an answer to our question."

Bohr explained that he was assuming that an atom is formed by the successive "binding" of electrons. One free electron after another would be drawn into the atomic solar system until the number of electrons equaled the charge of the nucleus and the whole system was rendered neutral. He cited the formation of a helium atom, which would be created if a nucleus with a charge of two captured two electrons into orbit.

Bohr assumed that a flash of energy would be radiated out as the captured electron was "bound" into the nuclear orbit. Then he assumed that the energy given off in this action would equal one of Planck's quanta multiplied by the frequency of the rotation of the electron. This gave startling results, startling because they agreed with the figures obtained in experiments.

In this way Bohr saw that he could determine the dimensions of the atom and the number of electrons in the outer ring. He also, he told Hevesy, obtained very close agreement with experiments on X rays, which "according to the theory is the radiation sent out during the rearrangement of a system if an electron in one of the innermost rings is removed," for instance by a scientist's bombardment.

Bohr's work was suggesting even more strongly than when he was at Manchester that the ring structure and its building up might not only make possible an understanding of the periodic system of the elements, but of how and why elements combine. It should be possible, Bohr added, to show why two hydrogen atoms will combine into a molecule "while two helium-atoms won't."

Bohr drew a clear distinction between the nucleus and the

electrons surrounding it. He emphasized to Hevesy too that he was not talking about radioactivity, which really was an explosion from the nucleus. But, he pointed out, the loss of the rays expelled in such an explosion would change the charge of the nucleus. And if the charge varied, the number of electrons bound to the nucleus would change. Hence the chemical and physical properties of the atom would vary.

"The latter relation is just the one you have found in your experiments and your results were therefore what I had expected and hoped," Bohr wrote.

Bohr was well aware that these were fundamental changes of which he was talking. And yet he was a young assistant in Copenhagen, unknown, and well removed from the great centers of science. He did not want Hevesy to think that he was suffering from illusions of grandeur. "I am sure," he wrote, "that you will understand, that I write to you in the same way as I spoke in Manchester, i.e., that I don't speak of the result which I mean that I can obtain by help of my poor means, but only of the point of vieuw—and the hope to and believe in a future (perhaps very soon) enormous and unexpected development of our understanding—which I have been led to by considerations as those above."

But the problems of trying to determine what the unseeable, unreachable atom really is, and why this inconceivably small universe produces visible matter and the visible world were nearly endless. Nicholson's paper, for example, had raised a second question—the possible relation of the atom's structure to the lines of the spectrum.

Bohr's friend H. M. Hansen had been bringing up the same puzzling problem. Hansen, a year younger than Bohr and destined to become a rector of the University of Copenhagen, had just returned from a year and a half of study at Goettingen. He had become a skilled spectroscopist. In Copenhagen he had been appointed an assistant in the physical laboratory of the Polytechnical School and the two young scientists met frequently. Bohr naturally told Hansen about his attempts to decipher the atom.

How, Hansen aeskd at once, can the atom you are proposing,

with its nucleus and orbiting electrons, account for the spectral lines an atom emits when suitably provoked? Bohr's first answer was a short one; "It can't." The bands of color in the spectrum seemed much too complicated for him even to hope to explain on the basis of the atom's structure. The problem was insuperable, he thought.

It was also a very old one. Isaac Newton (1642–1727) had sent a beam of sunlight through a narrow opening into a dark room. Where he let the light fall upon a glass prism, the light suddenly was broken up into bands of color, into a spectrum made up of all the colors of the rainbow, red, green, yellow, blue, and violet, in that order.

Considerably later, in the early part of the nineteenth century; Wollaston in England and Fraunhofer in Germany discovered dark lines in between the colors of the solar spectrum. The darkness could only indicate that certain colors were missing. Then it developed that some of the dark lines were double and that there were hundreds and even thousands of dark lines that could be seen against a light background.

Other phenomena appeared as scientists continued to study this colorful mystery of light. Light from a spirit flame colored with common salt produced a broad yellow line against a dark background. Other elements produced their own distinctive spectra. The spectra were in fact the signature of the elements. No two were exactly the same, and thus the appearance of sodium's special yellow or the distinctive rainbow of barium or calcium or any of the other elements was an infalliable sign of the presence of that element. Even a minute trace of an element could be discovered by these benchmarks.

Nor was this all. Beyond the visible light the human eye could see was an ultraviolet region at one end and an infrared one at the other, each having its own spectrum. Even X rays produced a spectrum when passed through the fine network of a crystal.

Hansen was not deterred. Look at the regularities that the Swiss physicist J. J. Balmer had found and the development of them made by J. R. Rydberg of Sweden, he urged. Hansen explained the Balmer findings to Bohr.

In 1885 Balmer had begun the study of the simplest of all spectra, that of hydrogen. It was made up of only the three bands of color, red, blue-green, and violet. Balmer found to his great surprise that the colored bands exhibited striking regularities; their frequencies (the number of vibrations per second) were proportional to $\frac{1}{2}^2-\frac{1}{3}^2$; $\frac{1}{2}^2-\frac{1}{4}^2$; and $\frac{1}{2}^2-\frac{1}{5}^2$. The number squared in the second term thus increased by one for each successive line of color—3, 4, 5, and so on.

No one could explain the clear regularities. Every attempt had failed. Nor did the Rutherford atom help. If the one electron circling the hydrogen nucleus radiated light as it went around, it should give off a continuous range of frequencies rather than well-separated ones connected by definite laws. Citing all of this, Hansen was quick to ask if Bohr's atom could throw any light on the spectral mystery.

Bohr was growing more and more abstracted as Hansen pointed to the orderly spacing of the bands of the hydrogen spectrum. He had numbered the stationary states 1, 2, 3, and so on. He knew the sizes of each. Suddenly he realized that the differences in energy between his stationary states were exactly the same as Balmer's figures for the spectral bands of color for hydrogen. In an instant of understanding Bohr saw that the spectrum's color might be produced if an electron, with a burst of energy, jumped from one stationary state to another lower one.

"As soon as I saw Balmer's formula," he said, "the whole thing was immediately clear to me.

Bohr had a whole new insight, a whole new clue. It was as though an automobile that could run only on, say, Highway 3, a major cross-country road, suddenly proved to have the powers of a helicopter and could hop across to cross-country Highway 2, a road designed only for driving at lower speeds. If the hop were accompanied by a flash of light, a spectral color might well be registered.

Bohr was not one to yield to such a startling new insight without testing it in every conceivable way. In the midst of all his work, he virtually started all over again. He worked feverishly, night and day, for the most complex calculations had to be made. But he

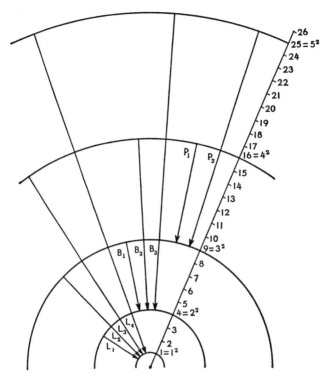

The first four orbits in Bohr's model of the hydrogen atom. The electrons might take any of the jumps indicated.

worked all during the month of February with high effectiveness and creativeness.

Bohr had first to establish his base. If there were no radiation, an electron in orbit around a nucleus would continue on its rounds, exactly as the man-made objects that were later rocketed into space continued on their orbits. On the other hand if the electron radiated energy according to the classic theory, it would soon—in about one hundred millionth of a second—collapse into the nucleus. Bohr then proceeded to demonstrate that atoms in their regular states have fixed dimensions. They do not collapse.

There was, however, a way out of the impasse. Planck had earlier established that other forms of energy are not given off in a continuous stream, but in distinctly separated packets. The same bit-by-bit change might apply here in the infinite world of the atom. "It seems necessary," said Bohr, "to introduce in the laws in question a quantity foreign to the classical electrodynamics, i.e., Planck's constant, or as it is often called the elementary quantum of action."

If the rules denied the actuality, was it not necessary to amend the rules? And when Bohr did amend the rules the amendment worked. "By the introduction of this quantity, the question of the stable configuration of the electrons in the atoms is essentially changed." With these assumptions Bohr could account for the line spectrum of hydrogen, and another massive problem disappeared.

If hydrogen's one electron had been driven out to a great distance from the nucleus, say by an electrical discharge in a vacuum tube, and if the atom were re-forming, Bohr showed that the electron would hop from the outer orbit to a closer one with a flash of light that would produce the bluegreen of the spectrum. If the electron then jumped to a still closer orbit, the quantum of light emitted would produce the red of the spectrum. Calculating distances and energies, Bohr found his theory in complete agreement with the actual measurements of the spectrum.

His theory also predicted exactly the series of lines that would be found in the extreme ultraviolet and infrared, if they should ever be observed.

In experiments with hydrogen in vacuum tubes scientists had found twelve black lines across hydrogen's bands of color. In the hydrogen spectra of some celestial bodies, 33 of the bands had been observed. Bohr's theory predicted that the condition for the appearance of a great number of lines is a very low density of gas. Thus once again theory was borne out by observation. A seeming obstacle further established Bohr's point, or at least opened it to confirmation.

Using his theory, Bohr could not account for a series of lines E. C. Pickering had recently found in the spectrum of the star Puppis. Nor could he explain the series Fowler had found in experiments with vacuum tubes containing a mixture of hydrogen and helium.

However, when Bohr ascribed the lines to helium, he could

account for them completely. It was a point that could be tested. In writing to Rutherford about the possible upsetting of the conclusions of Fowler and Pickering, Bohr suggested that an experiment might be made:

> I have tried to give reasons for ascribing the lines to helium. This point might, however, be tested experimentally. In a discourse in which I proposed my point of view and tried to explain (as I have done in my paper) that the presence of hydrogen in the experiments of Fowler might indirectly be the cause of the appearance of the lines considered, the chemist Dr. Bjerrum suggested to me that if my point of view was right the lines might also appear in a tube filled with a mixture of helium and chlorine; indeed it was suggested that the lines might be still stronger in this case. Now, we have not in Copenhagen the opportunity to do such an experiment satisfactorily. I might therefore ask you, if you possibly would let it perform [*sic*] in your laboratory, or if you perhaps kindly would forward the suggestion to Mr. Fowler, which may have the arrangement used still standing.

Bohr had to face the problem that radiation can be absorbed as well as emitted if his theory were to be valid. He had to assume that a system consisting of a nucleus and an electron rotating around it could, under certain circumstances, absorb radiation of a frequency equal to the frequency of the radiation emitted during the passing of the system between different stationary states.

Bohr was then ready to return to what he called "the main object" of his paper—the "permanent" state of a system consisting of a nucleus and bound electrons. According to his theory the only neutral atom containing a single electron would be the hydrogen atom. But there was very little experimental data on it; for a closer comparison experiment he had to consider more complicated systems.

"Considering systems in which more electrons are bound by a positive nucleus, a configuration of the electrons which presents itself as a permanent state is one in which the electrons are arranged in a ring around the nucleus."

If there were more than one electron on a ring, Bohr demonstrated that each would be at an equal interval around the circumference, like so many evenly spaced children playing ring-around-arosy. But would such a ring, made up of a number of evenly spaced "players" be stable? According to the old theory it would not be; according to Bohr's theory it would. And Bohr's theory accorded with the scant available experimental evidence.

Bohr was dealing constantly with the unseeable and the unknown. He had to reason from very few clues and the reasoning had to be of the most meticulous. He had laid the basis for a theory of the constitution of atoms and molecules, but he decided that he would have to present it in a second section of his paper. The first paper was already getting longer than the journals liked. And besides if he delayed, he might be superseded. Bohr had written to a friend: "I am afraid that I must hurry if it is to be new when it comes; the question is such a burning one."

Bohr reluctantly ended his first paper. The paper was typed and he sat down to draft the letter to Rutherford that would go with it. It was dated March 6, 1913. Bohr felt that he must add a few words of explanation:

Enclosed I send the first chapter of my paper on the constitution of atoms. I hope that the next chapters shall follow in a few weeks. In the latest time I have had good progress with my work, and hope to have succeeded in extending the considerations used to a number of different phenomena; such as emission of line spectra, magnetism, and possibly an indication of a theory of the constitution of crystalline structures.

I have, however, some difficulties in keeping it all together at the same time, and shall be very glad to have some of it published as soon as possible on account of the accumulating literature on the subject. And besides the paper is getting rather long for publishing at one time in a periodical I have thought it best to publish it in parts. Therefore I shall be very thankful if you will kindly communicate the present first chapter for me to the Phil Mag.

As you will see the first chapter is mainly dealing with the problem of emission of line-spectra considered from the point of view sketched in my former letter to you. I have tried to show that it from such a point of view seems possible to give a simple interpretation of the law of the spectrum of hydrogen, and that the calculation affords a close quantitative agreement with experiments.

"I hope," he added, "that you will find that I have taken a reasonable point of view as to the delicate question of the simultaneous use of the old mechanics and of the new assumptions introduced by Planck's theory of radiation. I am very anxious to know that you may think of it all."

Anxious days lay ahead while Bohr waited for an answer from Rutherford. But Bohr could not afford the time for fruitless worry. He had to put the two remaining sections into final form. While he watched for the red coat of the postman, he wrote and rewrote. Much polishing was necessary to make all the statements exact and not to slight any of the implications. Bohr also constantly turned back to the first paper to check and recheck, and to think further about the problems.

The letter from Rutherford came, dated March 20. As Bohr tore it open and scanned the first sentence, the tension of the past months gave way to a surge of relief and happiness. Rutherford did not keep him in doubt: "I have received your paper safely and read it with great interest, but I want to look over it again carefully when I have more leisure. Your ideas as to the mode of origin of the spectrum of hydrogen are very ingenious and seem to work out well; but the mixture of Planck's ideas with the old mechanics make it very difficult to form a physical idea of what is the basis of it."

Unlike many physicists who think largely in abstract terms, Rutherford visualized the structures and forms they were discovering. An electron was as visible to him as a bee buzzing around a flower and almost as alive. It was in these terms that he weighed Bohr's proposals.

"There appears to me one grave difficulty in your hypothesis, which I have no doubt you fully realize, namely, how does an electron decide what frequency it is going to vibrate at when it passes from one stationary state to the other? It seems to me that you would have to assume that the electron knows beforehand where it is going to stop."

Rutherford had hit the nail, or perhaps it should be said the electron, straight on the head, and Bohr laughed with pleasure both at the accuracy of Rutherford's point and its total lack of pretension. Bohr cherished the sentence for a lifetime and used it, as well as the remainder of the typically Rutherfordian letter, in the lecture that he gave at the Rutherford Jubilee in 1961.

A man who could sum up the whole case by asking how the electron would know where it was going also liked directness and conciseness. Rutherford thought Bohr's paper was too long and involved, and he made his point emphatically though with a heartiness that did not lacerate the man being criticized.

"There is one criticism of minor character which I would make in the arrangement of the paper," he continued. "I think in your endeavour to be clear you have a tendency to make your papers much too long, and a tendency to repeat your statements in different parts of the paper. I think that your paper really ought to be cut down, and I think this could be done without sacrificing anything to clearness. I do not know if you appreciate the fact that long papers have a way of frightening readers who feel that they have not time to dip into them."

Rutherford promised that he would go over the paper carefully and send it to the "Phil. Mag." He also said that he would make any corrections in English that might be necessary. There was more good news in the letter. Rutherford reported that he had talked to E. J. Evans about doing the experiment Bohr had suggested to test the spectral lines given off by helium.

Rutherford ended with a few friendly words about his own work—the mass of the alpha particle was coming out bigger than it ought to be—and with a postscript which gave Bohr some bad hours, for it said: "I suppose you have no objection to my using my judgment to cut out any matter I may consider unnecessary in your paper? Please reply."

Bohr was torn between elation at Rutherford's general approval of his work and uneasiness about the proposed cutting. He was even more embarrassed because some revisions he had just dropped in the mail would make the paper even longer.

To Bohr it would have been unthinkable to utter a shrill protest or to insist that his work could not be altered. He and Margrethe decided that the only solution was for him to go to Manchester. He tactfully wrote Rutherford that he would be "glad for any alterations you consider suitable" and apologized for giving him so much trouble. Almost as though it were happenstance, Bohr said: "I now have some holidays and I have decided to come over to Manchester." In the first draft of his letter he added "to spare you trouble in correcting the paper and to get the opportunity to speak to you about different questions." But even this seemed a little too insistent, and Bohr scratched it out. In the final copy that Margrethe wrote it did not appear.

Bohr was on the way to England before Rutherford's prompt reply reached Copenhagen. Rutherford had written that he thought the additions "excellent" and that he considered them "quite reasonable." "However," said Rutherford returning to the question of length:

> the paper is already long and full for a single paper. I really think it desirable that you should abbreviate some of the discussion to bring it within more reasonable compass. As you know, it is the custom in England to put things very shortly and tersely in contrast to the Germanic method where it appears to be a virtue to be as long-winded as possible.
>
> I should consequently be glad to hear what parts you think might be jettisoned or cut down. I think it would not be difficult to reduce the paper by one-third without sacrificing any of the essential points. As I mentioned in my last letter it is very desirable not to publish too long papers, as it frightens off practically all the readers.

Bohr was less concerned about the lazy reader than about his closely reasoned argument. He believed fervently that none of it could be "jettisoned." When Bohr reached Manchester and walked into Rutherford's office he was cordially welcomed. Rutherford suggested that they begin a section by section review, and this went on through several evenings. Bohr tenaciously defended every section and all the points. As they went along Rutherford yielded; he changed only a few words here and there to put them into better English. But it was high intellectual debate and Bohr said later that Rutherford's patience was angelic. Rutherford came to look upon the battle of Bohr versus brevity as a fine tale. With characteristic gusto he began to tell it to friends and colleagues, and to describe how he gave in.

As they finished Rutherford clapped Bohr on the back and

with a big grin told him: "I never thought you would prove so obstinate." And he sent the paper off to the *Philosophical Magazine* virtually unchanged.

Almost every one at Manchester that June of 1913 was hard at work on atoms. "Atoms hardly enjoy a greater patronage anywhere," laughed Hevesy. Bohr talked to Moseley and the others about the arrangement of the elements according to their atomic number. Moseley said that he would try to settle the matter by systematic measurements of the high frequency spectra of all the elements.

Bohr could not linger long in England; too much work still had to be done on the remainder of his paper. Soon after his return to Copenhagen the proofs of the first section arrived. Bohr excitedly pored over them, weighing every word and making as many changes as he dared. He returned them to Rutherford with a note: "I have altered very little, and have not introduced anything new. I have, however, attempted to give the main hypothesis a form which appears to me to be in the same time more correct and clear."

All the while Bohr was working constantly on the next two sections. The numerical calculations proved particularly troublesome, but on June 10 the second manuscript was mailed to Rutherford. A section of the third part was finished on July 1, but the whole was not completed until August 27. Bohr wanted to go farther. He felt remiss in not dealing with the magnetic properties of the atomic system, but he was afraid to make the paper much longer. He also feared that it might not be advisable to introduce too many new and hypothetical assumptions at one time.

Part II of "On the Constitution of Atoms and Molecules" was titled "Systems Containing Only a Single Nucleus."

Bohr again stated his position: "Following the theory of Rutherford, we shall assume that the atoms of the elements consist of a positively charged nucleus surrounded by a cluster of electrons. The nucleus is the seat of the essential part of the mass of the atoms, and has linear dimensions exceedingly small compared with the distances apart of the electrons in the surrounding cluster."

Bohr contented himself with this simple statement. Others later developed comparisons to give the layman an idea of the vast distances. Selig Hecht in his book *Explaining the Atom* calculated that if hydrogen's nucleus were enlarged to the size of a baseball, its electron would be a speck about eight city blocks away. Actually, atomic distances are almost incomprehensibly small. The diameter of a hydrogen atom is about 1,200,000,000th of an inch, and thus 200,000,000 of them, could be placed on an inch without squeezing the electrons at all.

Bohr continued, reviewing and reinforcing his base:

> As in the previous paper, we shall assume that the cluster of electrons is formed by the successive binding by the nucleus of electrons initially nearly at rest, energy at the same time being radiated away.
>
> This will go on until, when the total negative charge of the bound electrons is numerically equal to the positive charge on the nucleus, the system will be neutral and no longer able to exert sensible forces on electrons at distances from the nucleus great in comparison with the dimensions of the orbits of the bound electrons.
>
> On account of the small dimensions of the nucleus, its internal structure will not be of sensible influence on the constitution of the cluster of electrons, and consequently will have no effect on the ordinary physical and chemical properties of the atom. The latter properties, on this theory, will depend entirely on the total charge and mass of the nucleus; the internal structure of the nucleus will be of influence only on the phenomena of radioactivity.

From Rutherford's experiments in bombarding atomic targets with alpha rays and continued experiments with X rays, the evidence indicated, Bohr said, that the number of electrons surrounding the nucleus would equal the element's number in the table of elements. For example, oxygen as No. 8 among the elements would have eight electrons and a nucleus carrying eight unit charges. Tin, No. 50, would have fifty electrons and fifty unit charges in the nucleus.

Bohr assumed on the basis of an idealized model that the electrons would be arranged in coaxial rings rotating around the nucleus. But how many electrons would be in each ring, how would the rings be built up, and how would the configurations determine the chemical properties of the elements?

If a second electron were added to hydrogen's one, the two would circle round the nucleus, Bohr theorized and the element would be helium. The orbit then would be filled and would have no tendency to take on another electron. And in actuality helium is an inert gas with no tendency to combine with other elements.

But if a nucleus with a charge of three captured a third electron, the added electron would take up an orbit of its own beyond the orbit occupied by the first two.

Other electrons would be added, Bohr's calculations showed, until there were eight electrons in the second ring. With the two electrons in the inner ring and eight in the outer the atom would have ten electrons—it would be neon, No. 10 in the table of elements. At this point a third ring would begin to form and electrons would be taken into the third circle until it also had eight. The two, plus eight, plus eight, would be eighteen, and the result would be argon, No. 18, with eighteen electrons. The next, the fourth, orbit, Bohr continued, should, when it was completed with the addition of electron after electron, include eighteen electrons. And in elements of still higher atomic weight, he said, the configurations of the innermost electrons might again be repeated. "However," he

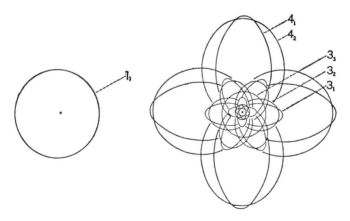

(left) The hydrogen atom with its one electron. (right) Main lines of the krypton atom, which has thirty-six electrons.

added, "the theory is not sufficiently complete to give a definite answer to such problems." In fact, the data were not sufficient to permit Bohr to go as far as he had. He had reached the right conclusions about atoms beyond hydrogen on the basis of arguments that in the end proved not supportable. Nevertheless at the time Bohr cautiously said, "It seems not unlikely that the constitution of the atoms will correspond to the properties of the elements."

Explaining the rings with their arrays of electrons made possible the solution of another major problem. If an electron was removed from one of the outer rings, the flash of energy given off would produce the ordinary line spectrum of the element—its own characteristic line-up of lines.

"In analogy," Bohr continued, "it may be supposed that the characteristic Roentgen X ray radiation is sent out during the settling down of the system if electrons in inner rings are removed by some agency, e.g., by impact of cathode particles."

This theory could be tested. The velocity of the cathode ray bombardment needed to produce X rays should be equal, Bohr pointed out, to the energy necessary to remove one of the electrons from the inner rings. Bohr made the calculations, and again the two fitted with nearly incredible exactitude—the energy needed to produce X rays equaled the energy necessary to knock out an electron from the inner rings.

Radioactivity was another matter. Emissions of energy from the electrons clustered around the nucleus could not explain it. Bohr maintained that only one conclusion was possible with the Rutherford atom; the alpha rays have their origin in the nucleus. It also appeared virtually certain to Bohr that the nucleus was the seat of the high-speed beta particles.

This view of the origin of alpha and beta particles explains very simply the way in which the change in the chemical properties of the radioactive substances is connected with the nature of the particles emitted. The results of experiments are expressed in the two rules:

1. Whenever an alpha particle is expelled, the group in the periodic system to which the resultant product belongs is two units less than that to which the parent body belongs.

2. Whenever a beta particle is expelled, the group of the resultant body is one unit greater than that of the parent.

This was exactly what Bohr's analysis of the ring structure of atoms had shown should be the case. Experiment and theory not only fitted, they fitted beyond all expectations.

Bohr was ready to carry his theory to the next step, the molecule, in "Part III, Systems Containing Several Nuclei," though here again he was reaching beyond the knowledge of the day. He again needed a clear definition: "the difference between an atom of an element and a molecule of a chemical combination is that the first consists of a cluster of electrons surrounding a single positive nucleus, while the latter contains at least two nuclei at distances from each other comparable with the distances apart of the electrons in the surrounding cluster."

A molecule thus was a combination of two or more atoms, and was made up of two well-separated nuclei and their electrons. But how, could two solar systems be fitted together? How is a molecule formed? One atom could not annex another as the nucleus does electrons, for two highly charged nuclei would repel one another. Bohr started his investigation with the simplest case—two nuclei each with a single ring of electrons. He showed that if the two approached one another in effect only the outer rims would touch. The inner part of the atom would be unaffected by the joining up.

"The greater part of the electrons must be arranged around the nucleus approximately as if the other nucleus were absent," said Bohr. "Only a few of the electrons will be arranged differently, rotating in a ring around the line connecting the nuclei. The latter ring, which keeps the system together represents the chemical 'bond.'"

Here was a premature, virtually intuitive explanation of the chemical "bond." Much work would have to be done before an actual explanation could be developed, but Bohr perceptively indicated the direction.

Molecules in Bohr's far reaching scientific imagination thus became ordered and explicable combinations of solar systems, of nuclei surrounded by electrons whirling through varied orbits on many planes. An odd base for the seeming solidity of the world.

In concluding the trilogy, as the three papers soon were called, Bohr summarized his main points. He had used Planck to explain Rutherford's atom, and in doing so had introduced new assumptions about the emission and absorption of radiation by an atomic system. Bohr summed up these assumptions:

1. Energy is not emitted or absorbed in the continuous way physics assumed in the past, but only during the passing of systems between different "stationary" states.

2. Though the equilibrium of the systems in the stationary states is governed by the ordinary laws of mechanics, these laws do not hold for the passing of the systems between the different stationary states.

3. The radiation emitted is one homogenous unit—one of Planck's quanta.

4. The stationary states are determined by the energy emitted and the revolving of the electron.

5. The "permanent" state of any atomic system is the one in which the energy emitted is maximum.

The two final sections also were long, according to Rutherford's ideas, but happily, he put aside his desire for terseness and sent both on to the "Phil. Mag." The first article appeared in the July, 1913, issue, the second in September, and the third in November.

The reaction was prompt and compounded about equally of shock and interest. To some physicists the idea that radiation emitted or absorbed by an atom did not come from its internal motion smacked of sacrilegion and madness. But Bohr could not be brushed off. As Hevesy wrote: "Everything is so clear . . . so truefull that nobody can avoid to be struck by reading it."

During the fall of 1913 the forces for and against the revolutionary ideas of the young Dane were lining up. At this point, however, events took a dramatic turn. At Rutherford's request Evans tested the question that Bohr had said was open to testing— his assignment of the Pickering and Fowler lines to helium instead of to hydrogen. Evans filled a tube with helium of extreme purity and obtained the Fowler lines. Thus they came from helium as Bohr had said and not from hydrogen as their discoverers had thought. It was exactly as Bohr had predicted.

Fowler, however, still hesitated. His measurements of the wave lengths of the disputed lines still did not exactly coincide with those calculated under the Bohr theory. Bohr suggested that the helium atoms Fowler had studied actually were ionized atoms or atoms stripped of their electrons. Fowler recalculated on this basis, the figures matched, and his last doubts vanished. He publicly announced the finding, and even pointed out that certain other spectra could be recognized as originating from excited ions rather than excited neutral atoms. The news of this dramatic confirmation of Bohr's theory raced through the world of physics.

At this moment Hevesy happened to meet Einstein in Vienna and tell him "the news." "When I told him of the Fowler spectrum the big eyes of Einstein looked still bigger," Hevesy wrote to Rutherford. "He told me 'Then it is one of the great discoveries.' I felt very happy hearing Einstein say so."

Hevesy was so excited that he also wrote to Bohr about his conversation with Einstein, explaining that he had asked Einstein about Bohr's theory.

He told me it is a very interesting one, important one if it is right and so on and he had very similar ideas many years ago, but had no pluck to develop it. I told him that this is established now with certenety that the Pickering-Fowler spectrum belong to He. When he heard this he was extremely astonished and told me 'Than the frequency of the light does not depend at all on the frequency of the electron.' (I understood him so??) 'And this is an *enormous* achievement. The theory of Bohr must be than wright.'

I can hardly tell you how pleased I have been and indeed hardly anything else could make me such a pleasure than the spontaneous judgement of Einstein.

The British Association for the Advancement of Science meeting in Birmingham on September 7, 1913, had scheduled a discussion of radiation. Rutherford suggested that the young Dane should be invited along with such celebrities of physics as Mme Curie, Lord Rayleigh, Sir Joseph Larmor, Lorentz, and Jeans. Bohr did not see how he could make another trip to England, but the opportunity was too great to be permitted to pass.

Jeans opened the discussion with a survey of the quantum

theory as it applied to the constitution of the atom. "Dr. Bohr has arrived at a most ingenious and I think we must add convincing explanation of the laws of spectral lines," he said. Turning to Bohr's fundamental assumptions about the structure of the atom and the quantum theory, Jeans commented that their only justification "is the very weighty one of success."

It was the first important discussion of the whole subject outside of the Manchester school. As Jeans presented Bohr's strange combination of classic and quantum physics and all the upsetting consequences that went with it, skepticism rose. Larmor solemnly asked Lord Rayleigh to express his opinion. The famed physicist who had contributed so importantly to the earlier understanding of radiation, took cover.

"In my young days," he answered, "I took many views very strongly and among them that a man who has passed his sixtieth year ought not to express himself about modern views. Although I must confess that today I do not take this view quite so strongly, I keep it strongly enough not to take part in this discussion."

Bohr himself was asked to give a short "explanation of his atom," and of course did so. As soon as he had finished, Lorentz put a skeptical question: "How is the Bohr atom mechanically accounted for?" Bohr had to acknowledge that this part of his theory was incomplete, but he argued "with the quantum theory being accepted, some sort of scheme of the kind suggested was necessary."

Thomson was another and a brusque questioner. He challenged Bohr's denotation of certain atoms as X-3. Thomson maintained that they should be identified as triatomic hydrogen molecules, rather than as atoms.

Bohr answered very modestly and quietly that, "Perhaps they might be ascribed to hydrogen atoms having three times as heavy a nucleus as hydrogen." Bohr suggested that the point could be tested by experiment, if a mixture of hydrogen and X-3 were diffused through palladium.

"Bohr had not been properly understood," said Hevesy, "and Thomson gave rather a quick answer saying after a brief consultation with Ramsay that Bohr's suggestion is useless, for not the molecules but the atoms of hydrogen diffuse through hot palladium.

Certainly, but it was just Bohr's point. The general appearance was that he, Thomson, had told something highly ingenious and Bohr something very stupid. Just the contrary was the case.

"So I felt bound to stick up for Bohr and explained the meaning of Bohr's suggestion in more concrete terms, saying that Bohr's suggestion is that X-3 is possibly a chemically non separable element from hydrogen. . . . Of course it is not very probable, but still a very interesting suggestion which should not be quickly dismissed."

Bohr's disputed X-3, or tritium as it was to be called, was discovered some years later. It was a heavy hydrogen atom.

Bohr was heartened and buoyed up by the discussion. The serious attention given the work of a young unknown outsider in the gathering of "lions" was itself an acknowledgment. He was also particularly happy about the support from Jeans. "I think he is convinced that there is at least some reality behind my considerations," he wrote Margrethe. Bohr never forgot this support at a critical moment. Throughout his life he could not hear Jeans's name mentioned without some expression of appreciation.

The doubt and opposition though were not soon dissipated. P. H. Zeeman, an authority on spectra, dismissed Bohr's papers in a line, and O. W. Richardson, a Nobel Prize winner, gave only passing note to Bohr's theory in a book he published at about the same time.

Lord Rayleigh was asked by his son if he had seen Bohr's paper on the hydrogen spectrum. "Yes," he answered, "I have looked at it, but I saw it was of no use to me. I do not say that discoveries may not be made in that sort of way. I think very likely they may be. But it does not suit me."

The Bohr atom did not suit many of the physicists brought up in the old school. However, they could not ignore it. Physics had been changed by the three papers of Niels Bohr whether anyone liked it or not. The change was comparable to that brought about two years earlier by Rutherford's discovery of the nucleus of the atom. Nuclear physics had been born then, atomic physics came to life now. Time would be required even to grasp the extent of the revolution Bohr had wrought, but the revolution had occurred.

5 /

Wartime Manchester and Orbits

A REVOLUTION in understanding had taken place. Though no physicist could foresee that it would ultimately alter the political structure of the world and the basis of life, many sensed that the change was profound and that science was entering a new era.

Few though could sense or believe that another kind of upheaval—war—was in the making. Neither Bohr in Copenhagen, nor Rutherford and the Manchester group with its members who had come from many parts of the world, were "ivory tower" scientists in the sense of feeling themselves separated from the ordinary run of events. Bohr began and ended every day with a careful look at the newspapers, and there were few days when discussion did not get around to the disturbances abroad and Danish policy at home.

There might be warlike incidents, all recognized, but surely, they argued, the clashes would not be permitted to pile up into the insanity of armed conflict.

In 1911 when Germany sent a warship into Agadir as a warning against French occupation of Morocco, the word "war" was bruited around Europe. At the time E. N. daC. Andrade was in Heidelberg studying for his doctorate in physics. As he later told Bohr to demonstrate the prevalent attitude: "I was sitting in a cafe with German friends, when one of them asked me if I was going back to England. 'Why,' I queried. 'There seems to be danger of war,' he replied, to which I answered 'Don't be silly. We are not living in the Balkans! You don't really think that the people sitting around here are going into the field to shoot at other people like them?' I thought that I was being very worldly wise in the face of an out of date attitude. No one in England took the threat of war seriously."

Ivan Bloch had argued in his widely read book *The Future of War* that war was an impossibility and would lead to universal bankruptcy and starvation. Physicists, as well as millions of others, were convinced by Bloch's logic or by Norman Angell who maintained in another book that influenced the times that war was pointless. Military victories and defeats were only illusions, he said; the essential economic life of a country went on regardless.

It was especially difficult in Denmark to believe that war might come. The country was prosperous and peaceful. Following the final loss of Schleswig-Holstein in 1879 and the equally disruptive loss of the Danish grain market to the growing flood of wheat from the American prairies, Denmark made an abrupt turn. Within a decade or two the country shifted into the production of "quality farm produce." With the establishment of cooperatives and of folk schools to guide the change, Denmark began to produce butter, eggs, and bacon of high and guaranteed quality. The world, and particularly England, bought all the Danes could supply and prosperity came to the land.

All the while the Germans with the reinforcement of the Triple Alliance were pushing Pan-Germanism and preparing for "Der Tag." A navy to challenge the British control of the seas was under construction. Huge armies were trained and honed to a pitch of readiness, and new railroad extensions stretched out toward the Belgian border.

No government could ignore the threat. Faced with the danger, the British and French had forgotten or suppressed old enmities. Behind a fortified frontier the French stressed *élan* and prepared for the possibility of attack. The army staffs had quietly agreed that if France were invaded British troops would move in to hold the frontier bordering on the English Channel. By the spring of 1914 British staff work was completed down to the billet for every battalion and even the points at which they would pause for coffee.

Russia, caught in the grip of a regime that even its defender

Count Witte called "insane, blind, crafty and stupid," still talked of marshaling an army of 6,500,000 against Germany if she attacked France and tried to turn on Russia.

Nevertheless the "armed peace" seemed to be holding. As the snow and ice of the winter of 1914 began to yield to the promise of spring, Niels and Margrethe Bohr were peacefully deciding their future. Bohr was only 29, but he and Denmark knew that his quantum theory of the atom had profoundly affected physics. His training had been thorough and complete. Bohr was invited to submit his application for appointment as professor of theoretical physics at the University of Copenhagen.

Bohr carefully and correctly drew up his letter and submitted it on March 4. On the day before he wrote to Rutherford, telling him of what he was planning, and asked for the "testimonial" Rutherford had long since promised whenever it might be needed. Rutherford's enthusiastic and impressive recommendation came back almost in the next mail. Rutherford's weighty praise was scarcely needed. Bohr obviously had all the qualifications and offered the university the theoretical physics that had not previously been taught as a separate science. There was actually no question of his approval; it was more like the university's claiming of its own. Only a month later Bohr wrote Rutherford that he had learned that the faculty would almost unanimously support his application.

Before action could formally be taken, however, another letter from Rutherford, on May 20, changed all Bohr's plans for the new post on the Copenhagen faculty: "I daresay you know Darwin's tenure of readership has expired, and we are now advertising for a successor at £200," Rutherford wrote. "Preliminary inquiries show that not many men of promise are available. I should like to get a young fellow with some originality in him."

Rutherford meant Bohr. The young Dane eminently fitted his specifications. Despite the uncertainty of the times, the opportunity to work for two years with Rutherford at the fountainhead of physics could not be refused. The £200 was no deterrent; in fact, Bohr considered it generous, and in 1914 it was not at all a bad amount. The junior scientists in the Manchester laboratory received

£120 to £150 a year and many graduate students could "manage" with a fair degree of comfort on £100 a year. Rutherford's own salary of £1,600 a year was considered munificent.

Bohr was filled with excitement at the prospect. It was what he wanted above all else. He found the University of Copenhagen understanding and willing to grant him a leave. He quickly applied for the two-year readership. Rutherford found his young fellow with originality.

Bohr was working feverishly to get everything in order to leave for England in September. He also had to answer several criticisms made of the trilogy. To regain his full energies and to have a few undisturbed days, Niels and Harald planned a walking trip through the Austrian Alps. Niels joined Harald at Goettingen, where Niels was invited to lecture. He also gave a lecture in Munich before the two brothers headed for the open road. They reveled in the mountains and the peace of the forests. War could not have seemed farther away, and the sudden outbreak caught them all the more by surprise.

On June 28 black headlines in the newspapers blazoned "Archduke Franz Ferdinand Assassinated." Austria-Hungary, with the foolhardy daring of a decaying empire, decided to use the assassination to seize Serbia. An ultimatum was issued.

Niels and Harald could wait no longer. They broke off their trip and caught one of the last trains to cross the border.

Bismarck had predicted some years before that "Some damned foolish thing in the Balkans" would lead to a European war. It did. A fateful march toward conflict began. Germany promised to support Austria-Hungary if Russia opposed the empire's "punitive action" against Serbia. Exactly one month later Austria-Hungary declared war on Serbia and the next day bombarded Belgrade. Two days later Russia mobilized, and on July 31 Germany issued an ultimatum to Russia, demanding her demobilization within twelve hours. A second ultimatum was dispatched to Paris, demanding French neutrality, and negating any possibility of acceptance, by further demanding that France hand over her fortresses of Verdun and Toul as a guarantee of that neutrality. Germany was asking for the key to France. Completing the prelude to war, a third ultimatum went to Belgium.

German troops moved further into position on the Belgian frontiers, for the decision long ago had been made to attack through that country. As Sir Edward Gray was soon to say on one of the nights of agony when all Europe was collapsing into the chaos of war, Belgium's neutrality was treated as nothing more than "a scrap of paper."

Two nights of anxiety and frenzy followed. Crowds jammed the streets, lights burned till dawn in chancelleries, staff cars raced through the streets. Last minute efforts to avert war failed.

At 8 A.M. on August 4, 1914, the gray German columns rolled across the Belgian frontier, making the declarations of war by France and Great Britain mere formality. Mobilization and long-laid military plans automatically moved into execution. The German plans were laid according to schedule—the German troops were to be in Paris on the 39th day after the attack began. Officers boastfully announced that they would be home "before the leaves fall."

In Copenhagen the Bohrs joined the anxious crowds in the big paved square before the high steepled town hall. The government met late into the night, though there was no question that Denmark would try to remain neutral. But the Danes knew from the experience of a thousand years the vulnerability of their position at the head of the Baltic. New reminders of it were not long in coming. Germany threatened to lay mines in Danish waters unless Denmark herself sealed them off to all belligerents.

To their anxiety about their country and about Europe, the Bohrs added their own worry about going to England. Bohr consulted with friends at the university and in the government. Could or should he still go to Manchester?

The immediate question rested on what would happen at sea. On July 28, when the outlook became ominous and the ultimatums were flying, Winston Churchill, the first Lord of the Admiralty had ordered the British Grand Fleet to sail to its war base at Scapa Flow, at the bleak northern tip of British land and at the head of the North Sea. The fleet had been engaged in maneuvers and

Churchill was convinced that it could not be permitted to disperse.

The moment war came, the fleet from its northern station set up a blockade of the continent and took up a ceaseless watch for the appearance of the German fleet. "The Germans have the strongest incentive for action," Churchill wired the fleet commander, Admiral Sir John Jellicoe. Neither the British nor the Danes knew that the Kaiser was unwilling to risk the fleet he had spent twenty years in building and that he had ordered it to remain in its base at Heligoland. The German war leaders thought the war would be too short to take any chances with their precious ships. Surprisingly, most of the German merchant shipping also withdrew into the safety of German or neutral ports.

The Germans announced that they were sowing mines. On August 28 a British destroyer flotilla steamed into Heligoland to attack the German base. The move was undertaken to divert attention from the landing of British forces at Ostend. When the day's confused fighting ended three German cruisers and a destroyer had been sunk, and more than a thousand Germans killed.

To travelers awaiting transportation to England the news was grim. The battle at Heligoland, however, proved the last of the year. The British strengthened their control of the seas and by September, trade sanctioned by the British and complying with their regulations against contraband was again moving.

Early in September the Bohrs embarked. To stay clear of the area of greatest danger their ship took a route far to the north and around Scotland. Storm and fog beset their way. To the Bohrs it seemed that nature was attuned to the awesome forces man had let loose.

The Bohrs were welcomed to Manchester with a warmth mingled with relief at their safe arrival. Bohr was also badly needed in the laboratories. The exodus to the armed services had already begun and was increasing. Though there still was talk that the war would be a short one, the idea was crumbling as the Germans pushed on through Belgium and into France. The French and British retreated before them in deadly, desperate fighting. Each day brought blacker, more discouraging news from the front. Only when the Germans finally were halted at the Marne could the Manchester group give more than a troubled attention to physics.

Bohr, Evans, and Makower had to take over most of the teaching during the fall and early winter, for Rutherford was away. Rutherford, his wife, and young H. G. J. Moseley had departed in June to attend the overseas August meeting of the British Association for the Advancement of Science in Melbourne, Australia. When they left they had no idea that war was imminent; their thoughts were about the new honor that had come to Rutherford, who had just been knighted. Despite his shock at the outbreak of war, Rutherford carried through his original plan to visit his home in New Zealand and to return via the United States and Canada for visits with American scientists.

By the time the Rutherfords sailed for England, German raiders were loose in the Atlantic and the danger of mines and submarines was increasing. The Manchester staff lived through anxious days while the Rutherfords were at sea. "Their safe return to Manchester was greeted by all of us with great relief and joy," Bohr said with understatement.

Moseley also had returned safely a little ahead of the Rutherfords. Bohr had a brief opportunity to see him before Moseley enlisted and left for the war. During the preceding year Moseley had not only made another of the great discoveries of this period of preeminent and revolutionary discoveries, but had both used Bohr's work as a starting point and had supplied striking additional proof of Bohr's theories.

Moseley and Charles Darwin—the grandson of "the right Darwin"—had started working at Manchester with X rays, another of the mysterious rays given off by the atom. Their work took a different turn when Bohr proposed an atom with electrons jumping from one orbit to another as energy was emitted or absorbed, and in effect making a record of these jumps in the visible spectra they created.

"There was a new stimulus giving a great impetus to the subject," Darwin explained. "This came from Bohr's work on the theory of spectra. For most of the time Bohr was working in Copenhagen, but there was good contact with him through visits and correspondence."

Moseley about this time went on to Oxford, where he missed Manchester's concentration on the atom and its excitements and close cooperation. "Here there is no one interested in atom building," Moseley complained. "I should be glad to do something toward knocking on the head the very prevalent view that Bohr's work is all juggling with numbers until they can be got to fit. I myself feel convinced that what I have called the H hypothesis is true, and that one will be able to build atoms out of e, m, and h [energy, mass, and quanta] and nothing else besides."

Bohr had conjectured that X rays were emitted by the innermost electrons only, and that the emissions would produce a typical spectrum, in a sense a self-produced picture. Confirmation might well be obtained, Moseley thought, if measurements could be made of the wave lengths of the X rays emitted by a sequence of elements. Moseley decided to begin with the elements in the sequence between calcium and zinc.

He had first to devise an apparatus. Darwin said that every time he went to Moseley's room a different set-up for the experiment was being tried. Moseley was a prodigious worker who was as likely to be found in the laboratory at 3 A.M. as at 3 P.M. Finally he hit upon exactly what he needed—a cylinder of glass about a yard long and a foot in diameter. Down the center of it he ran a toy railroad track. He loaded blocks of the elements he wanted to test on tiny cars and pulled them back and forth into the path of a beam of cathode rays. The X rays produced emerged from the large tube to fall upon a crystal of potassium ferrocyanide, which in turn reflected them upon a photographic plate.

The photographs were startling and revealing. As Moseley put them together, he found that he had a stair-step progression. "We have proof here that there is in the atom a fundamental quantity which increases by regular steps as we pass from one element to the next," Moseley reported. "This quantity can only be the charge on the central positive nucleus."

Bohr had hypothesized that the charged nucleus of hydrogen

had one electron orbiting around it; the next element, two; the third, three; and so on up to the ninety-two electrons of the heaviest known element. Moseley's photographs and measurements identified the number of each with the number of units of positive charge in the nucleus. The structure Bohr had proposed was thoroughly substantiated.

Moseley early pointed out that the system he was uncovering might lead to "the discovery of missing elements, as it will be possible to predict the position of their characteristic [spectral] lines." In between hydrogen with a weight of one in the first position and uranium with its weight of 92 and its ninety-second position, there were known elements for each number except 42, 43, 72, and 75. As Moseley had predicted, all four were found within a relatively short time.

In his paper published in the *Philosophical Magazine* in 1913, Moseley commented: "The results obtained have an important bearing on the question of the internal structure of the atom and strongly support the views of Rutherford and Bohr."

It was the third finding of paramount, enduring importance of those five years in which a science was remade. Moseley's clarification of the table of elements ranked close to Rutherford's discovery of the nucleus of the atom and Bohr's deciphering of the atom's structure. All three discoveries were closely related.

Bohr scarcely had a chance to discuss these findings with Moseley before Moseley enlisted. Despite the pleas of Rutherford that he accept a position in which his scientific abilities could be used (Rutherford had made determined efforts to see that he was assigned to such a place) Moseley insisted upon joining the Royal Engineers and was assigned as a Signals Officer.

The laboratory was deep in another development. If each element was created by the addition of one more electron, as Bohr's work indicated, then each element should be a multiple of one. The problem was that the elements were not simple multiples. The atomic weight of silver, for example is 107.880, not an even 107. The question was why?

But research was becoming increasingly difficult. Pring went off to the Royal Fusiliers; Florance, Andrade, and Walmsley joined the artillery. Robinson was to go at any moment. Rutherford himself was drawn increasingly into war work. The old separation of laboratory and government was ending in the necessities of modern war. With German submarines working critical devastation on Allied shipping, the government turned to science for some method of detecting the hidden, unseeable weapon. Only the methods of science offered any possibility of countering this largely invisible menace.

The newly established tie between scientist and government was eventually to alter the method and support of science. The new day visibly dawned at Manchester with the installation of a large tank in the dark basement of the physics laboratory. Experiments in radioactivity continued, but naval officers as well as scientists now received the command to duck their heads as they came down the basement steps, and apparatus to detect underwater sound waves vied with the scintillations given off by radioactive sources.

The university wondered if it would have any students left. But still they came; the young had to be prepared for the heavy responsibilities that lay ahead, though they might be snatched from their training before it was finished.

Despite the heavy teaching load and all the other demands, Bohr and Rutherford managed to carry on some of their own research. Late at night or at any hour open between other duties they hurried in for a few hours of work.

At the end of 1914 the war of movement ended and the embattled armies settled down into bitter, bloody trench conflict. The battlelines stretched all across France. Despite this barrier and the embargoes and restrictions that backed it up, word came through that James Franck and Gustav Hertz had proved the stepwise transfer of energy in atomic processes. Exactly such a transfer in quanta or steps had been predicted by Bohr. It lay at the heart of his theory.

The two German experimenters had bombarded mercury vapor. They reported that a ray of 4.9 volts was sufficient to knock an electron out of the mercury vapor atom. Anything less than 4.9

volts produced no effect. Bohr's theory that an electron could be affected only by the right quanta was seemingly upheld. However, when a mercury vapor atom was hit it gave off an ultraviolet flash of "wave length 2536." It was on this basis that Hertz and Franck reported that the 4.9 volts were necessary to remove an electron.

Bohr questioned the latter assumption. His calculations indicated that more than twice as much energy, 10.5 volts, would be required to remove an electron. Bohr suspected that the 4.9 volts was only enough to kick the mercury electron from one orbit to another. As Bohr put it: "it seems that their experiments may possibly be consistent with the assumption that this voltage corresponds only to the transition from the normal state to some other stationary state of the neutral atom."

In long discussions in the laboratory Bohr argued that if 10.5 volts were necessary to remove an electron and 4.9 to jump it from one orbit to another, the Franck and Hertz experiments would give the strongest support to his theory. If, on the other hand, the electron could be removed by 4.9 volts as Franck and Hertz said, Bohr admitted that his theory would be in "serious difficulty."

Bohr was not primarily an experimental physicist, but he was eager to investigate the point. Makower agreed to work with him on a test experiment, and Rutherford urged them to go ahead. With this encouragement the two planned an intricate quartz apparatus with various electrodes and grids.

To construct it they enlisted the services of the laboratory glass blower, a German named Otto Baumbach. His skill was great; he blew the exceedingly thin-walled glass that permitted the unimpeded passage of Rutherford's "darlings," the alpha rays. Though he had become an enemy alien with the outbreak of war, Rutherford had obtained permission for him to continue his work. Between intervals of blowing glass for Bohr, Baumbach loosed a blast of denunciation of the English and made fiery predictions of what the Germans would do to them. Bohr, who of course understood the German, let it pass. "His temper was not uncommon for artisans in his field," he said calmly. "It released itself in violent super-patriotic utterances."

Andrade, who did not have Bohr's placidity and heard the same tirades, told the belligerent German that he had better keep his mouth shut or he would get into trouble. At this, the glass blower complained to the vice chancellor of the university that Andrade had threatened him. Andrade was lightly reprimanded for being so hard on "a poor defenseless German in our midst." Baumbach though was not there for long. His continuing violence and vehemence soon led to his internment.

The little episode would not have mattered except that Bohr and Makower's fine new apparatus caught fire. It was ruined and there was no skilled glass blower to make another. When Makower then volunteered for military service, the experiment had to be entirely abandoned.

"The problem was solved with the expected result by brilliant investigations of Davis and Gauthier in New York in 1919," said Bohr. "I have mentioned our fruitless attempts only as an indication of the kind of difficulties with which work in the Manchester laboratory was faced in those days. Our difficulties were very similar to those the ladies had to cope with in their households."

The year opened tragically, with a blow that went beyond the general tragedy of the war and the disasters it brought in its train. A dread telegram arrived at the Manchester laboratories: Moseley was dead. The scientist who had made the third world-shaking finding of the physics revolution had been killed at the Suva Bay landings at the Dardenelles, Churchill's attempt to force the back door of the enemy. At the age of 29 one of the geniuses of English science was gone. If he had lived he would have received the Nobel Prize.

"A terrible shock to all of us was the tragic message in 1915 of Moseley's untimely death in the Gallipoli campaign," Bohr recalled with new sorrow in after years. "It was deplored deeply by physicists all over the world, and not the least by Rutherford who had endeavored to get Moseley transferred from the front to less dangerous duties." Moseley's work, Rutherford said, "ranks in importance with the discovery of the periodic law of the elements, and in some respects is more fundamental."

For all of the sorrow, pressure, and anxiety of this war year,

Bohr was virtually compelled to adapt his theory to the new findings that so directly affected it and to reply to some of the accumulated criticisms. In time snatched here and there he worked on this important task. His paper was published in the *Philosophical Magazine* in 1915 under the title "On the Quantum Theory of Radiation and the Structure of the Atom."

There was no change in his basic picture of the atom as made up of a nucleus surrounded by electrons whirling around it in set orbits. But Bohr could be more specific about the atomic states:

1. An atomic system possesses a number of states in which no emission of energy or radiation takes place, even if the particles are in motion relative to each other, and such an emission would be expected in ordinary electrodynamics. Such states are denoted as the "stationary" states of the system.

2. Any emission or absorption of energy radiation will correspond to the transition between two stationary states.

Bohr willingly conceded, even urged, "the preliminary and tentative character" of his general assumption. But work on spectra as well as Moseley's work had also confirmed it.

Bohr also clarified his view of what might go on inside the tiny but mighty atom. If one electron is removed, he pointed out, its place might be taken by an electron coming directly from outside the whole system, or by an electron jumping inward from one of the outer rings. In the latter case the vacant place in the outer ring would be filled by another electron from outside the system.

As an electron was drawn to the innermost orbit a certain type of radiation would be given off, producing a certain spectral line; the removal of an electron from ring two would produce a still different type, and Bohr thought it likely that still a third type of radiation could be ascribed to ring three. If more than one electron were removed by some violent impact, Bohr held that the missing electrons could be replaced by electrons from other things. Such a violent change would give rise, he argued, to spectra of still higher frequency.

Not even war could eliminate the pleasures and interest of being with Rutherford and his stimulating friends. Rutherford for years had met for dinner once a month with Alexander, the philosopher, Tout, the historian, Elliot Smith, the anthropologist who was in the midst of the search for man's early ancestors and the dispute about Piltdown Man, and Chaim Weizmann, the chemist who thirty years later was to become the first president of Israel. Bohr was invited to join them. He was no longer the student, but the esteemed friend and collaborator, a man with ideas to contribute even to so erudite and delightful a group. "To me it was a most pleasant and enlightening experience," he recalled with his typical formal statement.

Bohr had obtained an extension to remain in Manchester for a second year, but his stay could not be further prolonged. The University of Copenhagen expected him to return. He was formally appointed to the newly created post of professor of theoretical physics. His stay in Manchester had to come to an end.

Early 1916 was one of the darkest parts of the war. The Germans with massive "hammer blows" were attempting to break through the wearying allied lines. The laboratory members moved the lines of pins with which they followed the fighting on a wall map. There was also a mighty attempt to break through at Verdun, but again the Germans failed.

Just before the British launched their counteroffensive on the Somme, Bohr remembered Rutherford quoting Napoleon on the British: it is impossible to fight them for they are too stupid to understand when they have lost. In these grim hours even this was a comfort.

As the Germans were checked on land, they intensified their submarine warfare. Detection devices developed by Rutherford and others in the laboratories of Britain helped in fighting the underwater boats, but could not halt their ravages. The seas were dangerous and every safeguard had to be taken to prevent the leakage of any information that would have aided the Germans in their preying on shipping. As Bohr prepared to return to Denmark, he collected his notes and calculations made during the two years. Rutherford knew they were exactly the kind of unusual looking material that certainly would alarm a military inspector, and so he equipped Bohr with a letter certifying that he had held the position of reader at Manchester and that the notes were to be used in preparing an article for an English scientific publication. Thus vouched for, the Bohrs embarked without difficulty. Their voyage home also was uneventful and safe, though many Danish ships were being torpedoed without warning. The moment the Bohrs landed in Copenhagen they cabled to allay the worry of their Manchester friends.

His first stay in Manchester had shown Bohr what a scientific center could be. The war years there had only reinforced this conception—the well-equipped laboratory, the dedicated search for truth, the fellowship of scientists from all parts of the world, and a director like Rutherford, both a research scientist and a leader eager to bring out the best in all of those around him and to aid them in discoveries that might otherwise have been unattainable. The whole took on a power and luminosity greater than the separate parts.

Such a center was not in any sense a sanctum, but a place which even before the war took a rounded interest in what went on in the world. It was a place with complete devotion to science but without solemnity or pomp, in contrast to the stiffness of the German universities. It was a place to play as well as work, and high intellectual achievement reached new peaks because there was relaxation in between. None of this had been lost on Bohr. In fact it was foremost in his thoughts as he returned to set up a new department of theoretical physics at Copenhagen.

Thomas Mann in his famous essay on Freud and Einstein maintains that everyone has a model after whom he models his own life. Bohr unquestionably had his—both consciously and unconsciously it was Rutherford.

6 /

The Institute and Einstein

THROUGH THE SUMMER OF 1916 the guns roared and the neutrals plied their busy though always perilous trade, but Bohr in Copenhagen concentrated as much as the tumultuous times allowed on more basic matters.

Two years earlier reports began to show that the red, blue, and violet lines of the hydrogen spectrum were not single lines but doublets. Bohr was both baffled and aroused. He had pictured the hydrogen atom as a nucleus circled by a single electron. If there was only one electron in orbit only a single line of color should have been produced as the electron flashed from one orbit to another. But if the lines were double, as many experimenters were confirming, there had to be some other structure in the minute, invisible universe of the atom.

Bohr worked over his calculations. Only with the long reach of mathematics and logic could one plumb the unseeable depths. There was no way to photograph the interior of an atom, or even directly to test its possible structures. Bohr had suggested in a letter to the *Philosophical Magazine* that there might be a second orbit, possibly an elliptical one in the same general area as the first ring around the nucleus. Each great highway might be made up of several different lanes or loops. This, however, was a likelihood exceedingly difficult to establish.

Bohr wrote to Rutherford on September 6, 1916, to say that he was sorry the paper he was preparing was not yet ready. "I have worked on it the whole summer and met with several difficulties in the teasing subject," he explained. "It is hard to get to the bottom of the problem."

Just when Bohr thought he was making some progress and had started to prepare his paper for publication, a group of papers arrived from Germany. German scientific publications came across the neutral Danish border. In one of the papers Arnold Sommerfeld demonstrated that Bohr's original quantum orbits had to be "generalized" or increased by adding quantum elliptical orbits to them, much as Bohr too had theorized. They became not single, but multiple, tracks. "It's one of the most beautiful papers I've ever read," Bohr exclaimed as he examined it.

Sommerfeld agreed with Bohr that the first orbit around the nucleus was a circular one. To Bohr's second orbit he added three elliptical orbits, any one of which the electron might take in the second "stationary" state. He also proposed the existence of eight elliptical orbits in association with Bohr's third circular orbit, or general path the electron might follow when it first moved into the ken of the nucleus and before it was drawn closer. The atom was no longer a simple planetary system. Drawn in two dimensions on paper it was an elegant, even beautiful, composition of loops and circles taking different though closely related ways around the nucleus, as though multiple space ships were whirling around the earth on various multiple tracked orbits.

Much of this Bohr had intended to say in his own paper. Once again he had been superseded. Bohr felt weary and overworked in this sudden break in a long and intensive effort. There was always, he had learned, one very good cure for such a state. He went for a walking tour in the country. Walking along country roads and through the sun-dappled Danish forests Bohr was quickly restored. He could come back with full energy for a "final effort" on whatever the work might be—in this case his quantum paper.

But when he returned to Copenhagen he found a new batch of German papers, among them a "very interesting" paper by Ehrenfest that also contained some of Bohr's proposed results.

"My paper now needs a great deal of re-writing," Bohr wrote to

Rutherford. "I will send it to you shortly for otherwise I am afraid that it will not contain much new when it appears. This whole field of work has indeed from a lively state got into a desperately crowded one."

Despite the war the Rutherford-Bohr atom had opened such rich possibilities that laboratories everywhere sprang to one promising investigation after another. Science tends to such rapid spurts when a new field suddenly appears. In the three years since its publication the trilogy had supplied such an impetus.

Three years and the growing mass of confirmation also were proving its towering and unique contribution. Bohr was singled out as a young man of rare talent, if not genius. The University of California invited him to spend a year there. At almost the same time a letter came from Sir Henry Miers, head of the University of Manchester, inviting him to join the Manchester faculty. Rutherford of course had a large hand in the latter offer.

I was so glad to receive a very kind letter from Sir Henry Miers about my possible connection with the University of Manchester," Bohr wrote to Rutherford on September 6. "I hope so much to get it arranged in some way. I am longing to be back in Manchester and hope already to come next summer.

After having spoken to some of my colleagues I am afraid that it would be very difficult at present because of the extraordinary times. I have had to give up going to California for a year. If it is convenient to Manchester University I should like to wait a few months in taking up the question with the university authorities.

Though scientific publications were arriving from Germany, the British were becoming increasingly strict about permitting the mailing of any information that in any conceivable way might give aid to the enemy. Kay, the Manchester indispensable, mailed a number of English scientific reprints to Bohr. They were returned by the post office with a statement that no printed matter could be transmitted to Denmark. The British feared that it might reach Germany across the neutral Danish border, and although the connection between science and the war was not close, they were unwilling to take any chances whatsoever. Letters, however, came through fairly well.

"I'm sorry," wrote Rutherford in a letter dated September 21, "that you have not had time to get your paper out before Ehrenfest & Co." Rutherford who had once urged Bohr to "keep it short and concise" this time reversed himself.

"I think I would make my paper as complete as before, especially for English readers, even though you have been anticipated in some of the arguments," he advised. "The subject is so new that I think it just as well to write it up very fully."

Rutherford assured Bohr that he understood the difficulties of making a decision about Manchester. "In these troublous times, there is no hurry about it," he added, "I am quite prepared to wait six months before I get any definite answer."

Rutherford reported on some of Bohr's friends in the armed services. Bohr had no other way of getting news about them and was always anxious about what might be happening. At this stage all the news was good. Marsden was back in England, training New Zealanders not far from Cambridge. All the others were safe in so far as Rutherford knew. The war also seemed to be going somewhat better; the new tanks were proving successful.

"Everyone here is fairly satisfied with the progress of the war," Rutherford added, "and there is much talk about the new war machines used on the Somme front. Apparently they were exceedingly useful in the last attack."

In December, 1916, atoms, the war, the German papers, and the recognitions from abroad all slipped momentarily into the background. The Bohrs's first child, a son, was born and named Christian. The rejoicing was great; and one of Bohr's first actions was to write the happy news to Rutherford.

The Rutherfords were nearly as happy as the Bohrs about the lively, blue-eyed baby. "My wife and I both send you and Mrs. Bohr our heartiest congratulations on the birth of a son," Rutherford wrote. "This will always be a remembrance of your stay in Manchester."

The war could not be more than momentarily put out of mind.

It was always in the foreground even in neutral Denmark. The Danes were carrying on a highly active trade with both sides, when Germany in January 1917 announced a policy of unrestricted submarine warfare. Danish shipping losses mounted, as did those of the Allies and the great neutral, the United States. The United States strongly protested the new policy. When American protests were unheeded, the United States, on April 6, 1917, declared war on Germany. New hope ran through the tiring Allied countries, though there was little expectation that the Americans would soon send troops to their support. England took encouragement too from the establishment of a vigorous new government headed by David Lloyd George. Lloyd George started, Rutherford said, with the almost unanimous support of the country.

Rutherford was being drawn ever more deeply into war work. His anti-submarine research took him to France, and then soon after the American entry into the war, to Washington to inform American scientific colleagues about what was being done and to enlist their aid. War was becoming more and more a scientific enterprise and the scientists were brought in as advisers on many phases of the conflict. The new pattern was being confirmed. Rutherford was serving too on the Board of Invention and Research and numerous other committees. He managed occasionally though "to find an odd half day" to try a few experiments.

"I have got, I think," he wrote to Bohr from Manchester on December 9, 1917, "results that will ultimately have great importance. I wish you were here to talk matters over with." Bohr and Rutherford had come to depend upon one another for the insight, the criticism, and discussion essential for probing into the deep where there were no guidelines and even no direct verifications. Such rigorous counterpoint can come only from one who knows the subject. Others cannot follow or provide the checking that alone indicates if the path is the right one.

"I am detecting and counting the lighter atoms set in motion by alpha particles and the results, I think, throw a good deal of light on the character and distribution of forces near the nucleus," Rutherford reported.

And then the scientist added a foreword on one of the great discoveries of all time: "I am trying to break up the atom by this method. In one case the results look promising. But a great deal of work will be required to make sure. Kay helps me and is now an expert counter. Best wishes for a happy Christmas."

Bohr knew instantly that Rutherford was attempting to do what had never been done before. For centuries the alchemists had labored to transmute the "base metals" into the "noble" gold and silver. Others had spent lives and fortunes searching for "the philosopher's stone" to accomplish the same magic. Even with the rise of science the hope of breaking into the atom and changing it had been called impossible. Lord Kelvin as recently as 1907 had put the idea beyond bounds with the declaration "the atom is an eternally formed, unbreakable unit." Rutherford, nevertheless, was saying calmly that he had taken up the task.

On the same day the letter arrived, Bohr and Margrethe, who shared the excitement with him, hurried off a reply, "I was exceedingly interested to learn of your new and important results."

In the long fruitful interchange with which the young Dane and the famous Englishman carried on some of the most critical phases of their work, Bohr was often the laggard in writing. Until the startling news came that Rutherford was attempting to disintegrate the atom, Bohr had delayed in answering Rutherford's last letter. The real reason was that he felt embarrassed to confess that his work on the quantum theory paper still was not finished.

"I feel so ashamed always to say that I have not finished my work on the paper which I have worked on for such a long time," he explained in the atom letter.

At last though, Bohr could report some highly important progress of his own. He had finally worked out a satisfactory explanation of the triplet bands of color P. H. Zeeman had discovered when he subjected the hydrogen spectrum to a magnetic field. Bohr also reported that in his "spare time" he was getting ahead on the quantum paper.

He was now trying to work out the system of rings and orbits corresponding to all the groups of elements in the periodic table of elements. It was a fantastically difficult undertaking, but now the end was within sight. Bohr told Rutherford that he thought the paper should be published in the *Transactions* of the Royal Danish Academy, because there might be trouble getting it to England for publication. The embargo was getting tighter and the necessities of the war also were limiting English scientific publication. If Bohr went ahead with publication in Denmark, he said he hoped the paper could later be published in the *"Philosophical Mag."*

Though Bohr had his paper well in hand by December, he did not finish the first section until May, 1918. He had been at work on it since his return to Copenhagen in 1916. The paper had, of course, undergone many changes in this time. Bohr felt that he had been "forced" to introduce new materials and to alter older ones. In addition there was nothing harder for Bohr than to end a paper. Every time he went over it he could see new possibilities or ways of making a point clearer.

Though he planned publication in Denmark Bohr also mailed the first section to Rutherford. In the letter that accompanied it he ruefully remarked: "I feel that you already a long time ago must have given up confidence in me as regards its appearance."

The Bohrs had other happy news that May to convey to the Rutherfords. Their second child, another son, Hans, was born. In telling Rutherford about the new baby, Bohr added that Christian, then about a year and a half old, was a fine, healthy child with a penchant for getting into his father's papers. Bohr enclosed a long-promised photograph of his first-born.

Bohr was a fond father who took a constant and detailed interest in everything that pertained to the children. Niels and Margrethe one day went shopping for a new baby carriage. Margrethe was drawn to one, a relatively simple model, while Niels liked a more elegant, shinier equipage. Each argued for his choice and neither could convince the other. Then Margrethe gave in and Niels's choice was purchased. Harald met Niels soon afterwards and found him looking very gloomy. Harald asked if they had not bought the baby buggy Niels wanted? Wasn't he satisfied? Niels typically answered that the way he won did not satisfy him at all.

He did not want a concession, but an agreement growing out of real conviction.

By the spring of 1918 all was not going so well with the Rutherfords in England. Lady Rutherford had contracted the influenza then epidemic throughout the world. She was somewhat slow in recovering from the serious illness. In addition war work was taking even more of Rutherford's time and he had lost his secretary. His letters now were scrawled in longhand. His attempts to break the atom also were hamstrung for lack of time, a situation that alone made Rutherford far from happy. What made the lack of time for experiment all the more frustrating was that on the rare days when he got into the laboratory he was obtaining "very striking results."

"I am sure something fundamental will ultimately come out of it," he wrote to Bohr. "I have been looking into the question amongst others of the range of atoms of different speeds carrying different charges—using your formula. All the work is heavy and slow and there is little time available. I wish I had a year to devote to it without interruption."

Rutherford commented as usual on the ever present war. But a new note entered, despite the English retreat at St. Quentin.

"Notwithstanding temporary setbacks we have reasons for being hopeful for the future. It is a long worm that has no turning."

Rutherford's inside knowledge, or intuition as it might be, was correct again. On November 11, 1918, the four-year holocaust came at last to a close. Delirium swept the Allied world. The release was beyond accepting without wild jubilation. The Bohrs, like those who had been directly involved, rushed shouting to embrace one another and all the friends they encountered. The end of the war was not only a cessation of an almost unbearable conflict and the release of their own country from a crushing strain, but victory and peace for their beloved England and for their friends there. Their hearts were filled.

More than this Bohr was convinced that the World War was mordant proof that there must never be another war. With his political acuity and awareness, he acknowledged that this would require a revision in the old relationships of nations. But he was also convinced that postwar anarchy in Germany would not contribute to a stable continuing peace and should be avoided at all costs.

Bohr poured this out to Rutherford in what was essentially a letter of thanksgiving:

> I need not tell you how often our thoughts in these days have gone to all our friends in England and how great a part we take in all the joy and happiness which everybody there must feel in the great result which the enormous effort in these critical years has achieved.
>
> Here in Denmark we are most thankful for the possibility which the defeat of German militarism has opened for us to acquire the old Danish part of Schleswig and at this time we feel an immense relief that the war is over.
>
> All here are convinced that there can never more be a war in Europe of such dimensions. All people here have learned so much from the dreadful lesson and even here in these small Scandinavian countries where for good reason there certainly was not much aggressive military spirit before the war, people have got to look quite differently than before at the political side of life.
>
> From all that we hear, we feel also quite sure that the men now in power in Germany take a real peaceful attitude, not for the occasion and not because they have always done so, but because all liberal-minded people of the world have understood the unsoundness of the principles on which international politics have hitherto been carried on.
>
> If therefore there does not come anarchy in Germany due to the great need . . . this time may certainly be looked upon as the beginning of a new era in history.
>
> Here I am longing to be in England now and to be able to talk to you about all sorts of things. I remember as if it was yesterday all the time I sat in your study and you developed for me your views on the different phases up and down through which the war went and how your unflinching belief in a happy end was always able to comfort me, however downhearted I could feel myself at times.
>
> Dear Professor Rutherford this letter is only meant as a greeting in these eventful days, but very soon I shall write all about my work.

The second part of my paper is coming from the printer in a few days. But I'm afraid it will meet with your disapproval because it is rather long (almost three times as long as Part One). My only good excuse is that I have tried really to put a good deal of work into it.

When Bohr accepted the chair of theoretical physics in 1916, he knew that he wanted to make it a post for research as well as teaching. With the end of the war the way now was at last open in Copenhagen to build the department into the kind of center Bohr believed it must be. The department itself was new. Previously there had been no professorship of theoretical physics and Bohr was unhampered by past obligations. Physics proper was taught by Professor Martin Knudsen who was known for his experimental verifications of the kinetic theory of gases. Knudsen had little patience with the quantum ideas Bohr was introducing. J. Rud Nielson, who later became professor of physics at the University of Oklahoma, was then studying with both Bohr and Knudsen. When he once asked Knudsen a question involving the use of quantum theory Knudsen snapped: "If we have to use quantum theory to explain this, we may as well not explain it."

At first Bohr's lecture classes were made up of six or eight advanced students. Gradually as he began verging more into atomic theory some of the faculty members drifted in. The discussion going on in his class drew them almost as if by osmosis. Bohr took his classes deeper and deeper into the subject, forcing them to extend their thinking beyond any point they had previously reached, and perhaps farther than they had ever thought it possible to reach. It was an exhilarating, extending experience.

As he talked Bohr filled the blackboard behind him with masses of figures, symbols, and sketchy drawings. Then he turned, palms resting on the dusty ledge of the board and with a piece of chalk still grasped in his fingers, to look at the class with an expectant smile that creased deep vertical furrows in his long, heavily boned face.

His thick, dark brown hair was neatly combed back into a pompadour that emphasized his high forehead and further elongated his face. Heavy eyebrows that grew horizontally with almost no arching broke the vertical lines of the face, but did not disguise them. Any heaviness or grimness that so ruggedly built a man might have had was denied by his litheness and the soft intelligent eyes. This was a pleasant, responsive face.

Bohr was dressed with all the formality expected of a university professor. His dark coat with its long lapels revealed excellent English tailoring. A watch chain looped through a buttonhole, and the knot of his dark, full tie was exactly in the center of his stiff, white collar. Short of wearing a morning coat Bohr could not have looked more the correct professor than he did. But his manner was far from the usual academic one. "He was always friendly and less remote and dignified than most Danish professors," said Nielson.

In class as the discussion waxed, Bohr tended to forget about time. He made up for it in between. Few moved more rapidly than this soccer-player professor. Bohr often took stairs two at a time. He also would come into the university yard pushing his bicycle faster than anyone else. He was a young man in a hurry.

In 1917, almost a year before the war ended, Bohr had approached the university authorities about establishing a small institute for theoretical physics as part of the university. Very soon after peace was restored, the university announced its formal approval of Bohr's proposal for the institute. The decision certainly was not hindered by the willingness of friends of Bohr's to raise "a big sum of money" to assist the university in building the institute. The "big sum" was about £4500 or $20,000. A former classmate of Bohr's who had become a leading Danish industrialist contributed generously himself and raised the remainder by general subscription.

The final necessary assistance came from the City of Copenhagen. A site was made available on the edge of what Bohr justly called "a beautiful park" not far from the center of the city. Great trees grew behind the allotted land, screening it from the open sunny meadows where soccer games nearly always were in progress in the summer months and where the Danes set up their array of carnival booths and a speaker's stand for the annual Constitution

Day celebration. Bohr was given authority to go ahead as soon as plans could be drawn.

"How glad I am," Bohr wrote to Rutherford. "I look of course immensely forward to it and we hope already that it might suit you and Lady Rutherford to come to Copenhagen to stay with us at the time of the festivities for the opening of the institute. I hope it will take place about a year from now."

A letter that came very close to upsetting "those splendid plans" was in the mails at the very time Bohr wrote his "end of the war," letter to Rutherford. The letter was written November 17, 1918, but in the still disrupted state of shipping, Bohr did not receive it until the second week in December. It came from Rutherford and was marked "private and confidential."

"Well you can imagine that everybody is very pleased and relieved at the dramatic issue of the great struggle," Rutherford began. "The country has been celebrating the great event in a *delirious* way this past week.

"As you may imagine lectures were unattended for a few days, and we were all glad to have a little relief from pressure. The whole nation is in a hopeful and energetic state and turning their attention to the problems of peace."

This brought Rutherford to the proposal that had moved him to mark the letter "private and confidential."

The universities are likely to see considerable development. You may have heard we have initiated a Ph.D degree. We also plan to make Manchester a research center for modern physics. You will remember our talks about a professor of Mathematical Physics in the department. Well, things are likely to develop rapidly. I would like to be sure you are still inclined to consider seriously an offer for a good post, probably at about £200 per annum.

You know how delighted we would be to see you working with us again. I think the two of us could try and make physics boom. Well think it over and let me know your mind as soon as you can. Possibly you might think of visiting us as soon as the seas are clear.

Rutherford added a few words about his work, and whether or not this was his intention, this was likely to be as persuasive as any other inducements he could offer:

"I wish I had you here to discuss the meaning of some of my results on collisions of nuclei," he wrote. "I think I have got some rather startling results. But it is a heavy and long business getting certain proof of my deductions. Counting weak scintillations is hard on old eyes, but with the help of Kay I have got through a good deal of work at odd times in the past four years."

It was an almost irresistible prospect that Rutherford held out—the matchless opportunity to work with him on the most advanced of research, a new center of physics, a professorship at Manchester, and England. Coming on top of the armistice and the go-ahead on the institute at Copenhagen, the Bohrs were again swept to the heights and at the same moment confronted with an impossibility. All the world was offered and all the world was taken away. Bohr and Margrethe almost ran to talk it all over with Harald. Nothing could be said to anyone else, because of Rutherford's stricture on secrecy. The discussion at the Bohr's apartment went on all day and through most of the night, and yet whatever the temptation Bohr knew deep within him what the decision would have to be. He was a Dane; he could not leave Denmark, his own nation which had given him every opportunity and now was presenting him with the institute of which he had dreamed. Such ingratitude and disloyalty were not in Bohr. He did not delay. On December 15 he wrote to Rutherford:

I don't know how to thank you for your letter of November 17, which I have just received and which besides being the source of the greatest pleasure to me has at the same time been the object of sorrowful consideration.

You know that it has always been my ardent wish to be able to work in your neighborhood and to take a part in the enthusiasm and imagination which you impart to all your surroundings and of which I have so much benefitted myself. At the same time I am not in a position to accept the splendid offer, and for which I feel more thankful than I can express because of the undeserved confidence in me that it contains.

The fact is that I feel I have morally pledged myself to do what I can to help in the development of the scientific physical research here in Denmark, and in which the small laboratory will play a part.

I should like so much to settle down in Manchester again and I know it would be of the greatest importance to my scientific work, but I feel I cannot accept the post you write about because the university has done all that they could to place all external means necessary for my work. Of course the pecuniary means, personal allowance as well as that to the running of the laboratory, will be far below the English standard.

I feel it is my duty here to do my best, though I feel very strongly the result will never be the same as if I could work with you.

Rutherford did not immediately give up. Soon after his return from a two-week Christmas holiday in Devon he wrote Bohr of his disappointment: "I hope you will not prejudge the question before you have a chance of coming to England to talk things over. I hope you will do so as soon as it is convenient to travel. There are many things in general I want to talk to you about."

Travel still was nearly impossible. All available shipping was tied up in returning the armed forces to their own countries and in reopening normal trade routes. The trains likewise were jammed. Everyone seemed to be on the move at the same moment.

In any event before he could think of leaving for a trip to England, Bohr had to finish the year at the university. He was also immersed in planning for the institute. Whenever the architect finished a drawing, Bohr would see an opportunity to make an improvement. Then the plans had to be redrawn. No sooner were they finished than another way to improve them would be revealed to Bohr. The architect alternated between collapse and explosion, but the plans were reworked and reworked until Bohr had the results he wanted. This did not make for speed of construction, and hopes for completion within a year faded rapidly.

By July, though, as travel again became possible, Bohr broke away from the drawings and excavations long enough to visit postwar England. Bohr and Rutherford had scarcely greeted each other with a heartfelt grasp of the hand when they plunged into the two questions of paramount import to them both.

The first, of course, was the incredible possibility that Rutherford was breaking into that redoubt of redoubts, the nucleus of the atom, and all the baffling problems this raised. The other was a new matter. An invitation of a lifetime had come to Rutherford. He was being asked to become Cavendish professor at Cambridge to succeed J. J. Thomson and his illustrious predecessors Clerk Maxwell and Lord Rayleigh. The world held no more distinguished a post in physics or in science generally.

In March Thomson had been appointed Master of Trinity College, Cambridge, and had decided to resign the Cavendish professorship. Rutherford was so eminently the man to carry on the incomparable succession that he was quickly "invited to stand for" the great honor. Rutherford had written this highly interesting news to Bohr and assured him that he was all the more anxious to see him in England to talk to him about it.

Rutherford would of course accept the Cavendish. A man does not reject the world's highest post in his chosen field, or certainly not a man of Rutherford's undiminished energy and large outlook. At the same time it was not easy for him to leave Manchester where everything—complete freedom to work as he wished, generous support, unusual students, and cooperative colleagues—had combined to give him almost an ideal setting for the work he wanted above all else to do.

There was also one troublesome question about Cambridge. Could J.J. keep his hands off a laboratory he had directed for almost thirty-four years? Rutherford talked the situation over with Bohr, who knew J.J. of old. Typically Rutherford decided to meet the problem head on. He wrote to Thomson: "Suppose I stood and were elected I feel that no advantage of the post could possibly compensate for any future disturbance of our long-continued friendship, or for any possible friction, open or latent, that might possibly arise if we did not have clear mutual understanding with regard to the laboratory and research arrangements."

Rutherford showed Bohr the reply he received. Thomson said

that he was glad Rutherford was considering coming to Cambridge and he added "if you do so, you will find that I shall leave you an absolutely free hand in the management of the laboratory." Complete honesty had settled a potentially disturbing conflict.

Bohr, though his own course was always more diplomatic, delighted in the exchange. He remembered all too well that his own tactful carefulness had failed for months to move a paper from the bottom of a stack on J.J.'s desk.

The Dane and the future head of the Cavendish were equally anxious to discuss the atom and its new mysteries. Until Bohr arrived, Rutherford had had almost no one he could consult on the amazing but still puzzling results. Rutherford showed Bohr how he had first bombarded the hydrogen nucleus with alpha particles. The alpha particles, four times larger than the nucleus and traveling at a speed of 18,000 miles a second, sent the hydrogen nucleus careening when they hit. A charged hydrogen nucleus also emerged from the violent collision. Rutherford reported this significant result in a paper published in 1919.

But what happened when he bombarded nitrogen proved both more surprising and puzzling. The nitrogen nucleus was heavier than the alpha particle colliding with it, and it should not have been dislodged, or disturbed. It was like a heavy car being hit by a child's toy automobile. Nevertheless, a hydrogen nucleus burst forth from this impact too.

Rutherford was forced to the nearly incredible conclusion that the nitrogen nucleus had somehow been shattered by the impact and that the hydrogen nucleus emerged from the broken nitrogen nucleus.

"It is difficult," he said to Bohr, "to avoid the conclusion that the long-range atoms arising from the collision of alpha particles with the nitrogen nucleus are not nitrogen atoms, but probably atoms of hydrogen, or atoms of mass two. If this is the case we must conclude that the nitrogen atom is disintegrated under the intense forces developed in a close collision with a swift alpha particle, and that the hydrogen atom which is liberated formed a constituent part of the nitrogen nucleus."

What did Bohr think? Nothing could have been more startling or less likely. Maxwell, who seldom erred or misjudged a situation, had argued that the origin of atoms went back, not to the establishment of the earth or of the solar system, but to the establishment of the existing order of nature, and he had declared "until these worlds and the very order of nature itself is dissolved we have no reason to expect the occurrence of any operation of a similar kind [the formation of any new atoms]." But the discovery of radioactivity had demonstrated the spontaneous decay of atoms and Maxwell's successor-to-be was forced to think that he was breaking up an atom and creating new atoms. The bastion of bastions seemed to be yielding, the irrefragable base of bases. Rutherford was a man of confidence who accepted the evidence of nature, but he hesitated. He and Bohr reviewed all the evidence in one long session after another.

"I learned in detail about his great new discovery of controlled or so-called artificial nuclear transmutations," said Bohr. "He had given birth to what he liked to call 'modern alchemy.' In time it was to give rise to tremendous consequences in man's mastery of the forces of Nature." However far out such conclusions might be, Bohr completely agreed with Rutherford's conclusion that the atom was being broken and new atoms formed from it.

To leave in the midst of all of this was a wrench, but Bohr had to get back to Copenhagen. He reluctantly had confirmed his regrets about accepting a post that might in time have made him Rutherford's successor at Manchester. He could do no more than tentatively accept an invitation to return to England the following January for a series of lectures.

The end of the war had once again opened Europe to science. Soon after Bohr's return to Copenhagen he was invited to Holland to lecture at the University of Leyden. He and the Dutch scientists talked "atomic problems from morning to night." Actually almost nothing else was talked in physics circles anywhere in Europe. Hevesy, who came to Copenhagen to visit Bohr, reported that in Berlin they spoke of nothing else.

Sommerfeld also came to Copenhagen. He was one of the first

of the German scientists to venture out again into the neutral world. "We had long discussions on the general principle of the quantum theory and the application of all kinds of detailed atomic problems," Bohr said. "I liked him exceedingly." Bohr held no rancor against Sommerfeld and his associates.

The busy comings and goings following the long closure of the war, the struggle to work on a new paper on the constitution of atoms, and the supervision of the construction of the institute finally wore Bohr down. He had to get away to the country for a little rest.

One worry lay in the increasing costs of construction. Prices were rising precipitously in Denmark as well as in virtually all other countries, and the money originally set aside for the institute was proving wholly inadequate. New sources of contributions had to be found. Bohr asked Rutherford and Sommerfeld to write to the Carlsberg Foundation, emphasizing the contributions the institute might make and its indispensability if physics were to go forward in Denmark. The foundation made a grant, and the necessary money was raised. It looked as if the new institute buildings could be completed and made ready for dedicatory ceremonies in late September or early October, 1920.

Bohr asked Rutherford to choose the exact date, a time when he could come to Copenhagen. He also optimistically invited the Rutherfords to stay in the Bohrs' new "flat" at the institute. In this way September 15 was chosen for the great day. The buildings were under roof and the prospects looked reasonably good as Bohr daily inspected the work. He climbed all through the structure, inspecting every detail and questioning whether it looked right. Was that window frame being installed so that it would be tight? Was another nail needed here, an extra brace there?

The Council on Physical Societies in England agreed to publish Bohr's papers on the quantum theory. At this happy, busy time with everything well on its way, the Bohrs' third son, Erik, was born.

All summer Bohr pushed the work on the institute. Everything though seemed to take twice as much time as he had hoped. By July it was indubitaly clear that the Bohrs' seven-room flat in one of the buildings would not be completed by September 15. The Bohrs were living in the spacious apartment of Bohr's mother, and changed their invitation to the Rutherfords to stay with them there.

Despite the new baby and the aborning institute, Bohr unquestioningly and eagerly accepted an invitation from Planck to lecture on spectral theory at the Physikalische Geseltschaft (Physics Society) in Berlin. An invitation from Planck, the discoverer of the quantum, and thus the man who had provided the first basis for the whole modern structure of physics, was a high point for Bohr. He rushed to show the letter to Margrethe and Harald and talked it over with all his close friends and associates. It would be his first meeting with the scientist he had so long venerated.

As Bohr arrived at the Berlin physics building, Planck and Albert Einstein came forward to greet him. Planck, very much the German professor in appearance, was shaking his hand with the quick German pump stroke. His bald head bobbed in a little bow, but his eyes, behind the small rimless glasses, were smiling warmly at Bohr, and the mustache drooping at the corners seemed to belie the formality of his black coat, his stiff, straight collar, and the black bow tie.

Bohr also was looking for the first time into what Hevesy had called "those great big eyes" of Einstein. His hair flared out in the already famous halo, and Einstein's clothes, whatever they were, never managed to achieve any degree of formality.

Three of the four men who were shaping and would shape the world for at least the next century—only Rutherford was missing—stood talking together, exchanging the very real pleasantries of their first meeting. Their talk stayed only briefly at this level. For almost every minute, except when they had to be in meetings, they settled down to a continuing and intense discussion of physics. "We spent the days discussing theoretical physics from morning to night," Bohr said.

Not long before Einstein had formulated the general statistical rules for the radiative jumps of electrons from one of Bohr's stationary states to another. He maintained that the jumps would occur, not only when the electrons were exposed to radiation and collision, but even without external disturbance. "Einstein emphasized the fundamental character of the statistical description in a most suggestive way, by drawing an analogy between the occurrence of the spontaneous radiative transition and the well-known laws governing transformations of radioactive substances," said Bohr.

Bohr in his lecture on spectral theory maintained that an exact determination could not be made. Einstein did not like this at all. As the heir of the scientific tradition of the last century, Einstein objected to any theory that left to chance "the time and direction of the elementary processes." He was fundamentally convinced that if all the laws were known all action could be predicted. At their first meeting this basic difference between the two men clearly emerged.

"These fundamental questions formed the theme of our conversations," said Bohr.

The discussions to which I have often reverted in my thought, added to all my admiration for Einstein a deep impression of his detached attitude. Certainly his favoured use of such picturesque phrases as 'Ghost waves guiding the photons' implied no tendency to mysticism, but illuminated rather a profound humour behind his piercing remarks.

Yet a certain difference in attitude and outlook remained. . . . With his mastery for coordinating apparently contrasting experience without abandoning continuity and causality, Einstein was perhaps more reluctant to renounce such ideals than someone for whom renunciation . . . appeared to be the only way to proceed with the immediate task of coordinating the multifarious evidence regarding atomic phenomena.

The "someone" willing to renounce in order to get on with the job, of course, was Bohr. It was a theme he was to develop much more fully as time went on. At the first meeting with Einstein, however, the two men raised questions that neither could answer to the satisfaction of the other. Planck seemingly stood outside the polite, though determined fray.

The meeting only whetted the interest of the two principals. Each looked forward to other meetings, where with more thorough preparation, either might be able to prove the fallacy in the other's point of view, and possibly even persuade the mighty adversary. Neither was ever quite able to give up about the other. The intellectual stakes were too high.

With the discussions and lectures taking this exotic bent, and teeming with unfamiliar words and concepts, even the most learned physicists found themselves in difficulties. The students were at sea. "I must confess," said Lise Meitner, many years later, "that when James Franck, Gustav Hertz and I came out of the lecture we were somewhat depressed because we had the feeling that we had understood very little."

Some of the young physicists in a half playful revolt against the esoteric proceedings, decided to ask Bohr to spend a day with them, explaining things to them, at the Kaiser Wilhelm Institute at Dahlem, a suburb of Berlin. All professors, all "bigwigs," were to be barred. Bohr was staying with Planck, and Lise Meitner, who was delegated to extend the invitation to him had to explain that only the guest was invited. Planck, as a professor and bigwig, was disqualified.

Professor Haber, the head of the Dahlem institute, had to be told because his cooperation was needed to obtain food. Food still was very short in postwar Berlin. Haber did not mind at all. He not only offered the food, but the use of his villa and asked only that he and Einstein be invited to lunch.

The young physicists spent the day firing questions at Bohr, who took great delight in answering them, or rather in stirring up discussions about the answers. At lunch, Bohr asked in return only that they explain the word "bigwig" to him.

Upon his return to Copenhagen Bohr had little time immediately to ponder the basic issues raised at Berlin. The institute would not be finished in time for the dedication ceremonies in September, though the buildings would be far enough along for the visitors to see what was being built. A thousand details had to be taken care of—invitations, hospitality for the guests, the arrangement of the seating, and even the placement of the red and white Danish flags that would fly over the buildings and ceremonies. Arrangements also had to be made for the lectures Rutherford would give for an audience that would include most of the "bigwigs" of Denmark. Bohr and Margrethe worked to the point of exhaustion.

September 15, 1920, dawned bright and clear and the *Dannebrog* waved crisply against the bluest of skies. Bohr, his weariness forgotten, stirred the audience with his vision of what the institute would mean to science and the world: "In scientific work no one man can make definite promises for the future," he said, "for obstacles can occur that can only be overcome by new ideas. It is important therefore not to depend on the abilities and powers of a fixed group of scientists.

"The task of continually having to initiate new young people in the results and methods of science leads to discussions and through the contributions of the young people themselves new blood and new ideas enter the world."

Bohr was forecasting much of the mission of the institute. He showed the many distinguished guests through the new buildings and happily expounded on every carefully thought-out feature.

The new four-story building, with the lowest floor partly below grade, formed a gracious, distinctive addition to a row of public buildings and hospitals along the park frontage. "Institut For Teoretisk Fysik" was carved above the double front door, though before long the lettering was only a formal designation. All Danes knew the institute by sight.

A few steps up from the entry a double door opened to a hall and to the lecture room, which with its terraced seating and big blackboard, would provide the setting for many famous discussions.

The library, on the floor above, was a modestly sized room looking out on the green of the park. In summer when the windows were open the shouts of the children playing in the park would drift pleasantly in.

On the top floor in later years was a lunchroom, where good Danish sandwiches spread with slices of cheese and red slices of tomatoes were always to be had.

The laboratories and offices filled the remainder of the building. In the beginning the equipment of the laboratories was not elaborate. Small offices were relatively numerous, for many of the students needed only a pencil and paper and blackboard for their work.

The Bohrs' still-unfinished flat at first occupied most of the top floor. A double door with etched glass panels opened into their seven rooms. Here in the Danish tradition the professor was to live at the institute.

In scale, in size, in the clear gray colors of the walls, and the flower boxes that were to come, the building was thoroughly at home in Copenhagen. It was Danish in spirit and design. No attempt had been made to copy on some lesser scale the majesty of Cambridge. The institute was fully itself.

It was January 18, 1921, before the buildings were sufficiently complete for Bohr to move his books and papers into his modestly sized office with its corner fireplace. As he put his books on the shelves, the real homecoming for him, he sat down to write his first letter at his new black leather-topped desk. The letter, of course, went to Rutherford. The institute was about ready to become the most renowned single center of study since Plato's academy.

7 /

The Nobel Prize

IF SOME LARGE FOREIGN BODY should someday flash through the solar system, the disturbance of our universe would be profound. The earth might well be thrown off its appointed rounds. At the very least the length of our year might be changed. After such an upheaval it would be most unlikely that the solar system would snap back into the ordered orbits men always have known.

But let some external body, perhaps a swift, powerful alpha particle, smash into an orbiting electron in some atom. Let the impact knock the electron into another and distant orbit. When the disturbance ended the electron would jump back to its original orbit and resume the rounds that it had followed from the beginning of matter and would pursue until the end, if end there is to be. A carbon atom remains a carbon atom, a gold atom a gold atom, regardless of what catastrophe it may suffer.

This stability, this timelessness surpassing that of the solar system and almost the imagination of man was explained for the first time in Bohr's atom. The atoms whose structure and workings had been discovered in the preceding eight years stood revealed as the most fabulous of all creations.

To find the answer as to why the atom was such an incredible structure, science would have to probe even deeper into its hidden structure and unfaltering workings. This was the nearly impossible, but exciting quest to which the new institute on Blegdamsvej turned. It was soon to produce an era that even the most conservative of historians and the most reticent of scientists called "heroic."

As the institute opened James Franck came from Goettingen to serve for several months as a visiting professor. To teach the institute fellows the fine technique of making the atom reveal its structure, Franck set up the apparatus he and Gustav Hertz had used in their famous studies of the effect of electric fields on atoms. The atom was excited by bombardment and the radiation given off was run through an elaborate system of prisms to separate the colors. Franck also gave several lectures on atomic collisions and the work that had provided such striking confirmation of the Bohr theory.

Lise Meitner was another early guest. She was invited to lecture on beta and gamma radiation. The invitations to the Germans were courageous acts in themselves, for the German scientists still were excluded from the major scientific congresses. Bohr was making every effort to have them readmitted, and demonstrated his convictions by inviting some of the leading German scientists to Copenhagen. He made them feel warmly welcome.

The Bohrs asked Lise Meitner to dinner at their home, and to put the pretty young physicist at her ease Bohr talked about some of his own war experiences in England, both serious and funny. "Even today," said Lise Meitner more than forty years later, "I can still feel the magic of that meeting."

Hans Kramers, a bluff, hearty young Dutchman, also was there. While the war still was raging Kramers had wandered into Copenhagen as a refugee. A trained physicist, he soon found Bohr, and at once began working with him. His title at the institute was lecturer, but actually the Dutch physicist was Bohr's assistant and he had a hand in almost everything that was done from his arrival until he left in 1926 to become professor of physics at the University of Utrecht.

Bohr also invited his irrepressible Hungarian friend of Manchester days, Hevesy, to come to the institute. Hevesy accepted and remained for twenty years, working on his noted physicochemical and biological research. He was soon to take up the search for the missing element, No. 72.

Oscar Klein came from Sweden and Svein Rosseland, who became an authority on astrophysics, from Norway. Both were young, but both already were known for their work on what were called "collisions of the second kind." These were the collisions in which electrons were transferred by bombardment from a higher to a lower stationary state in the atom.

The early visits of Franck and Meitner confirmed the ties of the institute with the German groups in Berlin and Goettingen. The stay of Klein and Rosseland forged close connections with physics in the other Scandinavian states. As might have been expected, the closest relationships and interchange also were established with the "great center" Rutherford was building at Cambridge.

Almost from the moment he went to Cambridge, Rutherford, as Bohr liked to point out, gathered a large and brilliant company of research workers around him. Aston was back from the war and continued the work that led to the discovery of additional isotopes. This was soon to bring him the Nobel Prize. James Chadwick, who had started to work with Rutherford at Manchester, had returned from a long internment in Germany to join Rutherford at Cambridge. Chadwick had been studying with Geiger at the time the war broke out, and before he could get out of Germany was arrested as an enemy alien. The war was long and the internment camp a tedious ordeal until C. D. Ellis was captured and assigned to the same prison. The two discovered many interests in common, and Chadwick began to teach physics to Ellis. By the end of the war Ellis was so skilled a physicist that he also could join the Cambridge group and contribute significantly to it.

Kapitza arrived from Russia, bringing with him ingenious ideas for the production of magnetic fields of previously unheard-of intensity. From the beginning of his work at Cambridge, Rutherford also was assisted by John Cockcroft, who also was to win the Nobel Prize.

Charles Darwin and Ralph Fowler, who married Rutherford's only child, a daughter, Eileen, taught theoretical physics. Among Fowler's students was P. A. M. Dirac, a mathematical genius who had distinguished himself from his earliest days by his unique powers of logic.

Physically and technically the Cambridge and Copenhagen centers were separate, the one in the old city of Copenhagen where it was establishing a whole new tradition of physical research, and the other in the ancient university town where it was heir to the world's most brilliant tradition of scientific research. And yet the two in many ways not only functioned as one but served as a nucleus to bind the world's physicists into a cooperating, close community of physics. The two centers quickly transcended all the old national and language barriers. They made physics truly international.

A steady procession of the Cambridge men came to Copenhagen. Bohr always was able to obtain fellowships for them. In this way Darwin, Dirac, Fowler, N. F. Mott, D. R. Hartree, and many others spent a year or two working at the institute in Blegdamsvej. In return Bohr and the other Copenhageners often were in England. Work was started in one laboratory and finished in the other. Papers went back and forth and a new idea, a gain, a problem, reached everyone in a time that seemed to rival the speed of the cables. What one learned was immediately open to all.

One of Rutherford's first acts after arriving at the Cavendish was to invite Bohr to England for a series of six to eight lectures. He was able to offer him a fee of £100 to defray expenses—payments for lectures and fellowships were the principal means of financing the close interchange and it is doubtful that any other comparable research was ever accomplished at so exceedingly modest a cost. The University of London joined on this occasion to ask Bohr to give several lectures there while he was in England. An invitation also had come in to lecture at Goettingen, and the first postwar Solvay Conference, was to meet in Brussels.

Bohr wanted to undertake all of these lectures and meetings, but before he could start on the crowded schedule he became ill from what was diagnosed as overstrain. He had to miss the Solvay conference and postpone the lectures.

"I dare hardly ask you if autumn would suit you as well for the lectures," Bohr wrote to Rutherford with obvious embarrassment at asking so august an institution as Cambridge to make a change in plans for his benefit.

Bohr thus missed the dinner that welcomed Rutherford to the Cavendish and to its hallowed though dusty and untidy precincts. One of Rutherford's first acts had been, he wrote to Bohr, to clean up the lab and buy "a great deal of" apparatus. The feat was celebrated in a song written to the tune of "I Love a Lassie."

> We've a professor,
> A jolly smart professor,
> Who's director of the lab in Free School Lane.
> He's quite an acquisition
> To the cause of erudition,
> As I hope very briefly to explain.
> When first he did arrive here
> He made everything alive here,
> For, said he, "the place will never do at all;
> I'll make it nice and tidy,
> And I'll hire a Cambridge 'lidy'
> Just to sweep down the cobwebs from the wall.

And then came the chorus:

> He's the successor
> Of his great predecessor
> And their wondrous deeds can never be ignored.
> Since they're birds of a feather
> We link them both together
> J. J. and Rutherford.

The atom was roundly dealt with in another verse:

> What's in an atom
> The innermost substratum?
> That's the problem he is working at today.
> He lately did discover
> How to shoot them down like plover,

And the poor little things can't get away.
He uses as munitions
On his hunting expeditions
Alpha particles which out of Radium spring.

Bohr's exhaustion proved a little beyond the mending of the usual walking trip in the country. For nearly six months he was unable to carry on his usual overwhelming schedule of work. Some articles were finished for *Nature* and the *Phil. Mag.*, but the lectures had to be postponed again.

To cheer Bohr, and in the high mood of the successful atom hunter, Rutherford wrote: "My work on the atom goes on in fine style. Several atoms succumb each week." Bohr's hearty laugh rang out at this. He read this as well as the Rutherford song to almost everyone in the laboratories.

Honors that normally would have climaxed the lecture series were under the circumstances awarded to Bohr in advance of it. In November, 1921, the Royal Society gave Bohr its highest honor, the Hughes Medal, in recognition of his work on the structure of the atom. Bohr also was made an honorary member of the Royal Institution.

"I feel I have deserved these honors by the English societies very little, but I am very thankful," Bohr wrote to Rutherford.

Bohr's normal health and vigor were returning and the postponed lectures were reset for March, 1922. Six were to deal with the general state of atomic theory and two with the special problems of the helium and lithium spectra.

When Bohr went to England he stayed with the Rutherfords at Newnham Cottage, the old neglected house they had found in "the backs" and had handsomely restored. Lady Rutherford was turning the once weed-grown garden into a Cambridge beauty spot. Bohr and Rutherford walked in the garden while they kept up their talk of physics. But even physics had to yield to the spring flowers; they stopped repeatedly to admire the tulips, the hyacinths, and the whole array of spring blossoms. Sometimes they also walked in the backs, under the great trees and across the close-cropped meadows where black and white cattle grazed in a scene of perfect tranquility.

The lectures went well. Rutherford wrote to Bohr after his return to Copenhagen that they were greatly appreciated. How well they were heard was another matter. Bohr's voice seldom carried to the back rows, and even those in the front rows sat with ears turned like the dish antennas of later years to try to catch the very low sound.

Bohr's sentences also tended to become long and involved, as he sought for the utmost precision and to encompass all essential qualifications. Bohr so frequently spoke and wrote in English and even insisted that he could think in it, that it was easy for the English and Americans associated with him to forget that it was a second language and to expect him to use the tongue as though he had been born to it. Rutherford had complained gently in a letter written shortly before the lecture series that Bohr's English was deteriorating because of his long absence from England. This was, of course, offered as another argument to induce him to come.

Soon after his return from England Bohr finished a group of essays, mostly made up of lectures that he had given, for publication by the Cambridge University Press. Rutherford and Ellis worked over his English, principally to shorten the sentences.

"I'm thankful for the admirable corrections of Ellis," Bohr wrote to Rutherford. "I also recognized your pencil and style. Here we are as usual in theoretical problems till over the ears, and some progress in elucidation of spectra is reached."

Bohr was also preparing for his long-delayed lectures in Goettingen. For this he was working in German. One of the penalties of having been born in a small country was that one was always expected to know and use the more widely spoken tongues. Bohr cheerfully and uncomplainingly accepted this additional burden.

It was in November when an inquiry came from Stockholm. Would Professor Bohr be ready to receive a telephone call from Stockholm the next morning? It could mean only one thing: Bohr was about to receive the Nobel Prize. The Bohrs' hearts beat quickly and seemed not to beat. Through the intervening hours they lived in a state of suspended animation, actually knowing what the call would be and yet not daring to admit that it could possibly be true. It was indeed true. A voice from Stockholm told Bohr that he had been awarded the Nobel Prize for physics for 1922, and congratulated him warmly. As the winner in 1922 he would succeed Albert Einstein, the recipient of the award in physics in 1921. At the moment the news was phoned to the Bohrs, it was released to the world at large.

From all of Denmark, it seemed to the Bohrs, the congratulations came. The King, and the man who ran the little ice cream stand at the corner below the institute, all rejoiced. For Denmark, as for Niels Bohr, it was the greatest of honors. A small state had outdistanced the great powers in one of the highest fields of human achievement. The students and fellows drank many a toast and serenaded their friend and mentor.

Among the stacks of telegrams from all parts of the world was one from England, from Rutherford and the Cavendish Laboratory. And it was quickly followed by a letter: "We are delighted that you have been awarded the Nobel Prize. I knew it was merely a question of time, but there is nothing like the accomplished fact. It is a well merited recognition of your great work and everybody here is delighted in the news. You will have received our telegram sent to you the morning we saw the news in the Times. I hope you will have a really good time in Stockholm."

Bohr was deeply moved. No other message of congratulations meant so much to him as this simple note from the man he revered. He answered: "I am sorry not to have thanked you for your kind telegram, but you will know how I have thought of you in these days. I have felt so strongly how much I owe you, not only for your direct influence on my work and your inspiration, but also for your friendship in these twelve years since I had the great fortune of meeting you for the first time in Manchester.

"I need not say that the award of the Nobel price which will mean immensely for my future conditions of life and work, and not least through the award of the price in 1921 to Einstein, has been such an unimpaired pleasure to me."

In the pressure of the moment, Bohr's English suffered. Prize became "price" and the sentence was a little tangled, but his pleasure in succeeding Einstein as the Nobel Prize winner in physics was abundantly clear.

In an earlier letter Rutherford had twitted Bohr about not writing more often. Bohr answered that he took Rutherford's complaint as "not the smallest sign of your friendship," and explained that he was overwhelmed with work.

Bohr was attempting to write a book on atomic theory, but was making no appreciable progress. Rutherford had advised him to put everything else aside and concentrate on this one undertaking. Bohr found this impossible. Besides administering the institute, carrying on his own work, trying to write the book, and talking to the students in discussions that defied the hours, Bohr was attempting to prepare his English lectures for publication. He was so far behind in the last task that he said he had a bad conscience. The Nobel Prize meant that he had to prepare a lecture, as well as to handle the huge volume of congratulations, the interviews, and all the attendant demands of suddenly having become the first and most noted citizen of his country.

The years, 1920–3, were years of stock-taking in physics, of weighing the accumulating proof about the atom, and of assessing what was unknown and incomplete. In a sense these three years ended the first period of discovery that had been ushered in with Rutherford's identification of the nucleus and Bohr's determination of the structure of the atom. Bohr paced his office and the halls of the institute debating with Harald, Kramers, and all the others the subject of his Nobel lecture. In the end the choice proved easy, perhaps inevitable—"The Structure of the Atom." The moment called for looking back and for appraisal.

On December 10, the anniversary of Nobel's birth and the day of the Nobel Prize ceremonies, the snow lay deep in the streets of Stockholm. Only the spires of the churches of the old town and the tower of the town hall stood out against the lowering winter skies. Stockholm in the low light of winter was a blending of snow and sky. And only the blue flags with their crosses of yellow that flew everywhere made a strong note of color in this chiaroscuro. They spoke too of the importance of the day.

At 4 P.M. in the early gathering Northern dusk, shining black cars and taxicabs converged on the Concert House. Two thousand guests were arriving for the presentation ceremonies. All wore formal dress. At 4:30 o'clock the commanding, stilling notes of a trumpet brought the brilliant assemblage to its feet. Down the aisle came the tall, lean King, ablaze with decorations and the Princess Ingeborg resplendent in tiara and jewels. They were followed by the laureates, Niels Bohr, Francis W. Aston who was receiving the award for chemistry, and the Spanish writer Jacinto Benaventa, the winner of the prize for literature.

The Royal Family took their seats in the gold armchairs of the first row. The laureates continued up a low flight of steps to the flower and crown bedecked platform.

"Neils Bohr"—the name rang through the large concert hall. As Bohr rose and bowed in acknowledgement, a distinguished Swedish colleague intoned: "The Nobel Prize is awarded for his services in the investigation of the structure of atoms and of the radiation from them." Proud words of praise followed.

Bohr gravely descended the steps, crossed the rich Oriental rugs spread over the floor and advanced to where the King stood awaiting him. The Swedish monarch ceremoniously handed him a leather-bound diploma, a gold medal in a leather case, and an envelope informing him of where he could receive the nearly $40,000 that went with the great honor. With a smile and a personal word of congratulation the King warmly shook Bohr's hand. Bohr bowed and, executing the required backward steps with his athlete's agility, returned to the platform, then and thereafter a Nobel Prize winner and one of the world's acknowledged greats.

Cars and chartered buses waited to take the guests to a gala dinner. A toast to the King! The company lifted its glasses. A toast to Alfred Nobel! A toast to Niels Bohr!

The next day, before an audience that was largely scientific, Bohr gave his lecture on the atom: "The present state of atomic theory is characterized by the fact that we not only believe the existence of atoms to be proved beyond a doubt, but we even believe that we have an intimate knowledge of the constituents of the individual atoms."

The quiet statement, so fundamental to all science, could not have been made a decade earlier. It marked the distance science had come in a few short years. The outlines now were clear. To describe the atom, Bohr outlined "our present conceptions": The atom is built of a nucleus that has a positive charge and is the seat of by far the greatest part of the atomic mass, and of a number of electrons of negative charge and negligible mass, moving at a distance from the nucleus, great compared to the dimensions of the nucleus and the electrons themselves.

"In this picture we at once see a striking resemblance to a planetary system such as we have in our own solar system," said Bohr. As the motions and relations of the solar system make the universe, so do the motions, the relations, and the structure of the atoms determine the properties of the elements that make up the earth, its occupants, and the universe.

It was now clear. All the ordinary physical and chemical properties of substances, their color, their chemical reactivity, in fact everything that makes them what they are is dependent, Bohr explained, on the electrons and on the way their motion changes under the influence of different actions. If this universe and all the parts of it have form and definition, and are more than a shapeless blob, it is because of this strange, ordered system.

At the same time, Bohr emphasized, the structure of the nucleus is responsible for another class of properties—the particle expelled with great velocity in the explosion that occurs in what is called radioactivity. "The inner structure of the nucleus is still but little understood, although a method of attack is afforded by Rutherford's experiments on the disintegration of atomic nuclei by bombardment with alpha particles," Bohr said. "Indeed these experiments may be said to open a new epoch in natural philosophy in that for the first time the artificial transformation of one element into another has been accomplished."

Bohr turned first, though, to stability, the ability of the atom to change and yet endlessly to resume the same form. The first two essential clues to an understanding of this unique quality came, he explained, in 1900 when Planck demonstrated that energy is radiated in little packets or quanta rather than in a continuous stream, and Einstein, six years later, established that light comes in quanta. Many workers tried to apply the quantum theory to atomic structure, but the effort was hopeless until Rutherford discovered the nucleus in 1911 and the atom suddenly was seen as a miniature solar system rather than as a "hard, gray or pink ball" seeded with electrons.

"One was led," said Bohr, "to seek a formulation that could immediately account for the stability in atomic structures and the properties of the radiation sent out from atoms." Bohr, the only one whose formulation succeeded, pointed out that he had proposed two postulates:

1. Among all the conceivable states of motion in an atomic system there exist a number of so-called stationary states, which possess a peculiar mechanically unexplainable stability. The stability is of such a kind that every permanent change in the motion of the system must consist in a complete transition from one stationary state to another.

2. The process of transition between two stationary states can be accompanied by the emission of electromagnetic radiation. Conversely, irradiation of the electron with electromagnetic waves of this frequency can lead to an absorption process whereby the atom is transferred back from the latter stationary state to the former. It thus can give out bursts of energy and absorb them.

While the first postulate can account for the stability of the atom, the second, Bohr noted, explains the sharp colors and lines the atom gives off when it is excited. To support his contention that a definite relation may be obtained between the spectra of the elements and the structure of atoms, Bohr cited the hydrogen atom.

Its red, green, and violet lines had been known for many years. Assuming that each is emitted as an electron flashes from one stationary state to another, Bohr, using this evidence, concluded that the hydrogen nucleus is surrounded by four stable states, or as they might be thought of by the layman as four tracks or highways, and only these tracks, in all the vast spaces surrounding the nucleus would be traversable by hydrogen's one electron.

If the transition occurred between the second and third or the second and fourth rings or states, Bohr pointed out, red or green lines would be produced. If the transition were between the inner rings it would yield the violet lines.

"This explanation of the origin of the hydrogen spectrum leads us quite naturally to interpret the spectrum as a manifestation of a process whereby the electron is bound to the nucleus," Bohr continued.

Thus he pictured a captured electron being bound into the outermost orbit. As it was drawn inward by the positive attraction of the nucleus it would jump to the next orbit, and then to the next, and then to the innermost or permanent state of the atom, where it would endlessly remain unless an encounter with some incoming electron should push it into an outer orbit or out of the atom entirely.

And so the general principles were worked out, the outlines of the amazing structure of the atom and its formation of all matter. But the strokes were big and broad. The fine lines that would fill in and support the outlines still had to be added.

The work of Sommerfeld and others had shown that each stationary state was not a single track, but a complex of many lanes or loops. The finding accounted for the finer lines that closer examination was disclosing in the spectra.

While the Germans had developed formal, mathematical methods for fixing the new complexity of the stationary states, Bohr had approached the continued study of the atom in another way. He told his Stockholm audience that he had decided to pursue the connections between the quantum theory and classical electrodynamics. In conjunction with some new work by Einstein and Ehrenfest, it led to the formulation of the so-called correspondence principle.

In careful, sophisticated language Bohr explained that the

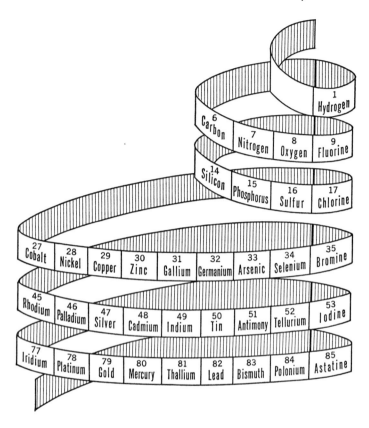

classical principles must be combined with the quantum principles if the atom is to be fully understood. Classical principles might be used as long as they serve, he indicated, but when they become inadequate to explain the stability of a solar system atom, different principles have to be brought in. This was what Bohr had done from the beginning.

Bohr proceeded to demonstrate that the two might be related in a very special way. He started with a quantum jump between two stationary states, a jump that could only be explained with quantum

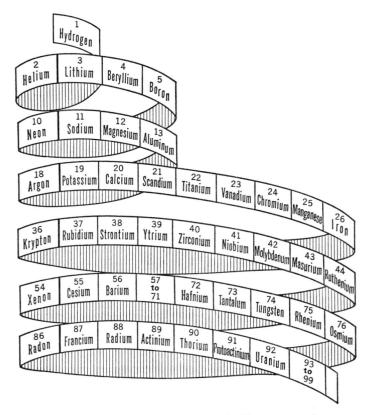

The table of elements shown as a coil. Elements in the same vertical line-up show many of the same characteristics: they may all be gases or all be metals.

principles. But, he said, this flash of energy from one state to another might be coordinated with the ordinary vibration of an excited electron in such a way that "the occurrence of the transition is dependent on the amplitude of the vibration." If an electron were vibrating with a certain violence, Bohr was arguing, it might be predicted that it would take the jump. Thus a vibration occurring according to the principles which physics had accepted for centuries could indicate the probability of a quantum action. It was a linking of the classical and the quantum in a way that brought the confirmations of experiment and the known into the quantum's remoteness and inaccessibility. It was also an answer to Bohr's critics who charged that he used one system when it suited his purposes and another when it did not.

Bohr had an exciting and newsworthy report to conclude his Nobel lecture. One of the gaps in the table of elements occurred at No. 72. According to Bohr's theory, the missing element would have to be similar to zirconium, No. 40, and titanium, No. 22, the two elements in the same column of the periodic table, and not like the rare earths which would stand next to it in the table.

Bohr was disturbed therefore when a report was published indicating that the element with the atomic No. 72 appeared to be identical with the rare earth family, Nos. 57 through 71. If this were true, Bohr knew, it would place an "insurmountable difficulty" in the way of his theory of how the atoms were organized.

It was a challenge. Hevesy and Dirk Coster, a Dutch fellow at the institute, had gone to work to test the matter. They began investigating zircon-bearing minerals by X-ray spectrographic analysis. But as Bohr left for Stockholm, the final checking had not been completed. Only on the night before Bohr was to speak could they wire him that the long-missing element had undoubtedly been found. After hastily revising his speech Bohr reported: "These investigations have been able to establish the existence in the minerals investigated of appreciable quantities of an element with the atomic number of 72. The chemical properties show a great similarity to those of zirconium and a decided difference from those of the rare earths."

The two scientists named the new element they had found hafnium for Hafnia, the ancient name of Copenhagen. It meant harbor or haven ("Copenhagen" means merchant's haven).

"I hope that I have succeeded in giving a summary of some of the most important results that have been attained in recent years in the field of atomic theory," Bohr said in conclusion.

However, he added a few final words of caution. All the knowledge gained, he warned, could scarcely be considered an explanation of the atom in the usual meaning of the word "explanation." "We are obliged to be more modest in our demands," Bohr said.

So far, Bohr declared, the new understandings of the atom had mainly contributed "to the classification of extensive fields of observation," and by their prediction had pointed the way to the completion of this classification. "It is scarcely necessary to emphasize," said Bohr, emphasizing it, "that the theory is still in a very preliminary stage, and many fundamental questions still await solution."

Full of the new honors and with the lecture completed, the Bohrs were ready to depart for the country for some rest and skiing. Before they could leave, however, word came that the children were ill with influenza. The Bohrs rushed back to Copenhagen, for the word flu still was a dread one following the pandemic of the war years. Fortunately, however, the flu was not a virulent form, and the children soon were well.

Niels and Margrethe made a second start on their long-delayed holiday. This time they went to Norway to ski. "Misfortune followed us," Bohr lamented. On their first day in Norway, Margrethe suffered a fall and injured her ankle. She was forced to remain in bed most of the time allotted for the vacation. Nevertheless, Niels said, they managed to get a holiday out of "being away from the household and work."

An auspicious year ended quietly.

8 /

A Heroic Time

BOHR WAS PACING BACK AND FORTH at one of the institute's irregularly regular colloquia. In his concentration his head was lowered, his hands clasped behind his back, his dark hair ruffled, and his eyes lost in thought. He was discussing the overwhelming problems that lay ahead in physics.

Students and fellows sat or sprawled intently in the amphitheater lecture room. "Professor Bohr . . ." came an interruption. To the amazement of students from the much more formal German universities, interruptions were always possible at Copenhagen. There was a question about the direction physics would take.

Bohr looked up, quite unperturbed, and answered with a quotation from Goethe. The words came from *Faust*, which Bohr, like his father, knew so well that he would recite long passages. This time he answered in a few words: "What is the path? There is no path. On into the unknown."

These were words Bohr was to use frequently in the next five years, when he was to make a deep penetration into the most unknown of unknowns. The previous decade with its brilliant light on the previously hidden structure of the atom had only raised other questions as intricate and baffling as those that had been solved. This always is the way of science.

Essentially Bohr had described the atom—electrons moving around the nucleus in certain well-defined orbits and occasionally making a jump from one orbit to another. But how and why the jumps took place, while the stability of this remarkable system was firmly maintained, was not known.

The old and seemingly naïve question of Rutherford remained: "How does the electron decide what frequency it is going to vibrate at when it passes from one stationary state to the other? It seems to me you would have to assume that the electron knows beforehand where it is going to stop." There still was no answer to this critical point.

The classification of the stationary states as great highways where the electrons made their rounds and were exposed to the action of external forces had helped only "temporarily," Bohr said. Models of the atom as a nucleus surrounded by circles and ellipses in many planes were becoming too complex for an artist to draw. They resembled strange flowers drawn in outline. Difficulties also arose in attempting to account in detail for the complexity of atomic spectra, and especially the duplex character of the arc spectra of helium.

Thus there were gaps and contradictions. Bohr had shown the structure of the atom and supplied the clue to understanding the chemical and optical properties of matter, but much remained to be done to incorporate these postulates into the framework of a general theory. There was a lack of the necessary logical and mathematical underpinning. As Bohr emphasized, a quantitative as well as a qualitative picture of the atom was needed.

At the end of one colloquium one of the students came up to Bohr and said that the discussion and the depths that had to be explored made him giddy.

"If anybody says he can think about quantum problems without getting giddy, that only shows he has not understood the first thing about them," Bohr replied with obvious personal agreement.

Bohr was the opposite of the scientist who arrives at a theory and spends the rest of his life defending it. Bohr rather sought out flaws and deficiencies. He insisted upon examining the problem from every possible side, and was never content to take the first solution that offered. Any problem under his study became almost unrecognizable from the one with which he had begun. This was the essence of the Copenhagen method. Together with the original, subtle mind of Bohr, it was to be a tool for bringing most of the world's physicists together in an interworking unity for the understanding of the universe.

In the fall of 1921 Bohr was invited to Goettingen for a "Bohr festival" at which he would give a series of lectures. Sommerfeld came from Berlin to attend and brought with him his two best students, Werner Heisenberg and Wolfgang Pauli. In the discussion following one of the lectures, the 19-year-old Heisenberg stood up and suggested that there was something wrong in some of Bohr's results; furthermore, he effectively supported his point. Bohr was delighted and promptly invited the youngster to go for a walk with him to continue the argument. On another walk Pauli joined them and jumped into the discussion. This was exactly what Bohr liked. He invited both young physicists to come to Copenhagen to carry on their work. Pauli came that year, and Heisenberg for his first long stay in 1924.

All during 1923 and 1924 Bohr with the particular assistance of Kramers worked to establish the coherence of atomic theory. At the same time he found himself involved in a very human fuss that often seemed as giddy as the atom itself.

Hevesy and his co-workers at Copenhagen had no sooner made their formal report on the discovery of hafnium, and popular excitement over something new in the world still was running high, than an elderly Irish chemist named Scott stepped forward to claim he had made the same discovery in 1913. He displayed a tube of material which he maintained was the hitherto missing No. 72. He had called it *celtium*, in honor of Ireland, and thus involved the national honor in his claim.

Hevesy, Bohr, and the institute were thrust into an embarrassing position. The new Nobelist and his associates were in effect accused of claiming credit for work previously done by another man. Bohr went around the institute telling everyone "It's terrible; it's a terrible muddle." The Copenhagen staff rushed to recheck their records. Bohr pointed out at once that it would have been virtually impossible for the Irish chemist to make certain of the existence of a new element without the most sophisticated X ray study of its spectra. There was no indication that such methods had been used. But Bohr, who was so scrupulous in giving credit to others that his papers became almost encyclopedic, could not take Scott's claim lightly.

As in every crisis, large or small, Rutherford was standing by. In a scrawled longhand note to Bohr he wrote: "Things are getting quite lively about the new element. Don't worry. I will see that you and your people get a square deal."

This was a partial relief, and Bohr quickly replied: "You can hardly imagine how great a comfort your kind letter was to all of us in the terrible muddle about the new element. We had not even dreamt about any competition with chemists in the search for new elements." Bohr looked upon the search for the missing elements as a problem of physics and structure.

In the end, it was clear to all that only a test, not letters, could settle the question. Through the intervention of some of the English scientists, a sample of celtium was sent to Copenhagen. Hevesy at once went to work. He soon found that the celtium sample contained not even a trace of element No. 72.

"I suspected it the whole time," said Bohr, returning to his point that any determination made without spectral analysis—without physical methods—was questionable at face. Bohr and Hevesy prepared a report for *Nature* and sent it on to Rutherford for preliminary review. As usual Rutherford corrected and livened up the English of the Danish and Hungarian authors.

"I'm sorry Scott was hoisted on his own petard," Rutherford wrote. "He's a good fellow, but elderly. The modern development of physics has gone over his head."

Bohr also was filled with regrets about the embarrassment of the Irish chemist. He was always willing to go to almost any lengths to avoid hurting the feelings of anyone. His harshest public comment on a speech or paper that he thought made no sense at all was "very interesting, very interesting." "I'm sorry for Scott's sake," he wrote to Rutherford. "We are continuing the investigation of his preparation in the hope of discovering some other new substance in it, which would of course give us much pleasure."

Unfortunately the sample did not yield anything that could be used to save face for Scott. But in the research stimulated by the dispute, Hevesy prepared a concentrated sample of hafnium. The colors and lines of the spectrum showed no resemblance whatsoever to the spectrum ascribed to celtium. Hafnium went into the table of elements as a lasting and self-designating tribute to Copenhagen. More than this, the dispute underwrote the correctness of Bohr's analysis of the periodic table and how it was formed. At Number 72, Bohr's theory indicated, hafnium should be homologous to titanium with 22 electrons and to zirconium with 40. When the elements were arranged in a helix, hafnium was in the column directly under titanium and zirconium (see pp. 122–3). So it would have some of their qualities rather than those of the rare earths next to it in the table. Hevesy's researches not only proved that hafnium was related to zirconium in all its chemical properties, but that it was contained in all common zirconium metals. It had not previously been detected, but it proved to be among the metals that are quite abundant in the earth's crust.

Even before the hafnium dispute had come to its natural death, Bohr had other and more pleasant duties to divert him from his work and the work of the institute. Sir Henry Miers, the head of the University of Manchester, wrote that the university senate was proposing to confer an honorary degree on him.

"I'm thankful that they remember me," said Bohr with great pleasure at such an honor from the university where the work of his life had been so auspiciously started. But Cambridge was not to be outdone, or perhaps it was Rutherford. On March 2 in a letter that had the sound of a whisper on paper, he wrote: "unofficially and quite confidentially your name has been submitted for an honorary degree in June."

As usual Rutherford knew exactly of what he spoke. Cambridge, preeminent among research centers in his eyes and the place where he had first ventured forth in the world of science, would make Bohr one of her doctors. Other than the Nobel Prize no greater honor could come to him.

For all of his deep gratitude and gratification with the honor Bohr did not think that he could manage to get to England for the June ceremonies. He was planning on attending the fall meeting of the British Association and then going on to the United States to accept a number of lecture invitations. His previously proposed trip to America had been canceled by his illness, and he had yet to see the American continent and meet its scientists. He knew that he could not afford the time for both a spring and an autumn trip. He therefore wrote to Rutherford that he was greatly honored, but did not think that he could be present. At this juncture he learned that Cambridge does not grant her degrees in absentia.

"The degree drops if the recipient does not turn up," Rutherford informed him, and added lightly and persuasively, "a sea voyage will be good for you and I'll be delighted to put you up."

Bohr did some rapid shifting of his plans, and made a reservation to sail on June 5. When he thus came back to Cambridge for her highest honor the university was at its most beautiful, with the buildings even more mellow than he remembered against the blue, cloud-flecked skies of spring and with the green lawns running smooth to the river's edge. Students punted along the quiet Cam, relaxing in the softness of the spring and the end of the university year.

At 38 Bohr was as lithe and muscular as when he had played soccer on the fields at Cambridge twelve years before. Clad in the traditional scarlet gown, he strode to the Senate House with a vigor that was not common to the generally older doctors of the university.

The solemn ceremony was followed by a whole succession of festivities. The venerable Cambridge Philosophical Society made Bohr a member at one of its formal meetings. Speaking briefly, he recalled that as a young student struggling with another language and the frustrations of research in unfamiliar precincts he had gone to the society meetings both to learn and to attune his ear to English. Bohr's feelings of awe at Cambridge had not entirely

changed. He still worried that Cambridge might, as he told Rutherford, find him not a "very representative person."

Rutherford gave a dinner in the "Old Combination Room" at
Trinity College. Most of the Cavendish staff in both physics and
mathematics was invited, and in this company Bohr felt at ease.
This was in part a gathering of friends and the talk soon turned to
physics.

The Bohrs had been thinking about a house in the country.
Niels's illnesses had served warning that he needed to get away
regularly from the pressures and tensions of the institute and the
city. In the summer of 1924, soon after Bohr returned from
England he and Margrethe found exactly the house they wanted,
Lynghuset (Heather House) at Tisvilde.

Tisvilde was less than thirty miles from the center of Copenhagen; in remoteness it was a century away. It was, in an oddly
literal sense, a fairy tale house set on the edge of an enchanted
forest. If all the illustrators of fairy tales and all the designers of
opera sets for *Hansel and Gretel* had seen Lynghuset, they could not
have created better copies of it. It was the kind of house country
people of an earlier time built on the edge of a forest—long and low
and small of scale, though with a generous number of rooms.
Thatch a foot thick covered the gabled roof and climbing roses grew
on the white-washed walls. The walls were thick and solid as they
are built in Denmark, and the thatch very effectively shed the rain.
Though the rooms were low ceilinged, they were not cramped, but
rather, the kind of rooms to offer shelter and comfort against
whatever the north winds or the forest might bring.

Lynghuset stood atop the low mound of a dune, though the
grass and trees had completely covered the sand when the Bohrs
first saw it. The forest, some fifty feet from the front of the house,
also grew on the dunes piled by sea and wind along the coast of
Zeeland. About two hundred years before the Kingdom of Denmark
had planted the forest to anchor the shifting sands. The beeches
and pines had not only sent their roots deep to hold the sand, but in
time had spread a high canopy of green over the undulating

seashore. Very little underbrush grew under the thick green roof of the trees. As the sun slanted through the leaves and fell in moving patterns of light across the tree trunks and the needled forest floor, the motes of dust caught in the light formed soft veils that draped the forest in a mystic glow. In the soft shadows it was easy to believe in the goblins and witches of story and tradition.

William Scharff, the artist, owned the house nearest Lynghuset. As a boy in the 1890s he had visited his grandparents at Tisvilde. He often told the Bohrs about those days when Tisvilde was completely isolated and the people lived according to the old traditions. "Witches and warlocks were actual facts to people who had contact with the outside world only when they drove with horse and cart either into Hillerod, or as very seldom happened, directly into Copenhagen."[1]

So it would always be, Scharff thought, as long as the forest stood: "The moon will always shine, big El Greco clouds will scurry across the dark sky and the treetops will cast their shadows over the moonlit earth, bringing forth the goblins who infested the neighborhood in our youth." The old stories had never died, but had lived on in the memories of adults and in the shivering delight and belief of children. Who could deny that the forest was enchanted?

The Bohrs' sons Aage had been born in 1922 and Ernest in 1924. Tisvilde was the right place for five lively boys and the Bohrs moved out in the summer as soon as school closed.

Bohr, however, had work to be done, even in the country. He generally was accompanied to Tisvilde by one or two of the institute fellows who were working most closely with him on some speech, or report, or piece of research. To obtain the quiet necessary for this work Bohr built "the pavilion," a small thatched room a short distance from the house. It was a retreat, but the work stopped frequently for a ball game, or a swim, or a walk with the boys. The men who stayed longest at the institute and collaborated with Bohr in his own projects generally became "uncles" to the boys. Over the years the children delighted in having a dutch uncle, Kramers, a

[1] See the chapter by William Scharff in *Niels Bohr His Life and Work*.

Swedish one, Klein, a Hungarian, Hevesy, and a German uncle, Heisenberg. They formed invaluable additions to all the games.

As Bohr and his assistants worked in the pavilion they often heard the whole troop of children going by on their way to Scharff's. The artist had established something of an art school for the Bohr youngsters and his own. They were all equipped with smocks and brushes. Sometimes they painted on paper, and sometimes, in what they interpreted as a search for realism and authenticity, they applied their paint directly to the trees and leaves.

Tree felling was another occupation in which they all engaged. Trees flourished on the old dune and to keep sufficient sun and light for the house and lawns some of them had to be cleared away.

"It gave father great pleasure to improve the grounds," Hans recalled. "It was his aim to encourage the natural development of every part of it and in order to protect its beauty and give it light and air some trees had to be cut down. This was not done until all had been thought out very carefully and it meant much to him that all interested parties fully approved.

"Tree felling was a great experience for us children and it was carried out according to the best precepts. Each one had his allotted task to perform, with axe, saw, or heaving on the rope to pull the tree down, and father preferably took part in everything from the felling itself to the stacking of the sawn and split logs. The trees which had suffered from standing in the shade had the withered branches sawn off and the bark cleaned of lichen with the gardener's gentle hand."[2]

The Bohrs also spent part of the Christmas holidays at Tisvilde. The country and forest in the glistening snow were the very picture of Christmas. A part of the tradition was a card game around the big open fire at the Scharff's. It was apt to go on late in the evening, and Bohr finally would appear to take the players home. Generally he himself then joined the game. Later still Margrethe would phone to ask if they ever were coming home.

[2] See the chapter by Hans Bohr in *Niels Bohr His Life and Work*.

During Christmas, too, Bohr would gather all the children around the fire and read Dickens to them. He was introducing them to Tiny Tim and the English traditions exactly as his father had drawn him into an early interest in England.

All the while Bohr and the workers who gathered around him continued the struggle to formulate the principles of quantum physics. The going was extremely difficult. Finally Bohr, Kramers, and J. C. Slater concluded that a consistent theory could be developed only if they renounced the conservation of energy in the single process. This was attacking one of the bastions of physics—the law that energy is never lost but is only changed into another form. Nevertheless it looked to the Copenhagen group as though the law did not hold in every instance. Bohr dared to say that energy might be conserved only "statistically" or on the average, rather than in every single process. "He was not afraid when the evidence led him there," said an associate.

However Bohr, Kramers, and Slater had scarcely published the heretical theory when it was shattered by a finding in America. Arthur Holly Compton reported that the frequency of X rays coming from a collision of a photon with an electron differs from the frequency of the wave that set up the action. Subsequent cloud chamber photographs of such a single collision showed that energy and momentum were in fact fully conserved. Bohr and his associates had been wrong. Their attack on the law of conservation of energy was short lived.

In the fall of 1923 Bohr made his long-anticipated visit to the United States. He went to Schenectady, Princeton, Washington, and Chicago. The schedule was crowded. At Schenectady Bohr met Irving Langmuir and was much impressed with his personality and ingenuity. The meeting with Albert A. Michelson at the convention of the American Physical Society in Chicago did not go so well. "I believe he found in me a more conservative scientist than he expected," said Bohr. "At any rate, decidedly more so than the younger school of American physicists."

Bohr also met Compton, and they discussed the "Compton effect" that had so affected the Bohr conservation loss theory.

The Danish scientist found the United States "strenuous but refreshing." If there was no paradox in this reaction to his busy trip, there certainly was in his view of the American future and past. Bohr told friends that he could not avoid "feeling how great the possibilities are for the future." At the same time he decided that he would not want to live there, for he would "miss the traditions that give color to life in the old countries."

Tradition had relatively little to do with the excitement and problems that gripped the institute all through the next four years, 1924 through 1927. The dispute over hafnium had called attention anew to the problem of the arrangement of the electrons in the atom. Bohr had shown what the arrangement generally is, but there was no answer as to why each orbit had its own number of electrons and no more and no less. Pauli had long prodded Bohr about how neatly he had put all the electrons in their own orbits around the nucleus. If the electrons are going to play ring-around-the-rosy with the nucleus, he asked, why doesn't the inner ring, the permanent state to which the electrons work their way, become more and more tightly packed with electrons? Obviously this did not happen.

Pauli undertook to solve the problem himself. Since Bohr had originally brought him to Copenhagen, Pauli had been shuttling back and forth between the institute and his teaching post in Hamburg. An expert physicist, he had a logical, critical mind and a sharp, nimble wit. If a loophole had been left in a theory, Pauli would be sure to detect it and comment devastatingly on the lapse. In combination with his roly-poly figure and jovial disposition, his criticisms were relished by Bohr and the institute.

"The witticisms amused Bohr," Rosenfeld once said. "The sarcasm left him undaunted, and the underlying criticism he took very seriously."

Stories about Pauli always were making the rounds of the institute. Bohr liked nothing better than to hear a new one, and lost no time in hurrying off to tell it to whoever might be working next door. The "Pauli effect" was thus well known in Copenhagen and throughout Europe. Pauli, like many theoretical physicists, was adjudged awkward and a menace in the laboratory. He had only to walk in the room, many stories could demonstrate, and the apparatus would break down. Once an elaborate apparatus set up in the laboratories at Goettingen blew up without any possible reason. Investigation then disclosed that the explosion occurred at the exact moment when a train bearing Pauli from Zurich to Copenhagen had stopped in the Goettingen station.

Pauli wrote in the same light vein in which he talked, and the arrival of a Pauli letter was an event at the institute. Under the guise of drafting a reply Bohr would carry on a dialogue with Pauli. It was almost as though his corpulent friend and critic were sitting there with his skeptical smile on his face.

Pauli succeeded at this time in showing that each quantum orbit, each of the lanes in the major highways or stationary states, can hold no more than two electrons. When the two places are taken, Pauli proved that the next electron must go into another orbit. And when all the orbits or lanes in any given highway were filled, the lanes in the next highway would begin to fill. Pauli descriptively and accurately called this ingenious idea the "exclusion principle." Once an orbit, or lane, had its two allowed electrons, all others were excluded.

As heavier elements are formed, more orbits are occupied by electrons, Pauli further demonstrated. Thus, Pauli argued, in person, in letters to Bohr, and finally in that star-struck year of 1925 in his paper, the lowest energy state or shell or highway is filled in the helium atom by two electrons. With all other electrons excluded from that ring the next electron must go into the next shell, consisting of one circular and three elliptical orbits, and the result is lithium, No. 3 in the table of elements. When the second highway is filled, ten electrons, two in the first highway and two each in the four lanes of the second highway will be whirling around the nucleus, and neon will be in existence, the element which has transformed American business districts at night into contorted patterns of glowing red.

The next electrons, two to an orbit, would fill the orbits of the next highway, and the others the additional highways and orbits until the 92 electrons of uranium, then the heaviest known element, were circling the nucleus in 46 different orbits.

Pauli thus completed the basic explanation of the periodic table of elements, the alignment of which was first seen and described by Mendeleef and made more understandable by Bohr and Moseley.

Bohr, delighted with this notable contribution from Pauli, predicted that it would prove fruitful for the understanding of some of the most varied aspects of the constitution of matter. His prediction was not long in coming true. Pauli's work and the discussions at the institute led two young Dutch physicists, Samuel A. Goudsmit and George Uhlenbeck, into a closer study of the electrons in their swift, two-by-two rounds in the atom.

Through an examination of the fine structure of the spectral lines they concluded that the tiny electrons revolve on their own axes as they make their circuits. They spin around themselves, much as the earth does on its own axis in its big loop around the sun. Gradually it was learned that the electrons, one revolving in one direction and the other in the opposite, produce weak magnetic fields that slightly change the orbits of the two electrons. Pauli's original concept was thus somewhat altered. Each electron was given its own orbit, though the two orbits of a pair are very close and similar.

In a few short months vistas had opened that would have been incredible even a year earlier. But always the questions were there, tantalizing and challenging: what is the exact relation of all the parts of the atom, and why does the atom behave as it does? The physicists wanted not only to describe the atom but to say why it is organized as it is. Unquestionably the rules were there, if they could be discovered and formulated.

As Bohr and Kramers worked at this overshadowing problem, Kramers concentrated his efforts on the dispersion or scattering of light as it hits atoms. By studying the effect produced, some information could be obtained about the stationary states of the atom. Kramers took the bold step of applying Bohr's correspondence theory to the phenomenon: perhaps there was an underlying correspondence between the way light was scattered in the tiny world of the quantum and in the ordinary world; perhaps the electron hit by a photon of light reacts in much the same way as a pendulum if it is struck a certain blow. At the time no one foresaw that this insight would lead rapidly to fundamental consequences.

At about this time young Heisenberg arrived from Germany. He went to work with Kramers on a further refinement of the dispersion theory. Suddenly Heisenberg perceived that the quantized vibrations of the electrons could be transposed into a mathematical law, or algebraic symbols. At first Heisenberg himself did not know the exact nature of the algebra. But the word of this momentous discovery was at once sent to Max Born at Göttingen. Born was familiar with a "theory of matrices" that had been developed at Göttingen and proposed that it could be applied to the Heisenberg idea. In what physicists consider an incredibly short time—three weeks at Göttingen—Heisenberg, Born, and Pascual Jordan wrote the famous paper that laid the foundation of what was called quantum mechanics, the basic working of the atom.

Bohr wrote to Rutherford: "Heisenberg is a young German of gifts and achievement. In fact because of his last work prospects have at one stroke been realized which, although only vaguely grasped, have for a long time been the center of our wishes. We now see the possibility of developing a quantitative theory of atomic structure."

Like a chain reaction the Heisenberg-Born-Jordan paper produced another advance at Cambridge. A young student there who went about his work so quietly that hardly anyone knew that he was around, developed an extension of the Heisenberg idea. P. A. M. Dirac, with the most elegant mathematics, established a correspondence between the quantum and classical mechanics.[3]

---

[3] Later, when Dirac was at Copenhagen, someone asked Bohr how Dirac was getting along. Bohr said that not much had been "heard" from him. J. J. Thomson, who was listening to the conversation, put in: "That reminds me of the parrot that wouldn't talk. When his purchaser went back to the pet store to complain, the owner had an explanation. 'So, you wanted a talker, I sent you a thinker.'"

At Copenhagen the discussion was fast and excited. Colloquia seemed to coalesce. Bohr would come bounding into the gray-walled lecture room, and they would start. Heisenberg argued that in the light of the new knowledge they should deal only in observable quantities. He wanted to throw out the whole concept of orbits as only an imaginary thing. This was too much for the ever skeptical Pauli. He exploded the new Heisenberg venture with some sophisticated mathematics and a sarcastic observation: "The moon, like the electron, occupies a stationary state. We all can see it following its orbit. So orbits must exist under some circumstances. There must be a place for them."

Bohr beamed: "How wonderful that we have met with a paradox. Now we have some hope of making progress." Was Pauli or Heisenberg right? Where would either idea lead? Both proposals would have to be examined to the fullest extent. By this time Bohr was pacing back and forth. His thoughts always came more readily when he was in motion. Tension mounted. Somebody threw open a window and leaned out for a little more air. And so the discussion went on late into the night.

Early the next morning Bohr was banging on Heisenberg's door at the institute—doors there always were kept closed. He had another idea. More figures went on the blackboard. At the end of another day or so of unrelenting effort and exhaustion, Bohr said to the group, as he often did in such moments, "Let's do something." Since a western movie was playing, this meant "Let's go to the show."

George Gamow unforgettably tells the story of the movies: "One of our duties was to take Bohr to the movies and explain the plot to him. He was a slow thinker and always asked questions 'Is this the sister of the cowboy who tried to steal the herd of cattle belonging to her brother-in-law?'

"He was an addict of American Westerns with such titles as *The Gun Fight at the Lazy Gee Ranch*, or *The Lone Ranger and the Sioux Girl*."

The Heisenberg thesis was scarcely dissected, tested, and extended, and put into association with the known facts about the atom when other reports again changed the whole situation: A paper from Erwin Schrödinger at Zurich and a newly discovered two-year-old paper by Louis de Broglie of Paris.

In 1924 de Broglie, a French aristocrat who bore the title of prince, had proposed in his dissertation at the University of Paris that matter—the particle, the unit of units, the concrete, the discrete—might also behave like a wave. It was a startling new concept.

A former student of medieval history who had turned somewhat late to physics—he was then 32—de Broglie took his own original way. His examiner, the scientist Paul Langevin, recognized that there might be something significant in what de Broglie had said. He called it to Einstein's attention, and Einstein mentioned the de Broglie idea in a footnote in his paper on the theory of gases at low temperatures. There Schrödinger saw it.

Schrödinger was instantly interested. But de Broglie had not been explicit about the nature of the waves he postulated, and Schrödinger began to examine the problem. Were the waves accompanying the electrons on their rounds real waves, like the waves rippling out across a pond if a stone is tossed into it? Or were they only mathematical fictions? Several years were to pass before the waves de Broglie had proposed were discovered experimentally.

Starting with the concept of waves, Schrödinger set out to build up a logical picture of the atom. He assumed that the atomic orbits are confined waves, waves that assume certain well-defined shapes and frequencies as do sound waves produced by violin strings or those on the surface of water in a vibrating glass. In this way the waves of each orbit would take on their own distinctive form. With this insight Schrödinger was able to explain the stability and identity of atoms and their astounding ability to regenerate themselves —all properties that had been strange and partly incomprehensible on the basis of the planetary model alone. Stability exists, Schrödinger demonstrated, because the lowest orbital wave pattern does not change unless considerable energy is added to it. In the absence of the energy necessary to move the electron to a higher state, the orbit holds; it is stable.

Because a confined wave pattern is always the same, Schrö-dinger showed, each atom must have its own distinctive identity. One hydrogen atom, for example, has, in this theory, to be like all other hydrogen atoms, for the electron waves are confined by the attraction of the nucleus and the electric effects of the other electrons in the atom.

"The identity of two gold atoms comes from the fact that the same number of electrons are confined by the same electric charge in the center and therefore produce the same wave vibrations," said Victor Weisskopf, explaining the Schrödinger theory.

The same confined wave theory, Schrödinger also found, ex-plains the atom's return to its original state no matter how it may be distorted or how many electrons removed. When the original condi-tions are re-established the electron vibrations must again assume the same pattern, and since the patterns are uniquely determined by the conditions in which the electron moves they are quite indepen-dent of what happened before. Whenever the atom returns to its original conditions the waves take up the same states they held before.

"It is remarkable that we actually find in the world of atoms what Pythagoras and Kepler sought vainly to find in the motion of the planets," said Weisskopf. "They believed that the Earth and other planets moved in special orbits, each unique to the planet and each determined by some ultimate principle independent of the particular fate and past history of our planetary system. There is no such principle in the motion of the planets, but there is in the motion of electrons in atoms—the wave principle.

"We are reminded of the Pythagorean harmony of the world: the atomic quantum states have specific shapes and frequencies that are uniquely pre-determined. . . . Here we find the 'harmony of the spheres' reappearing in the atomic world, but this time clearly understood as a vibration phenomenon of confined electron waves."

When Schrödinger first had the idea that the wave theory might be generalized into a concept that would explain all of the atom, he applied his equations to the behavior of the electron in the hydrogen atom. His results did not quite agree with the experimental data. Schrödinger was so disappointed that he abandoned his work for a few months. However, he could not stay away from it. In working at it again he noticed that when he applied the theory in a more approximate way it did agree with observations. Despite the small discrepancy, Schrödinger then went ahead with the publication of his paper. Only later did he learn that he was right and the observations wrong. Mathematics proved more accurate than experiment.

Schrödinger's paper and the discovery of de Broglie's work hit Copenhagen with explosive effect. A colloquium was immediately called. "Here is an exceedingly powerful and fertile method," said Bohr, and the examination began. After the long dearth of over-all mathematical determinations of the atom, the sudden appearance of two different systems, the Schrödinger-de Broglie system and Heisenberg's, was overwhelming. Were they in conflict or agreement? Both theses were intricate, dizzying even to the keenest intellects.

Physicists felt the need to consult with their colleagues about the confounding double breakthrough in 1925. Many came to Copenhagen. The debate went on in large meetings and small, in twos and threes, at dinner at the Bohrs' and in late-night sessions over trays of sandwiches Margrethe brought in.

How could both theories be so different and yet so inevitably correct in their major aspects? Nerves grew tense and some of the seminar participants were limp with exhaustion, when Bohr turned around from the blackboard and with one of his winning smiles, said: "This reminds me of the little boy."

Heads were raised; the weary physicists revived at the prospect of a story. Bohr went on: "The little boy went to the grocer's shop with a penny in his hand and asked 'Could I have a penny's worth of mixed sweets?' The grocer took two sweets and handed them to the boy, saying: 'Here you have the two sweets. You can do the mixing yourself.'"

The roar of laughter could be heard out in Blegsdamvej. The

one story brought forth a few others, and then they went back to work with even deeper concentration, for these were questions touching the very core of the universe and they all knew they were a part of one of the greatest of explorations.

The mixing of the two finally came with dramatic suddenness. Schrödinger's wave theory proved to be the equivalent of Heisenberg's quantum mechanics. Each was a way of describing the same reality, and in the end each clarified and supported the other. As Rosenfeld later suggested, one might go around the block in one direction and the other in the opposite, but they both circled the same area.

At this staggering moment in this *annus mirabilis* Bohr was invited to Cambridge. He was glad to accept, for the visit would give him an opportunity to consult with Rutherford about the problems that lay ahead and that were unresolved by all the brilliant progress. "I look forward to private discussions about our present theoretical troubles which are of an alarming character indeed," Bohr wrote to Rutherford as the arrangements were being made.

Rutherford took the great difficulties more philosophically: "Your letter sounded alarming, but you cannot expect to solve the whole problem of modern physics in a few years. So be cheerful over the fact that there is still a great deal to do."

Bohr was thoroughly delighted with this wry comment and prized the letter. He and Rutherford laughed about it when he arrived in England. The discussions were stirring and continuous, but the social demands of Cambridge were not to be denied. A tea was given in honor of Bohr, a tea immortalized, or made unforgettable, in the verse of Gamow.

> . . . that handsome, hearty British lord
> We knew as Ernest Rutherford.
> New Zealand farmer's son by birth,
> He never lost the touch of earth;
> His booming voice and jolly roar

Could penetrate the thickest door,
But if to anger he inclined
You should have heard him speak his mind
In living language of the land
That anyone could understand!

One day George Gamow, as his guest
By Rutherford was so addressed
At tea in honour of Niels Bohr
(Of whom you may have heard before).
The men talked golf, and cricket too;
The ladies gushed, as ladies do,
About a blouse, a sash, a shawl—
And Bohr grew weary of it all.
"Gamow," he said, "I see below
Your motorcycle. You will show
Me how it works? Come on, let's run!
This party isn't any fun."

So to the motorcycle Bohr,
With Gamow running after, tore.
Gamow explained the this and that
And Bohr, who on the saddle sat,
Took off to skim along the Backs,
A threat to humans, beasts and hacks,
But though he started full and strong
He didn't sit it out for long,
No less than fifty yards ahead
He killed the engine dead
And turning wildly as he slowed
Stopped traffic up and down Queen's Road.
While Gamow rushing to the fore,
Was doing what he could for Bohr
Who should like Jove himself appear
But Rutherford. In Gamow's ear

He thundered: "Gamow! If once more
You give that buggy to Niels Bohr
To snarl up traffic with, or wreck,
I swear I'll break your bloody neck!"

After his return from England, Bohr invited Schrödinger to come to Copenhagen, in September, 1926, for a series of lectures. They were planned to be a full-scale discussion of the whole atomic problem. Many of the institute alumni and other physicists came on for the seminars. Everyone was reaching out for all the information he could get and Copenhagen held out the highest possibilities of finding enlightenment. Bohr's guidance and appraisal were badly needed.

"Schrödinger gave us a most impressive account of his wonderful work," said Bohr. But in the Copenhagen style, Schrödinger did not get very far in a straight lecture. The seminar paid him the compliment of regarding his work as much too important to pass without the fullest examination and discussion. The questions flowed. The blackboard was covered with equations, erased, and filled once more.

Again it was admitted that Schrödinger's equations were mathematically equivalent to Heisenberg's quantum mechanics. But then Schrödinger daringly went on to argue for abandoning the whole idea of quantum jumps—the jumps of electrons from one orbit to another. The effects might be replaced, he maintained, by his three-dimensional confined waves.

Half a dozen physicists were shouting objections and questions. Bohr, his pipe forgotten, was pacing the room. Everyone was haranguing his neighbor. Bohr broke through the uproar to bring seminar back to the point. He started with an analysis. New figures went up on the board. But the commotion went on for most of the week.

"We argued with Schrödinger that any procedure disregarding the individual character of quantum processes would never account for Planck's formula of thermal radiation," said Bohr.

As fierce as the argument might be, it was always a search for the truth. It was not an attack on Schrödinger. Schrödinger was swept up in the attempt to give even greater scope and validity to his ideas if they were correct, and if not, to rebuild them in such a way that they would bring even greater honor to him and to physics. Nevertheless it was a harrowing week.

"If I had known that it [the wave theory] was going to be taken so seriously as to cause all of this discussion, I would never have invented it," Schrödinger remarked with a grimace as one long wearing session ended.

"But the rest of us are thankful that you did," replied Bohr. "You have contributed so much to the clarification of the quantum theory. You have succeeded in developing a wave-theoretical method which has opened up new aspects and proved to be of decisive importance for the great progress of atomic physics."

The two walked out arm and arm and set off for an excellent Copenhagen dinner. Was it possible though that Schrödinger could be right and that quanta could be replaced by three-dimensional waves? During the space of the next few months Bohr and Heisenberg and scores of other workers plunged into an intensive study of the whole problem.

"I remember discussions with Bohr which went on until late at night and ended almost in despair," Heisenberg later recalled. "At the end of one discussion I went alone for a walk in the neighboring park. I remember repeating to myself again and again the question, 'Can Nature possibly be as absurd as it seems to us in these atomic experiments?'"

Those who found relief in laughter, and this included Bohr, turned to light verse and even practical jokes. A newcomer from a stiffer tradition who found this a little disrespectful, said something about it to Bohr. Bohr drew on his pipe and with a smile that gave warmth to his words explained: "There are things that are so serious you can only joke about them."

George Gamow preserved some verses which he attributed to a Russian who was working at the institute during these baffling and exciting times:

HAIL TO NIELS BOHR

Fill up the tankards and brighten the fire!
Drink to him—drink to him—toast him once more!
Plucking the strings of our latter-day lyre
We sing to our hero and idol Niels Bohr!
    Prosperous days to you,
    Honor and praise to you,
Bohr the colossus we fear and adore!
Endless your merits are; none can deny it.
What if your theory is madly obscure?
You have proclaimed it; we dassent defy it.
Your words, like God's words, are totally pure.
    Niels you Apollo, you,
    Humbly we follow you!
Laws you devise are devised to endure.

Quantum reveres you. At edicts you utter
Each tiny electron examines his plight;
Confined in his orbit, he flies in a flutter,
Rushes and radiates, estimates height,
    Prejudges the motion
    And dreams up the notion—
Mixing up Cause and Conclusion in fright—
    To leap from his mother orb
    Straight to another orb,
Seeking, if vainly, a refuge in flight.

Hail to Niels Bohr from the worshipful Nations!
You are the Master by whom we are led,
Awed by your cryptic and proud affirmations,
Each of us, driven half out of his head,
    Yet remains true to you,

Wouldn't say boo to you,
Swallows your theories from alpha to zed,
Even if—(Drink to him,
Tankards must clink to him!)
None of us fathoms a word you have said!

In all the ferment of desperate work and light relief, some answers began to develop to some of the questions. The continuing study disclosed that the spin of the electron could explain the very discrepancies that had once stopped Schrödinger. Mathematics—art—had prevailed after all.

At Cambridge, George Thomson, the son of J.J., then proved the existence of the waves. Shortly before in America C. J. Davisson and L. H. Germer directed a beam of electrons at a metal crystal and obtained diffraction patterns analogous to those produced when light is passed through a pinhole. Experiment was providing proof.

The lights were burning as late at Goettingen as at Copenhagen and the other physics laboratories. Early in 1927 Max Born, who was studying collisions between atoms and free particles, reported that electrons should be regarded as "real" particles, though entirely different in behavior from the particles of classical physics, whether those particles were billiard balls, or the moon.

If enough particles are taken into consideration, he showed, their behavior can be predicted much as can that of a human population. Though it was impossible to predict the behavior of a single particle, in large numbers the problem was manageable. It was the old principle of statistics and insurance.

Born also found that where the Schrödinger and de Broglie wave motions are intense, more particles are likely to be found. By using the wave theory or wave mechanics as it was coming to be called, he could calculate the probability of electrons flashing from one orbit to another.

More sessions were held in Copenhagen. Letters and papers and visiting physicists coming to the vortex poured into the institute. The new concepts were beginning to shape up, to fit together.

"In the characteristic vibrations of Schrödinger's wave equation we have an adequate representation of the stationary states of an atom and an unambiguous definition of the energy of the system," Bohr concluded about this time.

Bohr was even becoming optimistic. In a letter to Rutherford, written on January 27, 1926, he made a prediction with relatively little qualification: "It looks now at any rate . . . that we shall be able . . . by the use of the correspondence principle supplemented by the new ideas to account comprehensively for the building up of atoms and for their properties." Bohr added one proviso: "Of course in so far as they are independent of the structure of the nucleus."

But once more the splendid progress only revealed still deeper problems and insights.

Heisenberg was back in Copenhagen and while the institute group watched with stunned, amazement he put on the blackboard equations picturing a "quite different" atomic world. Stationary states or the great highways circling the nucleus were set, not by measuring their distance from the nucleus, but by calculating the transitions—the flashing jumps of the electrons from one orbit to another.

"This led," said Bohr, "to quantitative calculations of the transition probabilities and to energy values for the stationary states which differ systematically from those obtained by the quantization rules of the older theory. The classification of stationary states is based on a consideration of the transition probabilities which enable the states to be built up step by step."

In newly classifying the stationary states (to the layman, the great orbiting highways) Heisenberg made an even more radical proposal. The position of the electrons racing around their orbits like the earth around the sun could never be absolutely defined, he said. Their position and velocity could only be approximated. It was equivalent to saying that the position of the earth could never be fixed, regardless of how much might be known about its rounds.

The trouble in the past, Heisenberg argued, was that the physicists applied their ordinary methods of observation to the atom. Their beams of light, exciting electrons and knocking them out of their orbits, destroyed the very condition they were attempting to "see." Despite Bohr's still radical proof that the ordinary rules could not be applied to the jumping of the electrons from state to state, quantum ideas had not been applied to the experiments on which physicists relied for their evidence of atomic structures.

To take the Copenhagen group with him into the subtle depths of the atom that he was penetrating in a new way, Heisenberg proposed an ideal experiment—a device often used by physicists. All the conditions of an experiment are imagined and it is rigorously carried out in thought even though it might be impossible to perform in actuality.

Imagine, said Heisenberg, a gun that could shoot a single electron into a dark chamber completely evacuated of all other atoms, even those of air. Imagine, also, he said, a scientist equipped with an "ideal" microscope that would enable him to see what went on in the chamber, if he directed a light beam into it.

What would happen, Heisenberg asked, when the light photon hit the electron? The answer was obvious to his sophisticated listeners: the electron would be knocked askew.

By the very act of lighting up the electron's movement, that movement would be disrupted. It was impossible both to see the electron and to measure its exact momentum. One action ruled out the other, Heisenberg argued.

But was there no other way? Heisenberg invited his audience to consider what would happen if the photon's energy were reduced by using a light of lower frequency. In that case, he pointed out, the weaker light encountering the stronger electron would be diffracted or deflected. There would be only a blur around the electron, and its exact position at any given instant could not be seen. The photon would not change the electron's course, but its position would be lost as the light was turned back or bent. Again there would be uncertainty.

Heisenberg scarcely needed to show that there could be a middle course. The light might be adjusted so that it would not knock the electron off its course yet would show its general path to a close approximation. Though its exact position could not be determined, it would be possible to obtain a general idea of its position, and the scientist could calculate from this data the probability of its being in a certain approximate area.

Heisenberg, presenting this experiment with the most formidable mathematics—and not in the words used here—was saying in effect that it would be impossible in principle to obtain an exact measurement of both the electron's position and velocity. But he was also proving that the probability of an electron's being in one place or moving at a certain speed could be known.

Bohr greeted the dazzling exposition with a fire of questions and arguments. The intensity of his interest was a certain indication of the importance he attached to what had been said. His reaction was the very opposite of his "very interesting, very interesting" and the indulgent smile that accompanied this equivalent of dismissal.

Once again another theory about the hidden inner workings of the atom was put to every test. And the Heisenberg principle of uncertainty passed the examination in so far as they could see. Bohr himself had been moving in the same direction for a number of years. However, Bohr was not the consummate mathematician that Heisenberg was, and the historic theory was Heisenberg's.

The new insights went so deep and so far that they were both exhilarating and terrifying to contemplate. Robert Oppenheimer caught the feeling and characterized this nearly matchless period with one word—"heroic."

"It was a heroic time," he wrote. "It was not the doing of any one man; it involved the collaboration of scores of scientists from many different lands, though from first to last the deep creative and critical spirit of Niels Bohr guided, restrained, deepened, and finally transmuted the enterprise. It was a period of patient work in the laboratory, of experiments and daring action, of many false starts and many untenable conjectures, of debate, criticism and brilliant mathematical improvisation.

"For those who participated it was a time of creation; there was terror as well as exaltation in their new insight. . . ."

A heroic era! A path had been cut into the unknown.

9 /

Opposites Meet

Iᴛ ᴡᴀꜱ ᴀ ʜᴇʀᴏɪᴄ ᴛɪᴍᴇ in the full sense of the word. Old worlds, or their foundations, were being destroyed. New worlds would have to take form. At the moment though there was only the crumbling and sapping of the old, honored, and established precepts. It was of such upheavals that the old Greek myths had told. The word heroic thus was accepted by nearly all who lived through the period.

Human knowledge and its meaning were at stake. Quantum physics, in the light both of Heisenberg's quantum mechanics and Schrödinger's wave mechanics, was decreeing that man through science could never know the final, ultimate answers to the universe, if that answer meant absolute certainty about each action. Thus the base itself was shaken.

Since the days of the Greeks and most particularly since the time of Galileo and Newton the goal and method of science had been clear. The physical universe was determined in its course by all-embracing laws. If the laws could be discovered they would lead to the answer. This was the certainty and the method; in it science found unfailing guidance. Laplace even held that if there were a great enough "intelligence" the whole state of the universe in any one instant might be comprehended and its history forecast for all time to come.

The doctrine called causality—the certainty that cause produces effect—underlay the whole of science. It was the base of knowledge and even the assumption of the child—if I count out five pennies I will know how many I have.

The abrupt discovery that this rule of the ordinary world did not hold in the atomic world, and that after counting out there would not be certainty, but rather uncertainty and indeterminacy, shook even the most resolute of minds. Heisenberg spoke of his "despair"; Bohr, of "alarm." To have made one of the foremost of advances and to have produced only a hazy, unalterable uncertainty at the very base on which the earth was built was almost more than the scientist could stand.

Bohr, who had led physics step by step into this dilemma, was most acutely aware that a way out must be found. Everything could not end in a blur. But Bohr could glimpse a solution of the impasse and, in fact, the way toward a greater understanding than ever before. At first, however, every attempt at this new approach met with defeat and frustration. In the cold, dark months of early 1927, he reached the point of exhaustion. He had to get away for a rest.

He and Mrs. Bohr went off to Norway to ski. Bohr was an excellent skier. As he zoomed down the snow-covered mountain slopes and felt the stimulation of the cold, crisp air, the concepts with which he had been struggling began to fall into place. The blockages disappeared.

By the time Bohr returned to Copenhagen two weeks later he had the basic answer to the paradoxes quantum physics had posed for the scientific world. Like the problem itself, Bohr's answer had implications that would extend far beyond the strict confines of science. It would in the end propose one solution for the bitter political and cultural divisions of the world. This, though, was for the future; in 1927 Bohr simply named his concept *complementarity*.

On the first morning of his return, Bohr raced down the steps of his apartment, across the graveled drive, and two at a time up the steps of the institute to find Heisenberg. He poured out his ideas. Heisenberg listened, disputed, and affirmed. It was the beginning of another of their intense work drives.

It was fortunate that Bohr felt rested, for during the next two months he and Heisenberg worked constantly. They would stop for a few walks in the park and Margrethe would drag them away for meals, but otherwise the work continued almost without interruption. Every attempt at interpretation was tested with every possible real or imagined experiment. It was the only sure-footed way through the immense subtleties and profundities with which they were dealing.

They also kept in the closest touch with Pauli. They needed his keenly critical view of what they were attempting. The detection of flaws or discrepancies was essential. Bohr prized nothing more than an objection that permitted him to clarify a point or to make a correction.

By spring the theory was ready, the theory that Dirac said "led to a drastic change in the physicist's view of the world, perhaps the biggest that has yet taken place"; that Oppenheimer called "the inauguration of a new phase in the evolution of human thinking"; and that John A. Wheeler, professor of physics at Princeton University, described as "the most revolutionary scientific concept of this century."

Essentially the theory to which Bohr and Heisenberg came was that there are two truths rather than one alone. Or, in other terms, that there can be two aspects, both true, and furthermore that the two together offer science and man a more complete view and understanding of the atomic world than either could offer separately, and thus in the end a clearer view of the visible world built out of the invisible substratum of the atom. Instead of division, Bohr showed, the parts, the divisions, the components can be combined into a harmony greater than that of the sections. Each separate, even contradictory, aspect complements the other.

Bohr began with one premise and two paradoxes. The premise was the well-documented one—the atomic world is not like the visible one. As we step through the looking glass, a new order appears, different in kind as well as degree.

This led immediately to the first paradox. Since Newton, science had rested on the assumption that a particle, be it electron or the planet Venus, can be observed in a certain place, and if its speed of movement is known, tracked certainly to the place it will reach in a specified time. The astronomer thus predicts a future eclipse of the moon, or a motorist driving to work at the permitted number of miles an hour can be certain if there are no upsets of arriving "on time." Almost every human move, if it is only reaching for the pencil lying on the table is consciously or unconsciously calculated in the same way, and Einstein in the theory of relativity emphasized that every observation or measurement rests on the coincidence of two independent events at the same space-time point. So profoundly is this general certainty built into the macroscopic universe that its existence is taken for granted, like breathing.

In the atomic world, however, an eclipse—if there were such an event in that world—could not be predicted to the moment; or if an electron started out on its rounds at a certain speed it would be impossible to say that it would arrive at a certain point at a certain time. An atomic hand reaching out—again if such a fantasy could be imagined—might only come close to the atomic pencil on the table. There might be a dismaying, disconcerting miss.

The reason, Bohr emphasized, is that the electron cannot be observed without a disturbance of it. The observer's beam of light, as Heisenberg had demonstrated, will buffet the electron around and alter its velocity. "Any observation regarding the behavior of the electron in the atom will be accompanied by a change in the state of the atom," said Bohr.

If on the other hand there is no disturbance, if no light beam or colliding electron touches the atom, there can be no measurement and no knowledge. And, Bohr pointed out, all knowledge concerning the internal properties of atoms is derived from experiments on their radiation and collision reactions. "This situation has far reaching consequences."

Equally staggering was the second great paradox with which Bohr had to deal—the wave-particle problem. By the strongest evidence the electron is a particle—a "concentration of energy and momentum at a single point of space and at any single instant of time." When an electron hit one of Rutherford's screens it gave off a scintillating flash of light. Something had struck. Or when it was hit by another electron the collision very much resembled a collision between two automobiles. Energy was transferred, and one or both were thrown off their course. Such a reaction unequivocally bespoke a particle, a thing of substance. But de Broglie and Schrödinger had shown that the electron also behaves as a wave. The two pictures were mutually exclusive and almost unthinkable and yet both indubitably existed. Physics was confronted with a fact and an impossibility.

The test experiments and equations devised by Bohr and Heisenberg showed that the velocity of an atom could be determined with precision. Or its position in space could be calculated with finality. It was only that both its velocity and position could not be found with accuracy. About its position at any particular point at any particular time there always had to be some degree of uncertainty.

It was the same with the wave-orbit problem. If an imagined, ideal microscope could be trained on the electron, and if it had sufficient resolving power to show the electron moving in its orbit, the same trouble would develop. In the act of observation at least one light quantum would pass through the microscope and would be deflected by the electron. The instant the electron was affected by the light quantum its momentum and velocity would change.

Thus there was no way of precisely observing the orbit of the electron around the nucleus. The waves that would normally be set off by the electron and that would be making their own orbit would be altered in their turn if the electron were jarred out of its movement by the light ray. It would never be possible to observe more than one point on the orbit of the electron.

But Bohr did not see helplessness in the quandaries that barred the way forever in the electronic world where the dart of light was as large as its quarry. "From the point of view taken here," he said, "just the renunciation forms the necessary condition for an unambiguous definition of the energy of the atom. We must consider the very renunciation as an essential advance in our understanding."

Both pictures could be used, Bohr saw. By employing both, by going from one to the other, a right idea could be obtained of the "strange kind of reality" in the atom. One, said Bohr using the key word, was complementary to the other.

The knowledge of the position of the particle thus might be considered complementary to the knowledge of its velocity, and the knowledge of the electron as a particle complementary to the knowledge of it as a wave. By knowing both with the greatest accuracy possible, a more complete description of experience, a new synthesis was possible.

"However contrasting such phenomena may at first sight appear," said Bohr, "it must be realized that they are complementary, in the sense that taken together they exhaust all information about the atomic object which can be expressed in common language without ambiguity."

By knowing both, Bohr maintained with revolutionary impact, a better description of experience—a new harmony—is attainable. Predictions could be made and checked by any scientist, Bohr further pointed out. They would not be predictions of exactly what would happen at any instant, but of the probabilities that certain things will happen. In the same sense no one can predict that some one individual will die at a certain moment, though there is no difficulty in predicting how many in 100,000 will succumb in a set period.

Thus Bohr resolved the appalling contradictions by synthesis. He concluded that complementarity is a requirement of the laws of nature and, far from being a hindrance, is an indispensable logical tool. Bohr was working out a subtle doctrine. It took words and attitudes out of their familiar contexts and re-used them in new strange ways. The staff at the institute, the first to be exposed to the new doctrine, found it almost impossibly hard to understand.

Bohr, happening upon a debate on the problem in the institute library, draped a leg across one end of the library table, and looking at the struggling students with an understanding grin, comforted them with a quotation from the Abbé of Galiana (1728–87.) "One cannot bow in front of somebody without showing one's back to somebody else."

The problem, Bohr noted, is as old as the Greeks. When Leucippus and Democritus proposed the atom as the indivisible, smallest unit of matter, the substance of which all else is made, they took it for granted that the coarseness of the sense organs would forever prevent the direct observation of individual atoms.

C. T. R. Wilson's construction of the cloud chamber, in which shooting electrons leave the marks of their passing in fine lines of water droplets, and the Geiger counter, clicking out the passage of electrons, had altered the early inaccessibility of the individual atom. Then and many times later Bohr was to say that the description of ordinary experiences "presupposes the unrestricted divisibility of the course of the phenomena in space and time," and the linking of all steps in an unbroken chain of cause and effect.

In the ordinary world too, Bohr emphasized, it is assumed that an ordinary measurement can be made without any effect on the object measured. The rough tearing of the cloth from the bolt does not impair the measurement of a yard. In the same way the physicist can turn his microscope on a piece of iron without in any way affecting it. In the eerie world of the atom it was different.

"The crucial point implies the impossibility of a sharp separation between the behavior of atomic objects and the interaction with the measuring instruments which serves to define the conditions under which the phenomena appear," said Bohr. Bohr argued that the new concept should not be difficult for humans. After all, he noted, "we are both onlookers and actors in the great drama of existence."

When the pressures of this demanding work became too heavy and there was no time for a weekend at Tisvilde, Bohr often went sailing. Like most Danes, living so close to the sea, he loved boating. In 1926 he and two friends, Niels Bjerrum, the chemist, and Holger Hendriksen, the xylographer, bought *The Chita*. During the next eight years they went on many trips, short or long.

With Bohr on board the discussions began swiftly. Even the sight of the moon on the water would prompt Bohr to ask why the reflection appeared as a streak rather than as a large patch, and in short order all of them were drawn into the debate. Bjerrum always said that Bohr had a remarkable ability to start his friends thinking and to make them feel much cleverer than they really were.

On one cruise *The Chita* stopped at Skagen. They all wanted to see a church that had been buried in the shifting sand. Bohr had always been skilled at throwing stones and skipping them on the water. He picked up a pebble and tried to throw it over the church. This proved easy. He then started trying to get the stone to fall outside the shutters over the peepholes in the first and second stories of the tower. This also was soon accomplished. Bohr then aimed at the small holes in the shutters and when he found that he could hit them, evolved the idea of throwing their walking sticks and making them stand in the sand outside the shutters. This worked. Bohr set the next goal—to knock the sticks down by throwing stones at them.

"Finally," said Bjerrum, "he got us to try to get the sticks to hang up there with the handles through the holes in the shutters and this really did succeed. The rest of us gave up the idea of getting them down again by throwing stones, but Bohr continued and was delighted when he finally succeeded."[1]

As Bohr rounded out his work on the theory that would give new directions to the whole of thought, he received an invitation to participate in an international congress of physics in Italy in September, 1927. It was to be held at the mystically lovely Lake Como in celebration of the 100th birthday of Volta.

Mussolini was at the peak of his power and Italy was determined to make the meeting an outstanding one. It was, in fact, planned as "one of the realizations of the Fascist regime," of which Mussolini and the newspapers were constantly talking. It was also to put Italy into the running in the suddenly important new area of theoretical physics.

Physics in Italy had been at something of a standstill in the preceding half century or more. Then, thanks to a physicist who was the only non-Fascist member of Mussolini's cabinet, a chair of theoretical physics was established at the University of Rome. A young Italian named Enrico Fermi was named to fill it.

To all who would listen, Fermi was expounding the quantum theory and the Bohr atom. As Laura Fermi relates in her fine book

[1] See the chapter by Niels Bjerrum in *Niels Bohr His Life and Work*.

Atoms in the Family Fermi was having trouble convincing the young group around him that matter and energy could be both particle and wave. They argued that such an outlandish contention was a dogma to be accepted only on faith. Fermi as the chief exponent of the "faith" thereupon was nicknamed "the pope."

Fermi, his friend Emilio Segré, who had just been persuaded to shift from engineering to physics, and Fermi's student Franco Rasetti, went to the Como conference. As the meeting opened, Segré whispered to his mentors: "Who is the man with the soft look and the indistinct pronunciation?"

"That is Bohr," they whispered back.

"Bohr, who is he?"

"Fantastic," Rasetti answered, "Haven't you heard of Bohr's atom?"

By the time the meeting ended, even an engineering student recently commandeered into physics knew about Bohr. Bohr had chosen as the title of his paper "The Quantum Postulate and the Recent Development of Atomic Theory." Behind this unrevealing title he made his first public presentation of his revolutionary theory of complementarity. It stirred the Congress as the Mistral sometimes roils the ordinarily calm waters of Como.

Bohr had begun cautiously and gracefully: "In a field like this where we are wandering on new paths and have to rely on our own judgment in order to escape from the pitfalls surrounding us on all sides, we have perhaps more occasion than ever to be remindful of the work of the old masters who have prepared the ground and furnished us with our tools."

His theory was presented in a language part physics, part philosophy. It was not, however, the language of any of the traditional philosophers, nor of their schools; it was not positivism, materialism, or idealism, though elements of each could be found within it. All of this only added to the growing unease. Bohr had not only destroyed the props of the whole structure of physics and science, but had done it in unfamiliar words and terminology. Anger and bafflement combined with the sense of shock. Schrödinger and Von Laue objected strenuously that Bohr's interpretation was neither convincing nor conclusive. They were entirely unwilling to sacrifice so much of the traditional base of physics. Others conceded that change was necessary, but disliked Bohr's ideas.

The widespread and vocal opposition was generally united only at one point—instead of turning to uncertainty, indeterminacy, and statistics, they insisted fervently that physics had to stand with "reality." The smallest part of the universe had to exist as objectively as a city or a stone, whether or not it was observed. They wanted a sharp, clear reality, not a haze or probability.

Einstein was not at the Como congress, and until the master was heard from, no one could be wholly certain in his own position. The question all over Europe was "What will Einstein say?" Would he demolish Bohr?

The fifth Physical Conference of the Solvay Institute was to follow almost immediately in October, 1927. Both Bohr and Einstein would be there, as well as nearly all others who were contributing to theoretical physics. Lawrence Bragg and Arthur Compton came from the United States. De Broglie, Born, Heisenberg, and Schrödinger all were to speak on the formulation of the quantum theory.

The subject was "Electrons and Photons." To leave no doubt that it was directed to the main question, the theme embroiling all of physics, discussion was centered around the renunciation of certainty implied in the new methods. There lay the rub. The stage was also set for a discussion of the possibility that wave mechanics—interpreting the structure of the atom primarily in terms of waves—might offer a way for keeping the cherished solidity of the old and combining it with some of the new. Such a course would have entailed a far less radical departure than Bohr's complementarity.

Against this background, Bohr was invited to give the conference a report on the epistemological problems confronting quantum physics. By asking him to speak on the science of knowledge and the grounds for it, the conference gave him full opportunity to present complementarity. There was no avoidance; the issue had to be directly faced.

Excitement mounted as Einstein rose to speak. He did not

keep them long in suspense. He did not like uncertainty. He did not like the abandonment of "reality." He did not think complementarity was an acceptable solution, or a necessary one. "The weakness of the theory lies in the fact that on the one hand, no closer connection with the wave concept is obtainable," he said, "and on the other hand that it leaves to chance the time and the direction of the elementary processes."

A dozen physicists were shouting in a dozen languages for the floor. Individual arguments were breaking out in all parts of the room. Lorentz, who was presiding, pounded to restore order. He fought to keep the discussion within the bounds of amity and order. But so great was the noise and the commotion that Ehrenfest slipped up to the blackboard, erased some of the figures that filled it, and wrote: "The Lord did there confound the language of all the earth."

As the embattled physicists suddenly recognized the reference to the confusion of languages that beset the building of the tower of Babel, a roar of laughter went up. The first round had ended.

What was euphemistically called the "exchange of views" was continued in smaller groups during the evenings. And here Bohr and Einstein met directly, face to face. Ehrenfest, who for many years had been a friend of both Bohr and Einstein, in effect served as mediator. His services were not needed in the sense of preventing personal combat. This was a battle fought with soft words and courtesies. Each held his opponent in a respect verging on awe. But the differences were as fundamental and as sharply defended as though the weapons had been guns.

Bohr had prepared carefully. He never left his words to last-minute inspiration, but compiled his material and tested it out against all possible loopholes in discussions with associates and students at the institute.

Bohr hoped that his proof would bring Einstein to his side. Complementarity, he argued, only carried on ideas that Einstein himself had ingeniously raised. Was it not Einstein who in 1905 had shown that the photon—a ray of light—is a corpuscle as well as a wave? Had he not explored novel procedures outside the classical framework of physics to do so?

And had not Einstein himself in 1917 formulated the rules indicating that the atom may spontaneously emit radiation at a rate corresponding to a certain prior probability? Only the probability of that disintegration could be calculated, and not the exact moment at which the radioactive material would give off another bit of radiation, in its long or short decay.

And did he not emphasize the dilemma still further by showing that the radiation will flash only in a certain direction, though in the wave picture of radiation "there can be no question of a preference for a single direction in an emission process?"

Einstein had even concluded his paper on radiation by saying: "These features of the elementary process would seem to make the development of a proper quantum treatment of radiation almost unavoidable."

A man who had laid this foundation should not quibble about the similarly drastic changes in foundation proposed in the theory of complementarity, Bohr argued.

Einstein also was prepared. He was ready to show by "ideal experiment"—one of his imagined but correctly set up experiments—that uncertainty would not be necessary if the right experiments were made, and if the interaction between atomic objects and the measuring instruments was more explicitly taken into account.

Einstein went to the blackboard. He drew one line with a small slit in it, and just beyond it another line representing a photographic plate. If a single electron or photon went through the slit, it would as it emerged fan out as a typical wave, in the wave's concentric lines. Einstein conceded that it would not be possible to predict with certainty at what point the electron would arrive at the photographic plate, say at Point A in the upper part of the plate, or at Point B, in the lower part. It could only be calculated that the electron *probably* would be found within a given region of the plate. But, Einstein argued, if the electron is recorded at Point A, it could never be recorded at Point B. Thus it should be possible, with control of the momentum and energy transfer, to determine precisely where and when the electron would strike.

Bohr objected immediately that the experiment Einstein had sketched was not comparable to the application of statistics in dealing with complicated systems. They quickly came down to the question—does the quantum-mechanical description exhaust the possibilities of accounting for observable phenomena? Einstein's answer was a decided "No."

Bohr challenged Einstein to examine "the simple case" of a particle penetrating through the hole if there were a shutter to open and close the hole. Bohr went to the blackboard. He added several lines to indicate a shutter. Could the shutter be used to control the momentum and energy transfer involved in the location of a particle in space and time? Could the collision of the light particle be controlled enough not to interfere with the course of the electron under study?

Bohr's answer was that the moment the light was weakened enough not to affect the electron, the electron could not be clearly found. It would become impossible to fix both its place and velocity.

Einstein was not in the least satisfied. "I am firmly convinced," he said, "that the essentially statistical character of quantum theory is solely to be ascribed to the fact that the theory operates with an incomplete description of physical systems." He was insisting that Bohr had still not gone to the bottom, and that he was taking an incomplete answer for the final one.

At their next session Einstein was ready with still another "imaginary experiment." Between the diaphragm with the slit and a photographic plate, Einstein inserted another diaphragm with two parallel slits.

Einstein suggested that proper control would make it possible to decide through which of the two slits the electron had passed before striking the photographic plate and registering its arrival with a black spot.

The single electron could go through only one of the two slits—after all it could not split in two. And the electron would strike the photographic plate at a predictable time and spot. Einstein argued that it would be possible to decide through which of the two slots the electron had passed.

This was not an easy one, but Bohr rose to it. After deep thought he demonstrated that in controlling the second diaphragm there would again be an inevitable interference with the electron and again indeterminacy would come into play.

"We are presented," said Bohr, "with a choice of either [and he emphasized the either] tracing the path of a particle or [again great emphasis] observing interference effects. We are faced with the impossibility of drawing any sharp separation between an independent behaviour of atomic objects and their interaction with the measuring instruments."

Bohr thought that he had proved his point. But Einstein was no more convinced than when the discussions began. If the arrangements suggested were not sufficient to obtain accuracy, he was certain that others should be possible. It was only lack of knowledge that blocked the way, not an inevitable barring of the way in. Einstein was not willing, because knowledge was lacking, to fall back upon uncertainty. He would not consent to be satisfied with saying "There's a chance that it will be this way," instead of "This is the way it will be."

"In spite of all divergences of approach and opinion," said Bohr, "a most harmonious spirit animated the discussions."

Einstein lightly jibed at Bohr and his support of chance and probability: "Do you really believe God resorts to dice-playing?" Bohr came back in the same spirit: "Don't you think caution is needed in ascribing attributes to Providence in ordinary language?"

Ehrenfest, though he was Einstein's collaborator, was shaken by Bohr's arguments. He teased Einstein, hinting that Einstein's attitude toward complementarity and the new development of the quantum theory was similar to that of opponents of relativity. Nevertheless Ehrenfest did not want to imply that he had switched to Bohr's side. With his next breath, he added that he could never feel any peace of mind about complementarity unless Einstein were convinced.

Despite Bohr's conviction that he had demolished Einstein, Einstein's insistence that there should be a way made him question. Again it was the kind of challenge Bohr loved. Could experiments possibly be devised that would enable the scientist to discover the particle at a particular spot and time?

He and the Copenhagen delegation worked most of the night trying to imagine other instruments and experiments that would control the momentum of the particles enough to make them fully measurable. But the more they tried that night—and in later years—the more impossible the task looked.

Einstein also persisted. In one of the general meetings Einstein again raised his general objections. The quantum attempt to solve the riddle of the double nature of all corpuscles had not found a final solution in the statistical quantum theory.

Einstein considered the impasse, as Poincaré had once said, "a measure of the depth of our ignorance," and he repeated his even greater dissatisfaction with another phase of the quantum theory, its failure to get down to reality. It did not give a complete description of the real, individual situation, irrespective of any act of observation or substantiation. Look at the radioactive atom, Einstein urged. At a certain time it emits a particle. The individual atom therefore has a definite disintegration time.

"One is driven to the conviction that a complete description of a single system should after all, be possible," Einstein declared. "But for such a complete description there is no room in the conceptual world of statistical quantum theory."

Bohr answered *yes*, but the consideration stands or falls with the assertion that there actually is such a thing as a definite time of disintegration of the individual atom. "The assertion of the existence of a definite time-instant makes sense only if I can in principle determine the time-instant empirically," he said. To determine, Bohr went on, would involve a definite disturbance of the system.

Einstein did not like this any more than when he heard it for the first time. He accused the quantum advocates of being "egg walkers," willing to go to almost any length to avoid the "physically real." Einstein was particularly irked by Bohr's general contention that a search for a complete description would be aimless, for the reason that the laws of nature can be completely and suitably formulated within the framework of the quantum description.

But, said Einstein, pounding as hard as his gentle disposition permitted, "for me the expectation of the adequate formulation of the universal laws involves the use of *all* conceptual elements."

Einstein hit hard again: "It is not surprising that, using an incomplete description (in the main), only statistical statements can be obtained out of such a description." Incomplete data, he was charging, produce incomplete results.

"To me it seems a mistake to permit theoretical description to be directly dependent upon acts of empirical assertion, as it seems to me to be intended in Bohr's principle of complementarity."

Solvay, 1927, one of the most influential conferences in the history of science, then ended. Physicists returned to their laboratories all over the world to continue the argument and to work on the problem.

The heroic period in physics, the unmatchable period, also was coming to an end.

10 /

Clash and Tragedy

AS THE TWENTIES ENDED and the thirties began, a feeling of confidence and elation swept the physicists. In a few short years the structure of the atom had been discovered and understood. The deepest problems of the structure of matter had been partly solved and the ancient question—where do the properties of matter come from—had largely been answered.

This illumination of the long hidden substratum had been achieved by going beyond all methods previously used in the exploration of the universe, beyond the observation that had showed the earth to be molded by the action of natural forces, beyond the analyses that revealed the evolution of life, beyond the experiments that had disclosed and put into man's service the electrical, chemical, and other forces of the universe.

Largely under Bohr's guidance, science had stepped into a new area, the atomic area where all was different and the old rules did not hold. But even this invisible and strange world had proved not beyond man's reach and comprehension. Its fundamental features had been identified and incorporated into a new conceptual structure. It was a heady sensation for the little band that had achieved this phenomenon.

"In a couple of years," said one exuberant young physicist, "we shall have cleared up electrodynamics; give us another couple of years for the nucleus, and physics will be finished. We shall then turn to biology."

"It is difficult for those who did not witness it to imagine the enthusiasm, nay the presumptuousness, which filled our hearts in those days," said Rosenfeld.

Heisenberg and Pauli, leaving the details to the lesser fry, turned to the electromagnetic field—the field to which all other forces are reduced in the end, to the forces of attraction and repulsion that control the organization of the atom.

The physicists were not alone in their assurance. The politically optimistic believed firmly that the Pact of Locarno had shown the way not only to the settlement of the problems inherited from World War One, but to future peace. Great confidence was placed in mutual guarantees against aggression. Sir Austen Chamberlain received the Nobel Peace Prize for his achievement of this high point in the search for peace and recovery.

Outside of Germany relatively little attention was given to the new Nazi Party or the rabid book called *Mein Kampf* which its *fuehrer* had written during a period of imprisonment.

The prosperity accompanying the peace and the intellectual exploits was hailed with the same confident exuberance. As the American stock market soared to new highs, the president of the exchange announced in September, 1929: "We are apparently finished and done with economic cycles as we have known them."

Bohr took it all more calmly. He was as elated as any at what had been accomplished, and Bohr was entirely capable of enjoyment of work well done. He laughed though at the idea that physics was about finished. In fact Bohr thought there was still so much to do on the quantum theory that he questioned the plunge into quantum electrodynamics. He distrusted mathematical fireworks unless some connection to the ground could be established, and he was afraid that in this case it could not be. Electrodynamics had few roots in experiment.

Bohr often would tease Rosenfeld, an exceptionally able mathematician and one of those rushing ahead. "In the sweetest tone of voice, but with an unmistakably malicious twinkle in his eye, he would ask me how I would handle some simple physical process with the learned methods of quantum electro-dynamics," Rosenfeld recalled.

Rosenfeld would start out, then the complications would become apparent. Bohr, delighting in his little stratagem, then would lecture his young assistant on the dangers of extrapolating too far beyond the realm of experience. Lofty abstractions, Bohr would repeat, should have a foundation in some aspect of experience accessible to observation.

For the next six years, until 1936, Bohr put most of his own efforts into the firming up and deepening of quantum mechanics. Physics actually was now moving in two directions. One was the further investigation of the quantum, Bohr's own choice, and the other was the exploration of the innermost part of the atom, the nucleus.

Much of the quantum work was done elsewhere, particularly by Heisenberg, Pauli, and Dirac, but the inspiration often came from Copenhagen; there the lines of investigation were hammered out in long discussions. The exchange and checking were constant and no one would have considered final publication without a critical review by Bohr. Bohr's name, however, did not appear on many papers of the period.

Bohr did not sit down and decide upon such a course. Nor was a special decision made that Copenhagen would concentrate on the outer structure of the atom, while Cambridge, Berlin, and Rome, in the early years of the thirties, delved into the riddle of the nucleus. It came about much more incidentally. Like Goethe, Bohr yielded "to the exigencies of the day."

Bohr could not resist any promising challenge. No matter how deeply he was involved in one problem, he would drop it and plunge with equal energy into some new matter if he thought that he could thereby clarify some point that was in doubt.

As Rosenfeld pointed out, the challenge might come in the form of a paper from abroad, a statement made in a seminar, an invitation to give a lecture or to amend some earlier paper. Any such occasion might launch Bohr on a new train of thought, "with an end as unpredictable as its beginning."

Many years before Hegel had made the fundamental distinction between this method of work and the strictly logical. In the one case the steps are arranged in a one-dimensional array, like the beads on a chain. In the other, in dialectic, a situation is argued out, a position taken, an error found in it, and the argument thus leads to a new position, only to have a new insufficiency appear, and to be drawn on to still another position.

"This is the way science really grows," said Wheeler. It was indubitably the way Bohr went, both in discussions and in the choice of direction.

Week after week might pass in Copenhagen in consideration of a problem. Time flowed by almost unnoticed. The institute was well equipped with clocks, but here they did not command. The continuum could be interrupted for institute business, for Bohr was scrupulous about the details of administration. But an interruption to sign papers or see a visitor or attend a public function was only a pause; it did not break the absorbing concentration until a clarification had been reached, or the flow was diverted into another direction.

Bohr had a sure instinct for selecting the fellows who came to the institute for a year or two of work. Their nationality or the political alignment of their countries mattered not at all, except that Bohr was happy to have as much of the world represented as possible. In any one year at least a dozen countries might have young physicists in Copenhagen, working under Oersted or other foundation grants.

Bohr was equally unconcerned about appearances. If young Placzek, a Bohemian from Bohemia, forgot to shave and was seldom awake in the day (his brilliant mind functioned best in the middle of the night) the institute had no objections, but enjoyed the stories that inevitably collected. Often visitors getting their first glimpse of the young "geniuses" thought them an "odd lot."

Disposition mattered no more than appearance. The fellows ranged from the silent and deeply studious to the unaccountables whose "mad" exploits and practical jokes produced tales that would ever after form a part of the annals of physics. If there was any one student Bohr sought above all others, it was the kind of young fellow Rutherford also had wanted—a chap with some originality in him. All of this drew the world's most gifted and perceptive young physicists to Copenhagen.

Among the most fabled of many fabled Copenhagen fellows was a young Russian, George Gamow, who arrived in 1928 and remained through 1931. Gamow had graduated from the University of Leningrad and obtained a fellowship for a summer of study at Göttingen. He bought his return ticket via Copenhagen, for he, like all young physicists, wanted to meet Bohr. Copenhagen was the Mecca.

Bohr met Gamow in the institute library where he frequently received visitors. When he talked with the young Russian about his interests and what he had been working on, Bohr's interest was at once aroused. Gamow had succeeded in showing by quantum mechanics how a decaying atom ejects an alpha particle and thus becomes a new element.

Bohr asked a few more questions and then with no delay for the usual checking of credentials said: "Listen, Gamow, would you like to spend a year here if I arrange a fellowship for you?" "I was in seventh heaven," said Gamow. Miss Schultz, Bohr's secretary, found Gamow rooms nearby and soon he was an integral part of the institute.

Ehrenfest introduced another of the fellows in what was to prove a remarkable period. He brought H. B. Casimir to Copenhagen. As they traveled together from the Netherlands to Denmark, Ehrenfest told the young physicist in words that he never forgot: "Now you are going to get to know Niels Bohr. Of all the things that happen in the life of a young physicist this is the most important."

Casimir also was invited to become a fellow. His parents doubted though that anyone could be as important and famous as their son told them that Bohr was. To test the matter, the senior Casimir addressed a letter to his son "Casimir, c/o Niels Bohr, Denmark." Despite the inadequacy of the address, it was delivered without delay. "After that," said Casimir, "I think my parents felt convinced that I was in good hands. They felt even more secure after they had met Niels Bohr and Mrs. Bohr."

As a young student watching with the absorption and acuity the young often focus on their heroes, Casimir was impressed that Bohr could grasp a subject without making detailed calculations and command an experiment without doing experiments. Others called it his "intuitive grasp."

"This was one of the characteristics of his work," Casimir said. "It made him not only the great theoretician he was, but also an experimental physicist without doing experiments, and an inventor without working on inventions."

Bohr's experiments tended to the simplest of the simple. For all of the complexity of his writing, he could when he talked and thought things out be almost as direct as Rutherford.

In one discussion Bohr used the library books as apparatus. He arranged the books on one shelf at the proper angles to represent two poles. With fountain pen in hand he then walked back and forth between the two illustrating his idea. Later a far more sophisticated version of the same "experiment" solved a troublesome problem. Casimir, who witnessed the Bohr "experiment," always regretted that he had not followed the lead it offered.

The institute was well equipped with laboratories, many of them at the well-lighted semi-basement level. Though Bohr rarely worked there himself, he often wandered in to watch the experiments under way. One day he offered to help, saying that he was not as awkward as he looked. With this he picked up one of the special thin-walled glass counters made for the study of rays with low penetrating powers. He did not know that the glass could barely stand the pressure of air. Instantly there was a nasty crackling sound, and it crumpled in his fingers. Bohr made a hasty, red-faced retreat.

The next Solvay Conference was to be held at Brussels in the autumn of 1930. Einstein would attend and both Bohr and the whole upper echelon of physics, the invited guests, anticipated that the celebrated controversy would be resumed.

It quickly was and under the most dramatic of circumstances.

With all of the conference looking on, Einstein undertook to destroy Bohr's position and his insistence on the ultimate indeterminacy of the universe. Einstein was ready with another of his "thought" experiments. He proposed to make use of relativity to permit the measurement of the exact time and energy of a single electron. If he succeeded, Bohr's theory of complementarity, his whole work, would be demolished, for it had to withstand every test. Its validity permitted no exceptions.

Einstein began to draw on the blackboard. He diagrammed, outlined, a box with a hole in one side. On the inside he placed some clockworks geared to open and close a shutter covering the hole. The box should then be weighed, said Einstein.

If, Einstein continued, the box contained a certain amount of radiation and the clock was set to open the shutter at a specific time, one photon, one light quantum, could be released. The exact time of its escape could be recorded with the greatest of accuracy.

The box, minus the one photon, would then be reweighed. Since under Einstein's famous formula, $E = mc^2$, mass is equivalent to energy and energy to mass, the loss in weight (mass) could be translated into loss of energy. Behold, there would be the exact energy of one electron at one instant in time. Precision, cause and effect, would be restored. The world would be right again.

And all of this could be accomplished, Einstein declared, without any interaction—any collision by an exploring light beam—to alter the energy of the electron and forever bar an exact measurement of both time and energy.

Bohr appeared to be in serious trouble. Einstein, with the aid of relativity and ingenuity, was waving away indeterminacy and the whole theory of complementarity—the new picture of the universe as a strange, but reconcilable, dualism.

Bohr and the other Copenhageners did not even think of sleeping that night. Bohr did not doubt that Einstein was wrong. Any experimental arrangement permitting the location of atomic objects in space and time inevitably implied a transfer of momentum and energy between the apparatus and the electron. But where did the error lie? Where? Only a few hours were available to find it.

All through the too short night, Bohr and his collaborators examined every phase of the Einstein experiment. Only after hours of struggle did Bohr put his finger on what was wrong with the seemingly irrefutable experiment.

When Bohr met Einstein the next morning, he went to the blackboard and drew the apparatus as a real apparatus, rather than as the simple diagram Einstein had used. He suspended the box from a spring coil. To one side he attached a pointer which would move up and down on the markings of a scale affixed to the support holding the box. The weight of the box thus could easily be read before and after the escape of the photon.

Then Bohr tripped Einstein on one of Einstein's own celebrated findings. Some fifteen years earlier Einstein in his theory of relativity had shown that when a clock is displaced in the direction of the gravitational force it will change its rate so that its reading in the course of a time interval will vary by the inevitable quantum.

After the initial weighing, Bohr demonstrated, there would be an unavoidable latitude in the knowledge of the adjustment of the clock. Here was the principle of indeterminacy all over again.

"Consequently," Bohr argued, "a use of the apparatus as a means of accurately measuring the energy of the photon will prevent us from controlling the moment of its escape."

The conference, or most of it, suddenly breathed easier. Bohr worked out his point mathematically, with Einstein and the others joining in. There could be no question that the pull of gravity would cause enough variation in the clock to prevent the precise timing of the photon's escape.

Einstein could not disagree. His own work was being used to disprove his contention, and Bohr's reasoning and mathematics were impeccable.

Once again quantum mechanics and Bohr had triumphed in this battle of the titans. The quantum view of the universe as a bafflingly dual one and yet one capable of synthesis into a higher

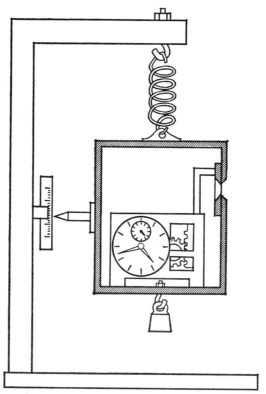

The Box. The imaginary box devised by Einstein was to disprove the reciprocal indeterminacy of time and energy in the atom and to confound Bohr. Bohr turned it against its inventor: he used it to uphold the theory of indeterminacy.

understanding was upheld by the very attack that had threatened it.

Einstein had lost. He had to concede that no exact time-space measurement could be made even with the aid of relativity. But he still did not like the world that resulted. When he and Bohr stopped to talk in the hall, Einstein was unhappy.

"He expressed a feeling of disquietude about the apparent lack of firmly laid down principles for the explanation of Nature," Bohr reported. "From my viewpoint, I could only answer that in dealing with the task of bringing order into an entirely new field of experience, we could hardly trust in accustomed principles, however broad."

Neither Bohr nor Einstein ever forgot or gave up in this olympiad of science, this struggle for the world, in so far as the world is comprehended in the minds of men. Bohr continued his studies of the box, developing more and more realistic drawings of it. He even put Gamow to constructing any actual model out of wood and metal. Einstein for his part began searching for alternative procedures to get around the fundamental objection Bohr had pointed up. The contest would be continued, though the 1930 clash was its highest point.

While Bohr swept to the quantum heights in the controversy with Einstein, the young physicists at the institute persisted, with Bohr's blessing if not his wholehearted participation, in the work on electrodynamics.

L. Landau from Russia and Rudolph Peierls from Germany found evidence that the relations between the different electric and magnetic fields set up by the nucleus could not be determined with arbitrary accuracy. Indeterminacy again was lifting its interesting head. If Landau and Peierls were correct another limitation would be imposed on classical concepts. Bohr's interest was irretrievably trapped. The subject had to be gone into after all.

"The challenge could not leave Bohr indifferent," Rosenfeld said. "He then cast himself heart and soul into the problem and it cost him (and myself who helped him in the task) many months of hard labor to solve it."

Though there were a number of interruptions, the harrowing struggle with electrodynamics went on from 1931 through most of 1933. Like all other Copenhagen pursuits of a difficult problem everyone was drawn into it. Everyone included Gamow, Rosenfeld, Casimir, Teller from Hungary, Bloch from France, Swensson and H. H. Nielsen from the United States. Never had the Copenhagen spirit, that blend of high intellectual endeavor, of venture, of profound involvement, and antics been more alive or yeasty. Nor had the concentration ever been harder or the play that offset it more high spirited.

One night at the movies an argument developed. The villain drew first, but the hero was a little faster and shot the scoundrel dead. Bohr insisted on the way home that this represented the difference between willful and conditioned reaction. The villain had, according to this theory of Bohr's, to decide to draw his six-shooter, while the innocent hero could react without thinking. Everyone else disagreed with this theory of the battle of Rattlesnake Gulch.

The next morning the irrepressible Gamow hunted up a toy store. He arrived at the institute bearing two cowboy guns. Bohr who was designated the hero, was armed with one. Gamow stored the other in his own equivalent of a holster. A test without warning was agreed to. As the afternoon wore on and fire power was forgotten in the heat of the debate, Gamow suddenly reached for his gun. But Bohr's draw—in the middle of a sentence—was even faster.

"We shot it out and he killed us all," Gamow certified.

The gun battle of Blegdamsvej, which has continued to echo through the tales of physics, was only one of several make-believe ventures into the realm of the outlaw and underworld.

For the annual September homecoming and seminar, the fellows decided to stage a production of *Faust*—a parody production. The Lord very obviously was Bohr and Mephistopheles was Pauli. Both were played with fine gusto by staff members:

BOHR ("the Lord"): Hast thou naught else to say?
Comest thou ever with complaint?
Is physics never to thy mind?

PAULI: Nay, 'Tis all folly! Rotten, as ever, to the core!
E'en in my dullest days it grieves me sore
And I must ever plague these physicists the more.

BOHR: (in mixture of German and English he often used when excited) Oh, it is dreadful! In this situation we must remember the essential failure of classical concepts . . . muss ich sagen . . . just a little remark. What do you propose to do with mass?

PAULI: What's that got to do with it? Mass? We shall abolish it!

BOHR: Well, that's very, very interesting, but . . .

PAULI: No, shut up, Stop talking rubbish . . .

BOHR: But Pauli, you must really give me a chance to finish what I have to say. If both mass and load are abolished what have we got left?

PAULI: Oh that's quite simple. What we've got left will be the neutron.

(Both pace feverishly back and forth.)

BOHR: It's not to criticize, it's but to learn. I take my leave now, later to return.

(Exit)

While Bohr continued to study the problems Einstein had raised and worked with the paradoxes of electromagnetism, he also thought more and more deeply about the extension of complementarity into nonscientific fields. Is not every human action changed and affected to some degree by the human thought that shapes it? Is effect not influenced by human participation, much as the atom is by the apparatus used in measuring it? "We ourselves are part of Nature," Bohr emphasized.

At about this time Bohr was invited to address the opening meeting of the International Congress on Light Therapy which was to be held in Copenhagen in August, 1932. Bohr decided to use the occasion to develop his thoughts on this further application of complementarity, and to call his address "Light and Life."

The subject had a special appeal to him. His father, Professor Christian Bohr, had long maintained that without a knowledge of the functions of an organism, there could be no hope of understanding its structure. At the same time Professor Bohr had insisted that the physical and chemical studies of organs should be pushed as far as possible. Otherwise, he warned, one would run the risk of deceiving himself with verbal explanations. Bohr was eager to continue the discussion of the problems that had so deeply concerned his father. The subject would honor his father, a thought that gave him pleasure.

Bohr was well aware that as a member of a science concerned strictly with the properties of inanimate matter, he was daring convention in discussing life. He was convinced, however, that the point had been reached where physics and life, the inanimate and animate, could no longer be entirely separated. He made the point to the surprised light therapists: the results reached in the limited domain of physics may influence our views of the position of living organisms within the edifice of natural science.

Bohr laid the basis for his parallels: Light is a wave, an oscillation. Physics also has established that it is a particle. A dilemma is thereby created of a character previously unknown in physics, for it is impossible to trace the path of an individual light quantum without essentially disturbing the phenomenon under investigation. Thus a complete account of light phenomena is unobtainable and must be renounced. We must therefore content ourselves with the knowledge that the phenomenon will occur within a certain range of probability.

"This situation may appear very uncomfortable," Bohr acknowledged, "but as has often happened in science when new discussions have led to the recognition of an essential limitation of concepts hitherto considered indispensable we are rewarded by getting a wider view and a greater power to correlate phenomena which before might have appeared as contradictory.

Bohr turned to his parallel: A plant's assimilation of carbon from the atmosphere or the formation of hemoglobin in the blood are similar to ordinary physical effects and to be accounted for by the ordinary methods of physics and chemistry.

But even complete understanding of such processes still would offer "no more satisfactory explanation of living organisms than would a comparison with such a purely mechanical contrivance as a clockworks.

"Indeed," said Bohr, "the essential characteristics of living beings must be sought in a peculiar organization in which features that may be analyzed by the usual mechanics (physics and chemistry) are interwoven with typically atomistic features to an extent unparalleled in inaminate matter."

Bohr explained further what he meant. The eyes are so constructed that the absorption of a single light quantum by the retina is sufficient for a sight impression. But this is the limit of the eye. No impression finer or smaller than the quantum is possible. There the limit is drawn.

The parallel was even more complete. To study the part played by a single atom in the vital functions of an animal would require killing the animal and thus destroying the very life the experimenter was attempting to study. "We murder to dissect," said Wordsworth. And if life is maintained, its final structure or flow cannot be completely analyzed.

"In every experiment on living organisms there must remain some uncertainty about the physical conditions to which they are subjected, and the idea suggests itself that the minimal freedom we must allow the organism will be just large enough, so to say, as to hide its ultimate secrets from us.

"In this view the very existence of life must, in biology, be considered as an elementary fact, just as in atomic physics the existence of the quantum of action has to be taken as a basic fact. Indeed the essential non-analyzability of atomic stability in mechanical terms presents a close analogy to the impossibility of a physical or chemical explanation of the peculiar functions characteristic of life."

How could any sharp line be drawn between living things and their environment? "We cannot even tell," said Bohr, "which particular atoms really belong to a living organism, since any vital function is accompanied by an exchange of material through which atoms are constantly taken up into and expelled from the organization which constitutes the living being."

The parallel extended most notably into the interplay of mind and matter. Bohr emphasized here that the interaction between measuring instruments and objects under investigation in atomic physics is essentially the same as that encountered in psychological analysis where the mental content is invariably altered when the attention is concentrated on any feature of it.

This was a strange and largely unheard of picture Bohr was painting. In fact it was about half a century ahead of its time. The men who used light in medical treatment were baffled. They applauded politely, and later complained that they had heard some farfetched, if not nonsensical, ideas. In addition to the difficulty and extraordinary character of his thought, Bohr's low-pitched voice and complex sentences—complex even when he was attempting to simplify them—made it difficult for even the expert to follow him.

Just before the lecture, Rosenfeld went to the railroad station to meet Max Delbruck, one of the institute fellows, who was returning from a short trip. The two went together to the Christianborg Palace where Bohr was speaking. They slipped into seats in the empty gallery.

It would make a fine story to say that Delbruck was so stirred by the speech that he resolved on the spot to become a biologist. But as Rosenfeld said later, that would be a "romantic exaggeration." Nevertheless the young German physicist was so deeply interested that he obtained a copy of the speech and pored over it. He was moved to turn to biology, the profession in which he was to make major contributions and to win international acclaim.

Thirty years later when Delbruck became the director of the new Institute of Genetics in Cologne, he invited Bohr to speak at the opening ceremonies. At Delbruck's request Bohr's subject was "Light and Life Revisited."

The poorly received speech of 1932 thus reached one person and produced one of the world's outstanding biologists. Not a bad score for one speech!

Regardless of the reaction of the audience Bohr himself was not satisfied with the speech. It was to be printed in English in the proceedings of the society, but he had promised a translation to a Danish journal and this offered a chance "to work on it a little." Bohr asked J. Rud Nielsen, one of the young men at the institute, to help him with the revision. When the first proof came from the printer they went over it word for word and sentence for sentence, reconsidering, changing, and adding. This went on through nine proofs.

One night while the work was under way Bohr and Margrethe had to attend an affair for a distinguished visitor to Denmark. Their presence on such occasions was a virtual requirement and a public duty. About 10 P.M. Bohr changed to evening dress and the couple set off. Nielsen rode with them to his room. As Nielsen got out of the taxicab, Bohr told him that he would be at the institute at 8:30 o'clock the next morning and would like to resume work on the proof. He was there promptly, looking again at a point that had troubled him. "Yes, we must change that; I've been thinking about it all night." Nielsen learned later from Mrs. Bohr that they had reached home at 2 A.M.

"I remember the last night of work on the Danish version," said Nielsen. "Bohr was overworked and nervous and paced back and forth while his secretary and I sat at a table with the proofs and the typewriter. When the last corrections were finally made and Bohr had to indicate to the frantic publisher whether the proof was okay or another proof was needed, his pen wouldn't write. His secretary gave him another one—and he signed in the wrong place."

The nine proofs and the prolonged struggle to make every sentence say exactly what Bohr wanted to convey were typical. No publication came out without such scrutiny and reworking. Bohr had a tremendous respect for the written word and its endurance. He wanted the words to be right, for the present and for the future.

The overweaning assurance and frenetic activity of the twenties had ended with shocking abruptness.

A few months after the president of the New York Stock Exchange had proclaimed the end of economic cycles, the market crashed. Banks tottered in the wake of the disaster, and many closed their doors. Prices fell and in disconcertingly short time plants cut back their production and reduced their work forces. Unemployment in the United States mounted to an appalling 10,000,000.

The ruin and paralysis spread throughout the world. In England unemployment increased from 1,000,000 in 1930 to 3,000,000 one year later. Denmark saw its vital agricultural exports decline and the whole country suffered from the impact.

The depression had a particularly disastrous effect in Germany. Many of the new factories on which the peaceful revival of the country had been based closed their doors, and millions were without work. The Allies tried to offset the mounting crisis by setting a terminal date for the reparations Germany had been paying, and by freeing the Reichbank and German railways from Allied control.

The substantial concession, however, only embolded "the dark, hidden forces" of Adolf Hitler and the Nazi Party, as they organized to destroy democratic government. In 1930 the former army corporal and his party elected 107 members to the Reichstag. By 1932 Hitler's party had captured 230 seats. A party that had garnered 13,000,000 votes was openly dedicated to war, to world-domination and to the eradication of the Jews. Hitler's storm troopers swaggered through the streets of German cities.

Bohr watched all of this with most acute unease. The world at large still generally dismissed the Nazis as of no importance, or disregarded their increasing brutalities. A Dane, living just across the border and with many friends and close connections in Germany, could not be so unheedful. Bohr was afraid of what was happening; he had no illusions about its seriousness.

On January 30, 1933, Hitler became chancellor of Germany, and the headlines "Germany under Hitler" awakened the unaware with a cold shock. What Nazi control meant rapidly became clear. On February 27 the Reichstag burned. Racial laws barring Jews from all public posts were quickly announced and the malevolent signs appeared—"No Jews," "Jews Not Admitted." The Nazis soon were in control of the military forces, labor, and business as well as the government.

Bohr knew that action would be needed to help the German scientists. Shortly after Hitler assumed power he went to Germany, ostensibly to visit the universities, but actually to check on the safety of the scientists and to find out how many would be dismissed under the racial laws. Bohr knew well that atomic physics would play a decisive part in the future of mankind, and he recognized that the endangered scientists must be brought to safety.

In Hamburg he met young Otto Frisch, the nephew of his friend and colleague, Lise Meitner. Frisch had just succeeded in measuring the recoil of a sodium atom when it emits a light quantum or packet. Bohr asked about this work with the warmest of interest.

He then took Frisch by the waistcoat button and with a smile that reminded Frisch of a kindly father told him that he hoped he would come to work in Copenhagen. To be confronted by a man who seemed quite legendary and to have him say: "We like people who can carry out such experiments," was almost more than the young physicist could imagine.[1]

The relief Frisch felt was expressed in a letter he wrote to his mother that night. She had been desperately worried about what the racial laws would do to her son, just beginning a highly promising career as a physicist. He told her not to worry. "The Good Lord himself has just taken me by the waistcoat button and smiled at me," he said to her. Many years later Frisch emphasized: "That was exactly how I felt."

Later that year Frisch went to London to work with Blackett. When his work there was finished an invitation was waiting from "the good Lord." It asked him to come to Copenhagen to work at the institute.

As Bohr traveled through Germany in those early forbidding days of Nazism, he unobtrusively laid the lines for the escape of the German scientists. The word was spread that Copenhagen would be standing by in case of necessity. With sensitivity and delicacy, Bohr invited Franck, Placzek, and Hevesy to continue their scientific work in Copenhagen, if they should want to come.

A few in the outside world still maintained that the ugly storm would blow over, and it became the thing to visit Germany. Reports came back of the bustling prosperity and confidence of the Germans. Again Bohr was not deceived. In 1933 he, Harald, Thorvald Madsen, director of the Serum Institute, and Albert V. Jorgen-

[1] See the chapter by Otto Frisch in *Niels Bohr His Life and Work*.

sen, barrister of the Danish High Court, formed the Danish Committee for the Support of Refugee Intellectuals. Bohr prepared for the reception of refugee scientists in Denmark and began writing to friends and colleagues in all parts of the world seeking positions for the men and women fleeing the Nazi terror.

All the while work on electromagnetism continued. It was arduous labor. Young Frisch arriving after his year in England went to his first colloquium. "The scene is imprinted on my mind," he said. Bohr and Landau were locked in a discussion. Landau lay flat on his back on the lecture bench gesticulating wildly. Bohr bent over him arguing and waving his arms. Neither showed the slightest awareness that there was anything unusual about this method of conducting a colloquium.

Frisch had spent three years in the formal atmosphere of Hamburg and a year in London where he had felt shy and had made few close associates. He was startled by his first glimpse of the "free-and-easy" atmosphere of the institute and the acceptance of a man solely on his ability "to think clearly and honestly."

Bohr at 48 was at the peak of his powers, physically and mentally. Few of the younger ones could keep up with him in the race up the stairs. A game of ping-pong was usually under way in the library, and Bohr was often one of the players. Frisch could not remember ever beating him, for Bohr's reactions and motions were fast and accurate. "He had tremendous will power and stamina," said Frisch. "In a way these qualities characterized his scientific work as well."

The concentration was intense, but for every step forward in the baffling study of electrodynamics there seemed to be another backward. It was with the greatest of reluctance that Bohr finally began a paper. The work of drafting went along even slower than usual. Change after change was made, until finally the last proof arrived. At this final moment Bohr suddenly had an inspiration; he saw the solution, at last. In fact, it fairly leaped out at him.

And it was such an anticlimax that the institute was thrown into an uproar of disbelief and joking. After three years of work, Bohr realized and proved that the field intensities at any point could be measured with complete accuracy by conventional, classical methods. The quantum did not have to be brought into it at all at this point.

Complementarity was required though for the relations between field intensities in different space-time regions. This took an extension of the methods developed in quantum mechanics and proved far from easy. Nevertheless Bohr brought it off in another of his works of rare imagination and thoroughness.

In a relieved and jubilant mood of accomplishment the group went off for a night at Tivoli, Copenhagen's enchanting amusement park. They had dinner in one of the restaurants and wandered along the tree-lined walks bordered by colored lights, through the amusement sections, gay with the Ferris wheel, games, and pink cotton candy, and stopped at the outdoor theater to watch one of the performances staged there several times each night. A magician circulated through the audience asking to be shown any object the giver might like to have his assistant identify. The assistant sat on the stage blindfolded and with his back turned to the audience. "What do I have here?" the magician would ask holding up a watch or a key, or anything submitted. After a little impressive fumbling, the answer would come back, clear and correct.

Bohr and Gamow insisted that it must be done with ventriloquism. Delbruck backed up his disagreement with a bet: "Ten Kroners it is not."

The problem had to be settled. Delbruck phoned all the ventriloquists he could find listed in the telephone directory, and when he had convinced them that he did not want to hire them for a performance, was assured that the magician did not accomplish his magic by any distant placement of the voice. Bohr and Gamow in support of their contention consulted the encyclopedias. In one they found a report that the Egyptian priests had made the images of the gods speak by ventriloquism. Bohr and Gamow claimed strong support from this bit of history and maintained that if the trick was not done with ventriloquism, Delbruck would have to prove how it was accomplished. This proved to be one of the Copenhagen impasses. The bet never was settled or paid.

Heisenberg came to Copenhagen at about the time of the electromagnetism solution. At a colloquium Heisenberg was skeptical; he did not believe a word of it. Seizing a piece of chalk he began to repeat some of the calculations, and to his amazement found them coming out as Bohr and Rosenfeld had said. "It was amusing to watch the growing expression of surprise on his face," said Rosenfeld. Thus convinced, Heisenberg changed into an ardent partisan of the new solution.

Pauli, who also was informed of the unexpected outcome, accepted it without disagreement, but pointed out that they would have to go on to some related problems. Bohr and Rosenfeld agreed entirely with this reaction, but warned that the additional work Pauli prescribed would be slow. Actually it took another twenty years.

Pauli had a very good idea of what "slow" might mean and he used the admission to twit the Copenhageners in a mockery of their own style: "Since you have managed to publish the work on the electromagnetic field it has become impossible to state with certainty that the other work will never be accomplished."

Those who were the butt of Pauli's pointed jibe prized and repeated it.

The institute had not only become the world center of physics; Bohr had become the undoubted first citizen of Denmark. The nation now made that high status official in the combination of honor and practicality that is part of the genius of the Danes. Bohr was invited in 1932 to occupy "The House of Honor"—Denmark's semi-official residence for her greatest citizen.

The House of Honor, a mansion built by J. C. Jacobsen, the founder of Denmark's famous Carlsberg Breweries, is made available for life to the man or woman who has brought the greatest honor to the country.

It stands at one edge of the brewery grounds. Only the tall trees and spacious gardens separate it from the railroad tracks on one side and the big buildings of the brewery on the other. The tallest trees, however, do not hide that Copenhagen landmark, the towering smokestack of the brewery. To reach the enclosed courtyard entrance to the House of Honor, visitors also drive by the remarkable entrance to the brewery. Its huge portal, bearing the inscribed words *Laboremus Pro Patria* rests solidly on the backs of two larger-than-life elephants carved from gray stone in every verisimilitude of life. Jacobsen was a man of imagination, as well as acumen and patriotism.

Jacobsen designed the House of Honor as his own residence in the flush of his enthusiasm after a visit to Pompeii. Some concessions had to be made for the difference in the Italian and Danish climates, but Jacobsen captured the grace and dignity of the Pompeian style, and built a house wonderfully adapted to the use to which he afterward dedicated it.

The spacious living rooms are flanked on one side by a Pompeian peristyle. The oblong arcade with its slender columns and tranquil outlook over the fountains and greenery of the central court were exactly the place for Bohr to pace in solitary meditation or frequently in the company of institute fellows or visiting physicists. In deference to the Danish weather the court was glassed over and formed a perfect setting for the big Christmas tree that was the center of the party the Bohrs gave each year for all institute fellows and their families who were away from home. It was also an almost unrivaled setting for any other large gathering.

In winter the guests were shown the adjoining garden room with its tropical trees and plants growing in a year-around profusion. In summer the guests were invited to wander in the spacious gardens. The wide central lawn was dramatized by one of the finest purple beeches in all of Denmark, and beyond stretched the rose garden. In later years the rolling side terraces were planted with rock plants the Bohrs brought back from the Orient.

As soon as the necessary painting and refurbishing was done, the Bohrs left their modest flat at the institute and moved happily into this unusual house. By any definition it was a mansion. Their first guests were the Rutherfords, and later they would entertain the King and Queen of Denmark, Queen Elizabeth and Prince Philip of England, presidents and premiers, as well as the greats of science. But it was also a house where the Bohr boys could freely pursue their own enthusiasms, and where the young physicists of the institute could come constantly as though they were coming home. Margrethe's warmth and graciousness and Bohr's cordial hospitality turned the House of Honor into a home rather than an institution. Those who came always remembered Carlsberg.

As they drank their after-dinner coffee, they would pull their chairs close to Bohr or sit on the floor around him in order not to miss a word. "Here I felt was Socrates come to life again lifting each argument onto a higher plane, drawing wisdom out of us which we didn't know was in us (and which of course, wasn't)," said Frisch.

"Our conversations ranged from religion to genetics, from politics to art, and when I cycled home through the streets of Copenhagen, fragrant with lilac or wet with rain, I felt intoxicated with the heady spirit of Platonic dialogue."

Then tragedy suddenly struck this happy, almost idyllic home. In 1934, Bohr, his eldest son, 19-year-old Christian, and two friends went sailing. A sudden storm whipped the gray waters of the Kattegat into lashing, white-crested waves that swept across the decks of the small boat. In the turmoil and crash of water Christian was carried overboard and lost.

The boat beat frantically back and forth scanning the waters in desperate hope against hope. When darkness forced them to acknowledge the finality of the tragedy, Bohr had to return to the House of Honor with the dread word.

In after years when the discussion sometimes turned to religion and Bohr expressed the opinion that consolation was one of the rocks on which the great religions stood, he often told the story of Buddha and the mother who had lost her only child. In her inconsolable grief she was brought to the great prophet. Buddha told her that he would assuage her grief under one condition—that she bring him six grains of mustard seed from someone who had never experienced a grief. The woman went from person to person and from city to city, asking, "Have you ever had a grief?" Many months later she returned to Buddha. She had no seed, but her inconsolability was cured. Bohr never explained why he told the story.

11 /

The Heart of Matter

A MYSTERY—another paradox—brought Bohr bounding into the lecture room of the institute. He was waving a paper in his hand. The discussion of electromagnetism was set aside for a "little development" that was unusually interesting. That it would foreshadow the splitting of the atom, no one could foresee.

Bothe and Becker, Bohr reported, had directed a stream of particles at beryllium, a rare light metal which in some of its forms produces the emerald and aquamarine. To their amazement the beryllium gave off a beam of rays of high penetrating power.

Bothe and Becker suggested that the mystifying rays might be gamma rays like those emitted by radium. Bohr, as he discussed the paper with the others, was immediately doubtful of this explanation. But if not, what could the rays be? Any radiation emerging from the atom was important. It might offer a new approach to a problem that had engrossed Rutherford and puzzled Bohr all through the late 1920s.

Rutherford, often assisted by Chadwick, had been disintegrating light nuclei by bombarding them with alpha rays. He had early demonstrated that the particles knocked out in this bombardment were, as he had suspected, hydrogen nuclei. And only hydrogen nuclei emerged when he chipped at the nucleus. It looked increasingly as though all matter, however hard, or soft, or white, or metallic, or gaseous, might be made up of multiples of the basic hydrogen nucleus. In testimony to the prime standing and significance of the hydrogen nucleus—the nuclear building block—Rutherford gave it a special name—the *proton*.

Rutherford then went on systematically to bombard and disintegrate additional light elements. Before long he had proof that twelve elements could be broken into; the once impregnable, invulnerable core by the atom was yielding, bit by bit.

In each instance one of Rutherford's newly named protons was driven out. The loss, he saw, could only mean that the nucleus had one less unit of positive charge, for the proton carried a charge of one. With one less proton and one less unit of charge, the atom would lose one of the electrons circling the nucleus. Its very nature would be changed; its characteristics would be altered; it would be transformed. When, for example, Rutherford broke off one of the eight protons of oxygen, the oxygen was transformed into nitrogen, with its seven protons. The dream of the ages and the unceasing quest of the alchemists of the Middle Ages had suddenly been realized in the Cavendish Laboratory in Free School Lane.

Rutherford had previously proved that atoms could spontaneously change their chemical nature through radioactivity. Now that other atoms could be transformed by subjecting them to radiation from a little plate coated with radium, no doubt remained that the nucleus, which had always been considered the solid of solids, was actually a complex system composed of many particles.

The question was "What particles?" Was there another particle in addition to the proton? There was every reason to inquire, for the protons alone could account for only about half the weight or mass of the elements. Lithium, for example, is made up of three protons each with a mass of one and therefore should have a weight of three. Nevertheless lithium's weight is seven.

It was clear that the other four units of weight had to exist in the nucleus, for the remainder of the atom is composed of the nearly weightless orbiting electrons and of great empty spaces. But if there were some other kind of unit in the nucleus, what was it? The reluctantly accepted theory was that the nucleus might also contain electrons.

Few, and least of all Bohr, were satisfied with the latter explanation. It was particularly unsatisfactory from the standpoint of the quantum theory. An electron bound close to the nucleus would have energies of billions of electron volts. Cosmic rays had such energies, but the energies involved in nuclear phenomena figured in the millions of volts, not in the hundreds of millions. When Bohr pointed up this major discrepancy, Rutherford had suggested that it might be necessary to assume the existence of a "chargeless proton." Almost a decade before, in 1920 in fact, Rutherford in his Bakerian Lecture had predicted the existence of nuclear particles of zero charge.

The existence of some such particle, Rutherford said, "Seems almost necessary to explain the building up of the nucleus of heavy elements. Otherwise how would a heavily charged particle ever reach the nucleus of a heavy atom?" How could the heavier elements, such as uranium, ever have been assembled?

Rutherford and the Cavendish went to work diligently to try to find the theoretical chargeless particle, the neutral particle of Rutherford's prescient imagination. But they did not succeed, and finally the effort was more or less dropped.

Thus when word came in 1930 that a new kind of radiation had been discovered coming from the nucleus, Bohr was excited and aroused. The time was ripe then to make use of any new opening. Advances in the understanding of atomic structure and in quantum physics offered a chance of success that had been lacking before.

When Bohr was invited by the British Chemical Society to give its Faraday Lecture in May, 1930, he devoted the last paragraph of his lecture to the nucleus. He pointed out that the new controlled disintegration of the nucleus offered a promising means of exploring the unquestionably great forces of the nucleus and concluded: "I have touched upon this mainly to emphasize that in atomic theory, notwithstanding all the recent progress, we must still be prepared for surprises."

No statement was ever more accurate. But could the outgoing radiation Bothe and Becker had discovered be a gamma ray? A paradox was posed. "How wonderful that we have met with a paradox," Bohr exclaimed to a colloquium. "Now we have some hope of making progress."

Bohr returned to the subject of the nucleus the next year when he went to England to speak at the centenary of James Clerk Maxwell. Rutherford had recently been made a peer, and a gala celebration was planned. Planck, Sir Joseph Larmor, and Bohr were to be the principal speakers.

Bohr reviewed the recent development of physics and spoke about the foundation Maxwell's work had provided for the new theory of electromagnetism. But in his conviction that a new promise was opening in nuclear physics—a conviction reenforced by what he was seeing in the Cavendish laboratories—Bohr added in conclusion: "I am glad to give expression to the great expectation with which the whole scientific world follows the exploration of an entirely new field of experimental physics, namely the internal constitution of the nucleus, which is now carried on in Maxwell's laboratory under the great leadership of the present Cavendish professor [Rutherford]."

Bohr was heralding the new day in physics. Ironically he was doing it at ceremonies honoring a scientist who had declared that atoms were unchangeable and always would be.

The breakthrough that Bohr anticipated came in the following year, 1932. The actual discoveries were made elsewhere, but as always the first step, even before publication, was to report them to Bohr at Copenhagen. His evaluation and criticism were essential. Within a few days the mail brought startling letters from Irene Joliot-Curie in Paris and Rutherford in Cambridge.

The daughter of the great Marie Curie, and her husband, Frédéric Joliot, had repeated the Bothe-Becker experiment of 1930. When they did, they found that the powerful and still mysterious rays emitted by beryllium could penetrate even the solidity of lead. The rays speeded through a lead shield that absorbed and stopped gamma rays. Thus it could be concluded they were not gamma rays. The Joliots discovered in addition that if a paraffin shield or a shield of any hydrogen-containing material were placed around the beryllium, the rays caused the ejection of protons of very high energy from the wax. The mystery was deepening.

Bohr was doubtful. "I suspect," he said in a letter in which he told Rutherford of the Joliot report, "that the beta ray tracks on her photographs were due to Compton effects in the walls of the Wilson chamber, rather than to nuclear excitation of the gas atoms within the chamber."

Exactly such letters helped to tie the physicists of the world together in the remarkable coalition that was solving some of the most fundamental problems of the universe. And generally Bohr was both the communicating center and the center where the disparate findings were given meaning, focus and direction. It was how this one man could influence science in almost every country around the globe and give direction to his era.

Bohr had scarcely rushed around the institute to show the Joliot letter to all of his colleagues when the Rutherford letter arrived.

James Chadwick, Rutherford's chief assistant and his close associate in disintegrating the lightweight atoms, also had repeated the beryllium experiment. He coated a disc with radium F (polonium), placed it behind a disc of pure beryllium, and hooked the two to a chamber in which every particle emitted could be recorded photographically on a strip of moving paper (an eye-saving improvement over the old method of darkening the room and counting scintillations by the naked eye). Like the Joliots, Chadwick tested the strange rays coming out of the beryllium with lead. When he slipped an inch-thick piece of lead between the beryllium and the counting chamber, there was no difference in the number of counts per minute. The rays came right through. They penetrated the relatively great thickness of lead as though it were not there. An electron or gamma ray would not have penetrated much more than the surface.

Then Chadwick tried the Joliot idea and placed a piece of paraffin between the beryllium and the counter. This time the ray count increased! Chadwick saw that additional particles were being knocked out of the wax. The scientist quickly established that the particles struck out of the paraffin were protons, or hydrogen nuclei.

The proofs piled up. The rays from the beryllium could not be deflected by a magnet. Here was conclusive proof that they were chargeless, or electrically neutral. And they traveled at a speed one tenth that of light, and thus much more slowly than gamma rays which are a form of light and as speedy as light. A new particle had been found.

Chadwick, remembering Rutherford's prediction of a charge-less particle, named the mysterious particles "neutrons." They were neuter, uncharged. And they had about the same weight—a mass of one—as the proton. But in significance the neutrons were large.

Chadwick and Rutherford saw that they had solved the old problem of the nucleus and its excess weight. The extra mass in the nucleus was made up of neutrons! Thus carbon, Number 12 in the table of elements and with a weight of 12, but with only six protons, unquestionably had six neutrons.

Bohr was elated. Copenhagen calculations immediately confirmed the correctness of Chadwick's discovery, and the neutron was welcomed into the world with a toast in good Copenhagen beer. Bohr hastened to write Rutherford about his enthusiastic agreement about the neutron and its name.

Rutherford answered almost by return post: "I'm pleased to hear that you regard the neutron with favour. I'll be interested to hear all about your theory of the neutron."

Then Rutherford continued, "It never rains but it pours. I have another interesting development to tell you about." Rutherford and Cockcroft had been examining the effects of bombarding lightweight elements with protons—charged particles. When they began experimenting with lithium, Number 7 in the table of elements, they found that it broke up into two "ordinary alpha particles."

"The simplest assumption," Rutherford wrote to Bohr, "is that lithium 7 captured one of the protons. It looks as if the addition of the fourth proton leads at once to the formation of an alpha particle and the consequent disintegration."

Rutherford added that he was glad "we're getting something out of high potentials," and predicted "you can easily appreciate that these results may open a wide line of research on transmutation generally."

This new startling glimpse of what might happen in the nucleus came, as Rutherford indicated, from the use of a new machine they had developed at Cambridge to step up velocity and direct a stream of particles. It was later to be called the accelerator. In itself it was to change the scale and methods of physics from that time on.

When Rutherford started his work on the nucleus he had broken into the nuclei of the lighter atoms with a little apparatus he could easily hold on the palm of one hand. It was lighter than a book, and not much larger—a small piece of board from which two bent tubes extended. A photograph was made of him holding it as though it were a child's toy.

The bit of radium Rutherford used in his "little gun" gave off several thousand million alpha particles a second. Rutherford thus had a prodigious supply of ammunition, but the trouble was that it shot off in all directions. He had no way of directing his bullets—his alpha particles—and only one particle in millions made a bull's eye hit on the small target offered by the nuclei in the material he was bombarding. Rutherford needed more accurate artillery and an energy of about 4,000,000 electron volts to penetrate the electrical barrier around the nucleus. The nucleus was the heart of matter; it was formidably well guarded.

John Cockcroft, who had served in the artillery during the war, set out to build some atomic artillery capable of shooting a stream of alpha particles through or over the barrier and into the nucleus. He succeeded. Compared to Rutherford's little hand-size "gun," the accelerator Cockcroft designed was a mammoth, with three-foot-long glass tubes and large pumps. It stood about 15 feet high and was considered appallingly expensive. It cost all of £100! Some later models were to be nearly a mile long and to cost many millions of dollars.

Cockcroft and his coworkers trained their new directable

stream of particles at a lithium target, with the astounding result that Rutherford reported to Bohr.

To lithium's four neutrons and three protons, the bombardment added a fourth proton. For one brief instant of time, Bohr learned, four neutrons and four protons were packed into the one nucleus. The former mass of 7 (4 + 3) was, with the addition of the proton (1) increased to eight. And then an alpha particle, bearing two neutrons and two protons (2 + 2) was given off. It was easy to illustrate and the blackboards at Copenhagen were covered with drawings.

The Joliot and Rutherford letters reporting all of these fundamental discoveries threw Copenhagen into uproar. Bohr at once wrote to Rutherford to congratulate him on "the wonderful results." He added that progress was so rapid "one wonders what the next post may well bring. . . . It should now be possible," he said, "to excite particle emissions from muclei by means of the recently discovered powerful agencies," and perhaps to settle "the fundamental problem of the constitution of the nucleus."

The "next post" was nearly right. Bohr's letter to Rutherford was dated May 21, and on May 26, 1932, Rutherford wrote again, with more astounding news. Working around the clock with the new accelerator, the Cantabridgians were investigating as many elements as possible.

"So far," Rutherford reported, "we have only had time to examine a limited number, but everything so far tested shows the same result. I may also tell you privately that uranium and lead also give a positive effect and we are able to increase the natural activity [nuclear transformations in which alpha particles shot out of the nucleus] several times.

"I am indeed inclined to believe that in these heavy elements the effect is likely to be due to the capture of the proton through one or more resonance levels, and this is a point I propose to examine using comparatively low voltages. Of course results so far are in the preliminary stage."

Rutherford spoke of his pleasure with the new accelerator: "The electrical counters completely confirm the scintillation method (counting the little flashes as the electrons struck a zinc screen) so there is no doubt we are on safe ground."

The two friends had a fuller opportunity to canvass the nearly fantastic events of 1932 when the Rutherfords came to Copenhagen to be the first guests in the House of Honor. Rutherford gave a lecture on "Transmutation of Matter" at a large dinner given in his honor. Both Bohr and Rutherford felt that nearly limitless possibilities lay ahead, though they could not imagine how limitless.

The pace continued into 1933. Early in the year Patrick Blackett in England and Carl D. Anderson in the United States independently found some unusual tracks in the cloud chambers with which they were working. Blackett was studying cosmic rays—rays from outer space—and had placed his chamber in a magnetic field to bend positively charged particles in one direction and negatively charged ones in the other. He found many typically curved electron tracks. But curving off in the opposite direction were tracks made by other particles of the same mass. He had discovered another particle that long had eluded detection, the positron or positively charged electron—the mirror twin of the negatively charged electron. The positron, however had the most fleeting of lives.

At this peak of achievement, the Bohrs went to the United States. Bohr was asked to lecture at scientific meetings being held in Chicago in conjunction with the Chicago World's Fair, known as "The Century of Progress." Bohr was weary from the intensity of the recent months, and he was certain that ten days in Chicago and several weeks in California would give him the rest and change he needed.

The change of scene materialized, almost too dramatically, but not the rest. In California, which he visited first, he was pressed to grant interviews to the newspapers. He refused, for he was fearful that "if the journalists get hold of one, he is lost." He was also asked to talk to the Trustees of the California Institute of Technology and give them his impressions of their rapidly developing school. "I said many nice things," he later told J. Rud Nielsen who met him in Chicago, "but I did not say anything I didn't mean, and it was not what they wanted to hear." Bohr had devoted much of his talk to his enthusiasm for the American Indian and his civilization.

Bohr and Millikan again found themselves at odds. On the other hand Bohr was delighted with a young physicist, Robert J. Oppenheimer, whom he met there, and with whom he formed a friendship that would continue under circumstances that neither could anticipate.

Then back to Chicago. Chicago was a study in contrasts—the fair on the lakefront with its crowds and frequent gaudiness, the ragged men standing under the elevated tracks in long lines waiting for soup and bread, and lavish private homes where the city's guest of honor was welcomed. One home had a swimming pool illuminated from below.

At one dinner Bohr ventured to observe that in Denmark the burdens of the depression were more evenly distributed over all levels of the population. He was surprised by the surprise with which his remark was greeted. Bohr looked at the depression in the light of his theory of complementarity—"one can't improve conditions for all levels of society without renunciation on the part of some."

Bohr had hoped too that he and Nielsen might start to work on a book he had in mind. "I should like to write a book that could be used as a text," he told Nielsen. "It would show that it is possible to reach all important results with very little mathematics. In fact, in this manner one would in some respects achieve greater clarity." The book did not progress beyond this discussion, and then it was time to return to Denmark.

The pace in the laboratories did not slacken. In January, 1934, Frédéric and Irene Joliot-Curie announced the discovery of artificial radioactivity. When they bombarded aluminum with alpha particles, a new substance was formed. It turned out to be phosphorus, but a phosphorus that vanished in half an hour. As it disappeared, it gave off gamma rays and the newly found positron. In the end it turned into silicon, the main constituent of sand.

The Joliots had barely missed the discovery of the neutron, and in fact might have found it, they later said ruefully, if they had happened upon Rutherford's prediction of a chargeless particle. But their discovery of artificial radioactivity in its turn brought them the Nobel Prize.

The discoveries were piling up—the proton, the neutron, the positron, artificial radioactivity, the transmutation of matter, and the powerful new artillery for smashing still further into that citadel of the atom, the nucleus. Physics was as alive as an excited atom.

The moment Enrico Fermi in Rome learned of the Joliot-Curie discovery, he decided to try to produce artificial radioactivity with neutrons. The neutron, he reasoned, would not shoot through the atom as rapidly as the fast alpha particles, nor would it as an uncharged particle be repelled by the charge of the nucleus. In this way, he thought, it would have a better chance of hitting a nucleus with full impact.

Fermi began systematically testing the elements in their order in the periodic table. Until he reached Number 9, fluorine, nothing at all happened and he was getting discouraged. Fluorine though was highly activated by the slow neutrons and so were the next elements in the list. Some of the radioactive isotopes created by the capture of the neutrons were so short-lived that he and his co-workers had to run full speed down a laboratory hall to get them to the counters before they disappeared entirely.

Neither could time be lost in publication. Fermi began publishing his results in an Italian journal that could get them out very quickly. The issues were awaited with the greatest eagerness in Copenhagen and elsewhere. "I was one of the few who could read Italian," said Frisch, "and I well remember how everybody crowded around me whenever a new issue of 'La Ricerca Scientifica' arrived."

But even greater surprises lay ahead. When Fermi and his collaborators reached uranium, Number 92 and the last on the list of the elements, the uranium became especially active. One of the products uranium gave off under the neutron bombardment was a complete mystery. It certainly was none of the existing elements close to uranium—generally the radioactive element created by the bombardment was only a place or two removed in the table of elements from the original material.

From preliminary calculations, Fermi thought that the strange element might, with its additional captured neutron, be a new element of Atomic Number 93, an unstable element that he knew could exist nowhere on the earth. When word of this possibility leaked out before Fermi was ready to announce it, the Fascist press seized upon it as a "Fascist victory," and the discovery of a new element was celebrated with great fanfare.

At this point accident entered, as it often does in science when the ground is prepared and work is intensively going forward.

Some silver cylinders were being irradiated by placing a neutron source inside of them and placing them in a lead box. Bruno Pontecorvo—who was later to decamp to the Russians—noticed that some odd things were happening to the metal cylinder. Its activity depended upon where it was placed in the lead box—in a corner or in the center—and apparently upon whether the box stood on a wood or a metal table. Everyone gathered around to see this oddity, and at this point Fermi suggested trying the effect of some light material, "paraffin for instance." Paraffin increased the radioactivity of the metal a hundred times.

Paraffin is largely hydrogen, and that afternoon Fermi proposed that they see what water—with its hydrogen—would do to the irradiated silver. A fountain and pool in the garden of an adjoining villa offered the best available supply of water, and so the silver cylinder with its radioactive neutron source was immersed amid the goldfish and lily pads. Again the radioactivity of the metal leaped upward.

Fermi theorized that the invading neutrons were slowed down by colliding with the similarly sized hydrogen nuclei, and that the slowed neutrons were much more likely to be captured and to remain in the nucleus than would a high-speed particle shooting through.

Word of these astounding results, with their indication that a slow neutron might more effectively break into the nucleus than a fast, powerful projectile, quickly reached Copenhagen. Christian Møller, one of the young physicists of the institute, was sent down to Rome to have a look at what was happening. "He came back with the details of the puzzling results of Fermi and his collaborators," said Wheeler, who was at the institute at the time. "Little was talked about except the news from Rome."

Who could have anticipated that neutrons would be slowed by passing through water or paraffin, or that they would be captured by the nucleus? Everything known so far indicated that slow neutrons should pass all the way through. But Bothe, in the United States, tried to calculate the chances of capture. During a colloquium held at the institute late in 1935 a report was given on his paper. "Bohr kept interrupting, and I wondered a bit impatiently why he didn't let the speaker finish," said Frisch, one of those taking part in the discussions. "Then in the middle of a sentence, Bohr abruptly stopped and sat down, his face suddenly dead. We feared he had been taken ill. After a few seconds he got up again and with an apologetic smile said, 'Now I understand it.'"

In those few seconds Bohr had grasped the idea that would lay the basis for the future of physics, and indeed the world. He had understood essentially what was happening in the nucleus and why, and thus he had the clue to the constitution and nature of the nucleus, to its transmutations and disintegration.

Bohr outlined his insight: The nucleus perhaps is a compound one and a nuclear reaction occurs in two stages. First an invading neutron would collide with a nuclear proton or neutron. The particle hit, however, would not be knocked out of the nucleus. On the contrary it would bump into another particle and that particle into still another particle until all the particles were in motion, and the original energy of the impinging neutron was distributed among them. And then would come an action in which the system rid itself of the excess of energy.

Once it was pointed out, the multiple collisions were easy enough to understand. The listeners stirred with excitement. This was a dramatically new picture of the nucleus. Again Bohr almost intuitively grasped the principle that underlay the phenomena the experimenters had discovered. It was the kind of comprehension that eludes most men, and in this sense it was his discovery of the structure of the atom and of complementarity all over again.

Bohr was in a fine fettle of creativity. He proposed the investigations that would have to be made to show whether he was right or wrong about the "compound nucleus" he had envisioned. The young physicists literally rushed from the room to begin.

Frisch and Placzek undertook to make measurements of the absorption of slow neutrons by gold, cadmium, and boron, or combinations of all three. They needed quite a thick layer of gold for experiment. Placzek suggested that they make use of several of the Nobel Prize medals Bohr's friends had left in his safekeeping. It would not harm them and scientific good might come of it. And so it happened. The Nobel gold showed a sharp "resonance" for neutrons of low energy; it was set all aquiver, as it should have been, if the nucleus were the compound one Bohr had suggested.

Bohr was greatly pleased about the outcome of the work and urged the two to write a paper at once. They did, though with difficulties. Placzek liked to work only at night and Frisch only in the day. However, after a few nights the work was completed and Frisch himself took it to the post office at four o'clock in the morning.

The full development of the concept of the compound nucleus took many more months. But sudden insight and a long marshaling of fact had again produced one of the great advances in science, as had the same combination in Newton and Darwin.

While Rutherford, Fermi, the Joliots, and many others were well equipped with radium—the first artillery of the atomic age, and with its second, the new accelerating machines, Copenhagen was without both. The emphasis in Bohr's institute was on theory, but theory needed backing by experiment and the new equipment was needed.

Bohr was to celebrate his fiftieth birthday on October 7, 1935. Hevesy, who was back at the institute, this time as a rebel against Nazi oppression, and a group of Danes decided to appeal to the Danish people for 100,000 kroners to buy half a gram of radium for

1. NIELS BOHR *(1885–1962)*

11. YOUNG BOHRS. *Niels at the age of four (right), with his mother, his brother Harald (in mother's lap), and sister Jenny (left). About 1889.*

III. ANOTHER GENERATION. *Bohr with his five sons, (left to right) Ernest, Erik, Christian, Hans, and Aage.*

v. Fifty years later. *Still hand-in-hand, the Bohrs walk down the terrace of their country home, Tisvilde. The thatch-roofed house stands on the edge of an ancient forest.*

iv. [Opposite] Engagement, 1911. *Margrethe Norlund and Niels Bohr happily hold hands at the time their engagement was announced. The photograph was taken at Margrethe's home.*

VI. SEVENTIETH BIRTHDAY, 1955, the patriarch. Niels and Margrethe Bohr with their four sons, their daughters-in-law, and the grandchildren.

VII. Margrethe Bohr

VIII. [*Opposite*] GREAT AND GOOD FRIENDS. *Lord and Lady Rutherford with Niels and Margrethe Bohr in the Rutherford's garden. About 1930.*

IX. THE ATOM, OF COURSE. *Bohr talks physics with James Franck, who had come to visit the Institute for Theoretical Physics, and the Danish physicist H. M. Hansen. The early twenties.*

x. a. CONFERENCE IN ROME, 1935. *Bohr turns to talk to a physicist behind him, while Guglielmo Marconi (left) and Frederick Aston (right), British physicist who won the Nobel Prize for chemistry in 1922, watch the camera.*

XI. THE INEVITABLE BLACKBOARD. *Bohr re-enforces the symbols on the board with a little manual demonstration of the problem.*

X. B. ROME AGAIN. *Another pause in the conference of 1935. Enrico Fermi (left), Werner Heisenberg (center), and Bohr (right).*

XII. BOHR WITH SIR JOHN COCKCROFT.

XIII. BOHR WITH ERNEST O. LAWRENCE, *American developer of the* *cyclotron.*

XIV. HOPES FOR A BETTER WORLD. *Bohr with Adlai Stevenson.*

xv. BOHR'S STUDY. On the wall are portraits of Bohr's parents and those closest to him.

XVI. BOHR WITH THE EVER-PRESENT PIPE.

Bohr's birthday. The money rolled in. The people of Denmark appeared grateful for an opportunity to express their pride in Bohr and to have a share in the support of the work that brought world-wide honor to their country.

Bohr accepted the gift with heartfelt emotion and appreciation. The tragic loss of his son the year before had only deepened the wellsprings of his feeling and made him even more sensitive to such an outpouring of sympathy and support.

At fifty, Bohr's hair was graying and his broad, heavy-set shoulders were developing a hint of a stoop. He was a little heavier, too, but these changes tended to be overlooked for his vitality and strength. He still zoomed down the ski slopes like a teen-ager and still took the institute steps two at a time. But his pipe was almost constantly in hand, and he seemed more reflective than before. The eyes were more deeply set under the thick beetling eyebrows.

In the dire months following the loss of his son Bohr had found surcease only in the construction of an adjoining institute of mathematics to be headed by his brother Harald. Planned on the same lines as the institute for theoretical physics, it formed a second wing. The new institute would use the lecture theater and other facilities of the physics buildings.

Niels and Harald had continued to be extremely close. The two institutes, standing side by side, also would enable them to coordinate mathematics and physics in a way that would contribute to progress in both sciences.

The plans for the new institute were made well before the family tragedy, but the construction of the building put Harald at Niels's hand at a time of great need. Thereafter when any important decision at the physics institute was in the making, Niels generally would say "I'll just go and have a word with my brother," and would disappear in the direction of the mathematics wing.

Or Harald would stroll over to Niels's office and, slowly sauntering around, both hands in his back pockets and head and shoulders thrust forward, would calmly and quizzically survey whatever might be under construction. With a puff on the cigar that he smoked almost as constantly as Niels did his pipe, he would give his valued opinion or produce an anecdote to make his point.

To make the birthday radium into a source of neutrons it had to be mixed with beryllium powder. Frisch was assigned the task of preparing the powder. Beryllium is a very hard metal and Frisch enlisted as many recruits as possible in the long tedious job of grinding it up in mortars. No one knew at the time that some people are highly sensitive to beryllium and could be killed by breathing traces of it. "Fortunately none of us were affected and no accidents happened," Frisch said with an evident sigh of relief.

Hevesy was pursuing his idea of using the artificially radioactive isotopes as "labeled atoms." In biological research, he pointed out, such atoms could be traced by their radioactivity all through the body and into any organ under study. He thought the "labeled atoms" might be useful to medicine. Bohr too was enthusiastic about the idea and the new radium would permit them to make the radioactive phosphorus with which Hevesy wanted to begin his research. At this point the institute for theoretical physics acquired a stock of rabbits, frogs, and sticklebacks.

To make Hevesy's labeled phosphorus atoms, some of the birthday radium was placed in a large flask containing 10 liters of carbon disulphide, a toxic, highly flammable liquid. As a safety measure the flask was lowered into a well in the basement of the institute. If protective measures had not been taken the radiation would have been strong enough to cause serious burns.

Every two weeks the liquid was drained off and the irradiated phosphorus, P 32, extracted. One day while the draining was being done the bottle slipped and broke. The assistant in charge bounded up the stairs, closely pursued by the fumes. If a spark had ignited them, the whole institute might have exploded. The largest pump was put to work to exhaust the dangerous fumes, but even so it took all night to clear out the building. That night Frisch, who was then living in a room on the top floor, remembered going to bed in a fatalistic mood.

"However, the next morning I was still there and the danger was over," he commented in the serenity of some thirty years later.

After this the well with its precious and dangerous radium settled down to uneventfulness. The story of the broken bottle survived principally at the annual Christmas parties held in the "well room." The lid of the well made a perfect table for the Christmas tree, the sausages, and the beer. Bohr always gave a report on the year's work and then the song burst forth. An ominous night might be settling over Europe, but at that moment there was only the unforgettable fellowship of the institute to revel in and to remember.

Shortly after the radium was presented to the institute, work was finished on an accelerator modeled upon the first one built at Cambridge by Cockcroft. It too could generate about 1,000,000 volts. Long sparks sometimes would leap from the machine as particles were hurled at their targets. A new wing was built to house the powerful machine.

Almost at the same time Cockcroft built the accelerator, Ernest O. Lawrence at Berkeley designed the cyclotron. In this second big machine of the new age in physics, particles were whirled through a magnetic field until they attained a high velocity and could be trained on any material under study.

L. J. Laslett came from Berkeley to design a cyclotron for the institute. Tall, bony, and taciturn, he sat with his legs propped up on a table and with a pad on his knees designing one component after another. Cyclotrons grew to great sizes in later years, and by comparison the Copenhagen machine was a small one. Even so, a wall had to be knocked out to get it into the institute when it was completed several years later. It remained in use for many years after some of its successors in other parts of the world had been retired.

The old days of putting apparatus together with a few tubes and lots of ingenuity were gone. The institute joined the move into the giant machine stage.

As 1936 began the huge machines were hurling their projectiles at the fastnesses of the atom's nucleus and the slow neutrons were insinuating their way into the still only partly known interiors of the nucleus, but Bohr continued to ask the prime question he had raised earlier—What is the constitution of the nucleus; what is the nature of the nuclear reactions? Though the accelerator and radium played their part in seeking the answers, the real weapon was a remarkable human mind supported by other trained, striving, human minds.

At lunch, as the institute staff ate their open-faced, Danish sandwiches, at tea, in the library, in the halls, and even out on the park playing field, but most particularly in the colloquia, the over-riding subject was the nucleus. An extraordinary dream also drew them on. In the nucleus lay energies vastly exceeding any then available to man. No one was so rash as really to think that this power could be released, but the knowledge that it was there lent a special fillip to this subject that so preoccupied them.

Bohr was further developing the picture of the nucleus that had come to him when he surprised the colloquium with his "Now I understand." In contrast to the outer part of the atom, where tiny, nearly mass-less electrons whirled their way through vast spaces, it was very likely that the nucleus was a tightly packed density of neutrons and protons. Because of their closeness a hit on one might set them all in motion.

All of this suggested a very realistic picture—a rarity in quantum physics. Everyone took to making drawings on whatever might be convenient—tablecloths, paper, or blackboards—and Frisch often added caricatures of the artists. Gradually ideas clarified.

At the colloquium Bohr went to the blackboard. Imagine, he said, that the nucleus is a shallow basin or bowl. He drew a big concave line on the board. If it is empty, he went on, pointing with his chalk, a billiard ball shot into it would go down one side and out the other with no change in its original energy.

But suppose the bowl was filled with balls. Bohr drew them in. Then if another billiard ball rolled in, it would not be able to pass through freely. It would strike the first ball it came to, divide its energy with it, and cause it to strike the balls nearest to it. Bohr traced the path with his chalk. The balls in motion would hit other balls, and the other balls, still others until in the end all the balls would be in motion, and the invading ball would be indistinguishable in the lot. It would be captured.

Under other circumstances, Bohr showed, the collisions might continue until enough kinetic energy happened to become concentrated on a ball close to the edge of the basin. Then that ball would be shoved out. Bohr showed it going over the edge of his bowl.

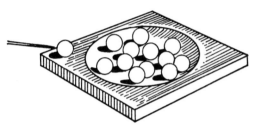

Billiard balls and the nucleus of the atom. If the bowl were empty, a ball sent in would roll down one slope and out the other. But if there are balls in the bowl, the ball sent in will not be able to pass through freely; it will strike one of the balls and divide its energy with it. The two then in motion will bump into others, and this will continue until all the balls in this multiple collision are in motion. Bohr used the balls in the bowl to illustrate the compound nucleus and the transmutations it may undergo.

The pushing out of the one ball would dispose of enough energy to reduce the motion of all the other balls, if the initial impact was not a heavy one. No other ball would then be able to collect the energy to climb the edge of the bowl and escape. "The comparison illustrates very aptly what happens when a fast neutron hits a heavy nucleus," said Bohr.

Artists more skilled than Bohr then drew the "bowl of balls" with full realism. Bohr used the drawing in speeches and it was printed in *Nature* of February 29, 1936, along with his formal paper, "Neutron Capture and Nuclear Constitution." The drawing proved highly useful in making his point.

Bohr emphasized "At the moment it is scarcely possible to form a detailed picture of such processes," but he could discuss the principles. The essential point, Bohr repeated, is the excitation of the particles of the nucleus and the formation of varied states of energy.

As the work went on, Bohr proceeded to show that the ejection of one ball near the edge is only one of the many things that could happen if the balls in a bowl were put into motion by an impact. The outcome might give no more indication of the cause, than does the pile-up of automobiles in a multiple highway accident.

Fermi had been changing one element into another and creating artifical radioactivity by bombardment with slow neutrons. The results, however, seemed to vary in a capricious way from one element to another.

Bohr worked at this case. If the energy of the incoming neutron was low, he pointed out, there would be less chance of escape than of capture. And if the combination of nucleus plus the new neutron were stable a new isotope of the original element might result. If, on the other hand, the addition of the neutron—the extra ball—to the balls already in the bowl created an unstable arrangement, the break-up might result in elimination of positrons, or electrons and gamma rays. In this case a stable isotope of a different element would be formed. Or, the added neutron might produce a radioactive element—one of the elements so short-lived that Fermi rushed to count them.

In the meanwhile Rutherford continued to bombard the nucleus, not with the chargeless neutron, but with powerful charged particles. The positively charged projectiles, though, were repelled by the positively charged nucleus. But, Bohr suggested, if the bullet particles were powerful enough to break through the "barrier" guarding the nucleus, a compound system, with all the particles in motion, would again be formed.

"The formation of such a state," said Bohr, "seems decisive for the explanation of the great variety of phenomena." "In fact," he concluded, "the essential feature of nuclear reactions, whether incited by collisions or radiation, may be said to be a free competition between all the different possible processes of liberation of material particles and of radioactive transitions, which can take place from the semi-stable intermediate state of the compound system."

Bohr looked ahead. If the nucleus could be bombarded with energies of 100 million volts, the energies of even this tremendous impact would in the first place be divided among the nuclear particles, and in such a way that a subsequent concentration of energy would be required to liberate any one particle. The principle would hold even with such a jolt.

If still more violent impacts could be conceived, say with particles of 1,000 million volts, Bohr said, "we must be prepared for the collision to lead to an explosion of the whole nucleus."

Bohr did not picture what would happen if an atom blew up in such a fashion. For a physicist the awful potentialities did not have to be filled in, but Bohr was not thinking of gargantuan explosions, for he continued: "Not only are such energies at present far beyond the reach of experiments, but it does not need to be stressed that such effect would scarcely bring us any nearer to the solution of the much discussed problem of releasing nuclear energy for practical purposes. Indeed the more our knowledge of nuclear reactions advances the remoter this goal seems to become."

Einstein was in complete agreement. When he was asked in an interview about the splitting of the atom, he conceded that it might be done, but added: "It would not be practical. It would be like a blind man in a dark night hunting ducks by firing a shotgun straight up in the air in a country where there are very few ducks."

In February, 1936, shortly before Bohr published his full paper on the nucleus, Breit and Wigner of the University of Wisconsin published the same concept of an exchange of energy between an invading particle and the particles of the nucleus. Their theory was developed in full mathematical detail. Bohr's was more general. The two papers neatly complemented one another.

"How does this balance of view come into being?" Wheeler

asked with wonder. As one who was in Copenhagen at the time the nuclear work was being done, Wheeler answered his own question. He attributed it in part to Bohr's methods of work:

> No one who struggled with a lively physical issue side by side with Bohr will forget the turbulence of the discussion, its ups and downs, the trial first of one extreme point of view and then another, and the forceful words with which Bohr so often summarized one or another conclusion reached along the way—"How can one possibly believe such and such a view, when such and such is so absolutely clear."
> One or another model, whether the individual particle model or some other model gets drawn on the board in a vivid diagram and pushed to the limit of its predictive power. But of all such details and special views there is often no trace to be found in the final paper. The extremes having been tried, one knows between what limits the truth must be found. Just because these limits are so wide, the wording may seem foggy to the uninitiated and unforewarned reader.

Ironically, this balanced, exploratory way of arriving at an answer, this testing, this searching out of errors in order to correct them, and the ceaseless struggle for the truth was developed into the fine science and art that Bohr helped to make it at the precise time when the opposite method in politics was pushing the world to the brink of disaster. Hitler, suppressing all disclosure of errors, and adopting, not rejecting, the unviable extremes, was feverishly preparing for war and hiding the brutal realities behind a screen of soft words. "Germany needs peace and desires peace" he proclaimed in 1937; "Germany neither intends nor wishes to interfere in the internal affairs of Austria or to conclude an Anschluss."

In the relative lull of 1937, Professor and Mrs. Bohr set out on a six-month trip around the world. It was the turn of their son Hans to accompany them. The Bohrs first went to the United States where Bohr had been invited to lecture at Princeton, Berkeley, and a number of other universities. Excitement about the nucleus and the baffling transmutations of the elements had led to an almost frantic search for whatever illumination could be gained. One conference followed another. Bohr spoke at most of the meetings and universities on "Transmutations of the Atomic Nucleus." He illustrated his words with drawings of the billiard balls clustered in their shallow nest.

From the West Coast, the Bohrs went on to Japan. The former Japanese students at the institute met them and, as Hans said, showed them "great and generous hospitality." Though Japan was about to launch her assault on China, the Bohrs were invited to a garden party at the Imperial Palace and Bohr had an audience with the emperor. As Bohr traveled around Japan to lecture at the universities, most of the former institute fellows went along, both to translate Bohr's lectures and to show him and his family all the scenic beauty of Japan.

In China the Bohrs were met with a similar outpouring of hospitality. In addition to lecturing at the leading universities, the Bohrs visited the pagodas at Hangchow and the Ming tombs at Nanking. Bohr was completely entranced by China, and thereafter always regarded it as one of his favorite parts of the world.

From China the Bohrs took the Trans-Siberian Railroad to Moscow. Bohr delivered lectures in the capital and in Leningrad.

In late September, not long after the Bohrs had returned to Copenhagen, Bohr went to Bologna to attend the centenary celebration for Galvani. He had hoped to see Rutherford at the Italian meeting. Instead word came that Rutherford had been injured in an accident—in a fall from a tree he was pruning. His death came a few days later, on October 19, 1937. The shock to Bohr was profound. He hurried to England to attend the funeral and was one of the mourners as Rutherford was buried in Westminster Abbey, close to the grave of the great Newton.

Soon afterward Lady Rutherford sent Bohr Rutherford's cigarette case. She felt that he should have the battered case that seldom left her husband's pockets. "My dear Mary," Bohr wrote in answer. "I cannot say how thankful I am for Ernest's cigarette case, which every day will remind me of his fatherly friendship and the great kindness you yourself have always shown me."

Bohr missed the confident spirit and insight of Rutherford. But even Rutherford could not have foreseen what would lie beyond the penetration they were making into the hitherto impenetrable massiveness of the nucleus. No more did they foresee that the carefully smooth surface of 1937 would so soon explode into chaos and war.

Beneath the surface mighty forces, both political and atomic, were astir and about to reveal themselves.

12 /

Crisis & the Cracking of the Atom

THE LOFTY TOWERS AND HALLS of Elsinore Castle seemed to echo
with Hamlet's tortured:

> That is the question,
> Whether 'tis nobler to suffer the slings
> and arrows of outrageous fortune
> Or to take arms against a sea of troubles. . . .

The question echoed too in Bohr's mind as he came to Elsinore, standing at the north end of the Danish land, walled and fortified against the gray sea. At the south end of the Kingdom of Denmark troubles were deepening, like a roiled sea threatening to engulf civilization and culture.

In the late summer of 1938, Hitler was denouncing Czechoslovakia for alleged oppression of the Sudetens, and proclaiming the superiority of the German race and culture. At one of the monster Nazi party rallies, replete with flaming torches and delirious shouts of "Heil Hitler," Goering typically screamed: "A petty segment of Europe is harassing the human race. This miserable pygmy race [the Czechs] is oppressing a cultured people [the Sudeten Germans] and behind it is Moscow and the eternal mask of the Jew devil."

In Denmark, not very far away, Bohr heard of all of this with sickening clarity. He was already deep in rescuing as many scientists as possible, but he decided to speak out, as only a scientist could, against the Nazi's pseudo-scientific dogma of race and Nordic cultural superiority.

An invitation to address the International Congress of Anthropological and Ethnological Sciences at a special session in the old castle Shakespeare used for the scene of Hamlet, offered an opportunity. Denmark had a tradition of neutrality and could not afford a direct, public attack on Hitler. Bohr understood this well. But in measured tones he could make his point almost as effectively as in naming the German dictator and the Nazis. He could emphasize the applicability of science, with its recognition of relativity, to all human judgments, and could defend the existence of two opposing points of view.

"The main obstacle to an unprejudiced attitude toward the relation between various human cultures is the deep rooted difference of the traditional backgrounds on which cultural harmony is based and which exclude any simple comparison between such cultures. In this above all, the viewpoint of complementarity offers itself as a means of coping with the situation."

Bohr reminded the anthropologists of his audience and those who might be listening beyond the thick walls and moats of Elsinore that it is "entirely a caprice of fate," that the culture of another people is theirs and not ours, and that ours is not theirs.

"Using the word much as it is used in atomic physics to characterize the relationship between experience obtained by different experimental arrangements and visualizable only by mutually exclusive ideas, we may truly say that different human cultures are complementary to each other," he continued.

"Indeed each such culture represents a harmonious balance of traditional conventions by means of which latent potentialities of human life can unfold themselves in a way which reveals to us new aspects of its unlimited richness and variety."

In the Danish castle made famous throughout the world by an English playwright, Bohr declared with obvious truth that contact between different societies has enriched and advanced civilization. At this point the German delegates walked out.

"It is indeed, perhaps the greatest prospect of humanistic

studies to contribute through an increasing knowledge of the history of cultural development to the gradual removal of prejudices which is the common aim of all science," he said.

Despite the temperate cast of Bohr's words—in themselves a rebuke to the stridency of the Nazis—the speech was a powerful denial of the distorted doctrine the Nazis were preaching in justification of their racial and political atrocities. Bohr was taking scientific arms against the Nazis. His was one of the strong European voices raised against them. To the Nazis, compounding prejudice into hysteria, war, and persecution, a declaration that the goal of science is the removal of prejudice was provocative in itself. Bohr's colleagues were convinced that the Elsinore speech went into the anti-Bohr files the Nazis had even then started compiling.

The next month, in September, 1938, when the institute held its annual seminar, almost none of the alumni who lived in Germany dared to attend. Others had fled, and were in no position to travel to Denmark for a meeting with Bohr. Attendance in the forbidding, troubled times fell far below what it had been before.

Fermi managed to come from Italy, though conditions there were becoming increasingly intolerable. During the conference Bohr quietly drew Fermi aside and told him that he was under consideration for the Nobel Prize. Under normal circumstances such a disclosure would never have been made, but Italy had ordered all Italian citizens to convert foreign monetary holdings into lire. Bohr inquired if Fermi would like to have his name withdrawn until the prize money could be used without restriction.

Fermi then told Bohr that he wanted to leave Italy. The enactment of racial laws in servile imitation of the Nazis had placed Mrs. Fermi, a member of a Jewish family of Rome, in potential jeopardy, and had climaxed Fermi's long-standing dislike of Fascism. Fermi said that if he should receive the prize, he would not return to Italy, but would seek a position in the United States. Bohr suggested that, if he finally decided upon such a course, he and his family come to Copenhagen following the Nobel ceremonies in Stockholm.

As Bohr had hinted, Fermi duly received the Nobel Prize.

Under the guise of going to Stockholm to accept it, the Fermis successfully made their escape from Italy. As they had planned at the September institute conference, the Fermis, with their children and nursemaid, then went to Copenhagen until they could sail for the United States. Bohr had again assisted in getting one of Europe's great scientists to safety and new opportunities in another country.

While Hitler was cowing the West with his threats of war and bringing Mussolini to heel, the physicists of Europe were preoccupied with a riddle—a riddle that if it had been promptly solved might have changed the course of history. Neither Hitler, nor Mussolini, nor the scientists themselves had the faintest suspicion that the work being done in the quiet physics laboratories of Berlin and Rome, largely by scientists the dictators soon would drive out of their countries, might have given the two heads of state exactly the weapon they needed for the world conquest of which they dreamed and boasted. The riddle essentially was a simple one—what were the strange substances produced when Fermi's slow neutrons made their way into the uranium nucleus?

Despite the loud and unauthorized Fascist announcement that Fermi had discovered a new element, No. 93, Fermi himself was not sure. Bohr's work on the compound nucleus had established that when the nucleus captures a neutral particle an unstable isotope might be formed and a beta particle emitted. The loss of such a particle could transform the original element into the next highest element in the atomic table. It was therefore plausible to believe that a new element No. 93, or even 94, might be created beyond uranium, No. 92.

Nevertheless no such element was known in nature, and a hot dispute arose over the short-lived substance (it had a half-life of 13 minutes, which meant that half of it disintegrated in that brief time). Fermi himself described it as a trans-uranic (beyond uranium) element, and the term was generally adopted. Some physicists argued that the mysterious material might actually be an isotope of protactinium, element No. 91.

Otto Hahn, and the gentle sweet-faced woman physicist, Lise

Meitner, had discovered and named element No. 91 in 1917. In 1938 Hahn was director of the radiochemistry department of the Kaiser Wilhelm Institute and Lise Meitner headed the department of nuclear physics. The suggestion that Fermi's mysterious substance might be an isotope of their protactinium immediately interested them. They decided to repeat the Fermi experiment.

The two collaborators had the chemical means for putting the mystery substance to the test. Hahn had previously discovered an isotope of protactinium with a convenient half-life of 6.7 hours. The scientists added a certain amount of this well-identified element to Fermi's trans-uranic product and chemically precipitated the mixture. Almost none of Fermi's substance "came out" with the protactinium. Thus, without question, the Fermi substance was not element No. 91. One possibility could be scratched off the list. Hahn and Meitner also proved that the unknown stuff was not an isotope of either thorium or actinium. The list was further reduced.

Ida Noddack, a chemist, suggested that the puzzling material should not be called a "trans-uranic element" until it had been determined with certainty that it was none of the other elements. But this implied that the uranium nucleus might break into a number of pieces. "To split heavy atomic nuclei into ligher ones was then considered impossible," said Hahn.

No one gave much attention to the Noddack suggestion, and Hahn and Meitner's work in eliminating the protactinium possibility was accepted as support for the idea that the mysterious substance actually was a "beyond uranium" element.

By this time Hahn and Meitner were thoroughly engrossed in the search for its identity. They went on, and themselves found a number of "trans-uranic" elements. In fact, they discovered these substances could be arranged in a regular series: their chemical properties corresponded to known elements lower in the table of elements, rhenium No. 75; osmium, 76, iridium, 77, and platinum, 78. (See drawing of table of elements on pp. 124–25).

The riddle of the Fermi substances also had involved the Joliot-Curies in Paris. They too repeated the Fermi experiments but with thorium, No. 90, and found a puzzling product with a half-life of

3.5 hours. They concluded that it was a "trans-uranic" element, with a strong resemblance to lanthanum, No. 57.

At this point the rising tide of Nazism again intervened in the course of science, as indeed in every phase of life in Europe. In February, 1938, Hitler summoned the Austrian chancellor, von Schuschnigg, to Berchtesgaden. As von Schuschnigg's records revealed, Hitler began: "I need only to give an order and over night all the ridiculous scarecrows on the frontier will vanish. You don't really believe that you could hold me up for half an hour?" If he were held up in the occupation of Austria, Hitler went on, "after the troops will follow the S. A. and the legion! No one will be able to hinder the vengeance, not even myself."

Mussolini gave his consent to the coming rape of Austria, and on March 11, the long German columns—with monstrous traffic jams caused by break-downs of some of the tanks—crossed the border. A few days later Hitler declared the dissolution of the Austrian Republic and the annexation of the territory to the German Reich. All Europe reeled with the shock.

Up to this time, Lise Meitner, who had been born an Austrian, had been able, although she was a Jew, to continue in her position at the institute and to go on with her work with Hahn.

"There was really a strong feeling of solidarity among us [the faculty of the institute]," said Lise Meitner. "It was built on mutual trust and this made it possible for the work to continue quite undisturbed even after 1933, although the staff was not entirely united in its political views."

With the annexation of Austria, Lise Meitner found herself overnight a German citizen and subject to all the racial laws of the Reich. Word spread that scientists would not be permitted to leave. Meitner knew that she could no longer remain; thirty years in Berlin with esteemed associates had come to an end. She left ostensibly on a short summer holiday and did not return. With the aid of friends in Holland she made her way across the Nazi frontier. Bohr's committee had been at work on her behalf and in the fall she received an invitation to continue her work at the Nobel Institute in Stockholm.

Back in Berlin, Lise Meitner was badly missed. In the absence of his long-time associate, Hahn recruited Fritz Strassmann, as solid and square of build as Lise Meitner had been small and delicate, to continue the testing of the Joliot-Curie results. He and Meitner had just begun a study of the puzzling French findings when she had to flee.

When Hahn and Strassmann precipitated the new French "trans-uranic" element, some radioactive products remained in their test tubes. In trying to find out what they might be, the new German team used some barium as a "carrier." And then came some most unexpected and remarkable results.

The radioactive materials "came down" (were precipitated) with the barium. Hahn and Strassmann meticulously rechecked. Had some impurities crept into the barium? The answer was that it was completely free of any. Hahn could only conclude that the precipitates must either be barium, No. 56, or radium No. 88, both members of the same series of elements.

"There was nothing in the knowledge of nuclear physics at the time to suggest that barium could possibly be produced as a result of the irradiation of uranium with neutrons," said Hahn. "Therefore we could only conclude that the products must be isotopes of radium. Still it was a strange affair."

Hahn and Strassmann then decided to try to separate the "radium" from the carrier barium. Thirty years before Hahn had worked out the technique for exactly such a process; he had separated barium from another isotope of radium. This time, to his complete amazement, the experiment would not work. There was no separation whatsoever. The two scientists tried larger amounts of the isotope, but all in vain. By this time they were shaken in their confidence in their own methods; they then mixed another radium with barium and began the separation process. The radium separated perfectly.

Hahn was one of Europe's most skilled and thorough chemists, but nothing that he could do altered the result. There was no conceivable conclusion except that the "radium" actually was barium, and this was an impossibility. Barium could have been produced only if the uranium nucleus had split, another impossibility as everyone knew. In this moment of impasse and frustration Hahn poured out the whole story in a letter to Lise Meitner, then in Stockholm.

It was only a few days before Christmas and the letter arrived just as Lise Meitner left to spend Christmas with friends at Kungalv, near Göteborg. She had invited her nephew, Frisch, to join her there for Christmas. For a number of years they had spent the holidays together, and it was easy for Frisch, who was with Bohr in Copenhagen, to go to Göteborg. He had only to take the ferry across the narrow Kattegat. Bearing warm Christmas greetings from the Bohrs and other friends at the institute, Frisch took a night boat and checked into his hotel. "It was the most momentous visit of my life," he said later.[1]

The next morning when he met his aunt was a momentous moment in itself. She was safe, and they were meeting again for the holidays after a disruption of her life and work that had nearly destroyed all hope. The reunion in the ordered, calm atmosphere of southern Sweden was overwhelming. But there was more to make the day forever memorable.

Lise Meitner immediately showed him the letter from Hahn. She was excited and puzzled about it, and was impatiently waiting to talk to Frisch about the incredible results it cited.

"I wanted to discuss with her a new experiment that I was planning, but she wouldn't listen," said Frisch. "I had to read that letter. Its content indeed was so startling that I was at first inclined to be skeptical." Hahn had set forth all the circumstances—it was absolutely impossible to separate the Joliot-Curie type of radium from barium. Frisch's first idea was that a mistake had been made.

This, Lise Meitner waved aside. Hahn was much too good a chemist for that. "But how could barium be formed from uranium?" Lise Meitner mused. It couldn't be, Frisch answered.

Only fragments had previously been chipped away from nuclei. Frisch dismissed the idea that a large number could be broken off

[1] See the chapter by Otto Frisch in *Niels Bohr His Life and Work.*

at one time; there could never be enough energy for that. That the nucleus could be split was even more unthinkable. A nucleus was not something brittle or solid that could be cleaved or broken. Bohr had shown, on the contrary, that it more closely resembled a liquid drop.

To clear their heads and get away from all interruptions, the aunt and nephew went for a walk in the woods. The snow lay deep on the ground and Frisch put on the skis he had brought with him for a few sessions on the slopes. Then, as they walked in the clear frosty air, they debated the riddle.

Perhaps a drop could divide itself into two smaller drops by elongating until a constricted neck formed in the middle and then perhaps tear itself apart into two drops. Strong forces would resist, just as the surface tension resists such a break-up in a drop of water. But, a hopeful, dawning "but," nuclei are electrically charged and this might diminish the effect of the surface tension.

In this moment of high creativity in thought they almost automatically sat down on a log. It was as though their legs would not hold them, for they both anticipated what was coming.

Lise Meitner began to do some calculations on the back of the letter and on some scraps of paper. Frisch followed every move of her pencil. Yes, the charge of a uranium nucleus was large enough to destroy the surface tension. Yes, the uranium nucleus might be a "wobbly uncertain drop, ready to divide itself at the slightest provocation, such as the impact of a neutron."

Aunt and nephew looked at one another in disbelief and in triumph. But, a "but" of doubt this time, when the two drops parted would they not fly apart with tremendous energy, an energy of some 200,000,000 electron volts? Where could such energy come from?

Lise Meitner remembered at this point how to compute the masses of nuclei from the so-called "packing fraction." If the uranium nucleus divided, the two particles formed would be lighter than the original nucleus by about one-fifth of the mass of the proton. They were hot on the trail.

Lise Meitner did some quick arithmetic. According to Einstein $E = mc^2$. She multiplied the lost one-fifth of mass by the speed of light squared. The ⅕ of the mass × 186,000 × 186,000 miles per second. They were scarcely breathing. It came out almost exactly at 200,000,000 electron volts. The lost one-fifth would supply exactly the 200,000,000 electron volts of energy with which the drop would tear apart.

"So here was the source of all that energy," said Frisch. "It all fitted." In the quiet and forgotten cold of the Swedish woods, aunt and nephew again looked at one another, again with incredulity, and again with triumph. Never before had this insight been given to any humans. They were discovering a new continent, looking into a new world. After the awful anxieties of the last few years, they would have been less than human if this rare moment had not held special recompense and joy for them.

Frisch might have set out for Copenhagen with all speed. But scientists are not reporters trained to the instant telling of the news; on the contrary they live by canons of deliberation and care. For two days Meitner and Frisch went over and over the revolutionary concept at which they had arrived. Then Lise Meitner returned to Stockholm and Frisch set off for Copenhagen in what even that generally imperturbable young man conceded was a "state of considerable excitement." Both wanted to submit what Frisch called their "speculations" to Bohr at the first possible moment.

When Frisch reached the institute on Blegdamsvej, Bohr and his 19-year-old son Erik were just about to leave for the United States. Bohr was going to spend several months at the Institute for Advanced Study at Princeton, and it was Erik's turn for a trip. Each of the boys in turn accompanied his father on an important foreign visit. But only a few minutes remained before their departure. Frisch talked fast.

"I had hardly begun to tell him," Frisch said, "when he struck his forehead with his hand and exclaimed 'Oh, what idiots we all have been. But this is wonderful! This is just as it must be!' "

Margrethe, Bohr's secretary, the other boys, and most of the staff were trying to get the Bohrs under way. Bohr had a reputation of barely making trains and ships and sometimes missing them, and the *Drottningholm* would not wait. Bags, briefcases, reminders, and farewells all mixed with the historic news Frisch had brought.

As Bohr was hustled out, he was asking "Frisch, have you and Lise Meitner written a paper?" Frisch pressed two pages of rough notes into Bohr's hand, and said that he had not written a paper but would do so at once. Bohr called over his shoulder as he was pushed into a taxicab that he would not talk about their findings until the paper was ready. And then in a final flurry he and Erik were off for the ship.

Soon after Bohr left, Frisch acted upon his plea to hurry with the paper. In several long distance telephone calls to Stockholm Frisch and Lise Meitner worked out about a thousand-word report for the English journal *Nature*.

They pointed out that Hahn and Strassmann had been forced to conclude that an isotope of barium, No. 56, had been formed as a consequence of the bombardment of uranium, 92, with neutrons.

"At first sight," they said, "this result seems very hard to understand." The two authors then presented their explanation, and argued that it was possible, if the uranium nucleus were considered as a liquid drop capable of dividing into two. They gave their estimates of the energy involved, the 200,000,000 volts, and identified the second part into which the nucleus should break as krypton, No. 32. (92 − 56 = 32).

They also reported that thorium should split in such a way that it should yield barium and lanthanum. (Here was the Joliot-Curie's substance that resembled lanthanum. It was lanthanum! Again the Joliot-Curies had barely missed another claim to immortality).

Meitner and Frisch described the division of the nucleus as "fission." As they were writing the paper, Frisch had asked one of Hevesy's young biologists what biologists call the splitting of the cell. "Fission," he called out, and Frisch had the word he needed. The process was simply, accurately, and perceptively named for all time.

As Frisch and Meitner sat pondering in the Swedish woods they also had seen that it would be relatively easy to test their "speculations" by experiment, and one of the sheets of paper Frisch gave Bohr indicated this possibility.

In the uproar of excitement at the institute, when everyone else heard of what had happened, Placzek, Frisch's night-owl partner in the compound nucleus work, was skeptical. "If you're right," he prodded Frisch, "why don't you do the experiments to show the existence of those fast moving fragments you are talking about."

"Oddly enough that thought hadn't occurred to me," said Frisch. "But then I quickly set to work and the experiment, which really was very easy, was done in two days."

Frisch drew up a short report on it—about five hundred words. "By means of a uranium-lined ionization chamber connected with a linear amplifier, I have succeeded in demonstrating the occurrence of such bursts," Frisch wrote. "About 15 particles a minute were recorded when 300 mgm of radium, mixed with beryllium, was placed one centimeter from the uranium lining."

In check runs, Frisch removed the neutron source and then the uranium lining. No pulses at all were recorded. When he surrounded the source with paraffin wax, the pulses were enhanced by a factor of two. Thorium gave similar results.

Both papers were ready to mail to England on January 16, 1939. Frisch noted the date on them and modestly sent them off to the editor of *Nature* without a word of warning that they were of unusual significance. Thus the first article, the Meitner-Frisch paper on the splitting of uranium, did not appear until the February 11 issue of *Nature*, and the second, on Frisch's experiment, not until February 25.

In the meanwhile Bohr and Erik caught the *Drottningholm*, though with no margin of time to spare. Rosenfeld, who was taking the trip with them, was waiting on deck. As the liner's big whistles sounded and they waved farewell to Copenhagen, Bohr was whispering under the din to Rosenfeld: "I have something to tell you." It was only public talk that Bohr had promised to avoid.

The ship was scarcely out of the harbor before Bohr was

showing Rosenfeld the notes Frisch had given him and telling him about the "wonderful" discovery. "We will have to think about it," he said, a sentence that in Bohr vocabulary meant we will study it, examine it, turn it upside down and downside up, we'll see how it fits into all that we know, and spare no effort to test it and thus to establish its true significance.

As the ship rolled and shivered through the rough January seas, Bohr and Rosenfeld did nothing else. A blackboard was installed in Bohr's stateroom and the two talked, argued, and calculated. The days at sea gave them a rare opportunity to concentrate on the dramatic breaking of the nucleus Meitner and Frisch had proposed.

"We were accustomed to thinking of small fragments being broken off," Rosenfeld said. "To imagine that the nucleus could break in two was startling."

Bohr first had to satisfy himself that the division was possible in actuality. He agreed that as the nucleus became distorted a critical value would be reached in relation to the surface tension. The figures showed that the nucleus would break. But, Bohr asked, why in two and not into many smaller pieces?

After lengthy thinking and dsicussion, Bohr could demonstrate that there would be as great a probability of the energy of the invading neutron being concentrated as that it might cause a shattering into many pieces. "This was a point Bohr found on the ship," Rosenfeld said.

By the time the *Drottningholm* pulled into New York on January 16, 1939, and docked under the skyscrapers of the metropolis, Bohr was thoroughly satisfied that the ideas of Meitner and Frisch were correct. Close examination, thinking about it, had confirmed his immediate reaction in Copenhagen.

Like all travelers Bohr, Erik, and Rosenfeld crowded to the ship's rail to watch the spectacle of New York, the skyscrapers, the mass of the city, its winter blue-grayness. Despite the constant work, often far into the night, Bohr felt somewhat refreshed; for a few days at least he had been removed from the constant oppressive worry about the state of Europe, and the threatening dangers. The symbolism of the Statue of Liberty, and the sight of a land free of much of the turmoil of Europe brought European troubles flooding to the forefront of his mind.

Then among the crowd filling the long narrow pier the Copenhagen group saw familiar faces and hands waving in greeting—John Wheeler, then professor of physics at Princeton University, and Enrico and Laura Fermi.

Heartfelt greetings were soon being exchanged. Only three weeks had passed since the Fermis had left Copenhagen bound for a new land and a new life. Now Fermi was established at Columbia, and Rome was receding into the past. Wheeler had not been at the Copenhagen institute since 1934–35, and reunion was good.

In all the noise and confusion of the pier, Laura Fermi could hear only a few of the soft words of Bohr, but they were serious words—"Europe . . . war . . . Hitler . . . danger." She thought that he had aged a bit even in the short time since she had seen him. As he spoke of Europe the lines seemed deeper in his face, and his shoulders stooped a little more than she remembered.

While they waited for the luggage to be taken to customs, Bohr drew Wheeler aside. In a low whisper he told him. Nothing could be said, but the atom had been split! Hahn and Strassmann; Meitner and Frisch. Wheeler remembers the shock with which he heard the strange words in all the hubbub of the pier.

But this was all there was time for. The Bohrs and Rosenfeld had to go through customs and the whole party organized to go different ways. Bohr and Erik would remain briefly in New York, while Rosenfeld went on to Princeton with Wheeler.

That night, January 16, 1939, Wheeler invited Rosenfeld to an informal meeting of a little group of physicists known as the Journal Club. After dinner the inevitable question was put to Rosenfeld, as it would have been to anyone arriving from that center of news, Copenhagen and the institute, "what's new?" In all of the long discussions on the ship, Bohr had not remembered to tell Rosenfeld that nothing must be said publicly until Meitner and Frisch could finish their paper and have it published.

At the question, "What's new?" Rosenfeld, who was bursting with the new discoveries, outlined what he knew. "I felt I was free to speak," he said. He was cautious though and emphasized that these were ideas mainly "to be thought about." But he very well recalls, "It created great excitement."

Some of those present phoned colleagues or friends, and the letters went out. Isador I. Rabi of Columbia, who happened to be in Princeton, soon returned to New York to talk to Fermi.

When Bohr reached Princeton the day after, Rosenfeld remarked quite casually that he had "talked" to the Journal Club. Bohr almost exploded, and for the first time explained to Rosenfeld that nothing should have been said before the papers could be published. Bohr foresaw immediately that a race would be on to test the splitting of the atom, and he felt fervently that Meitner and Frisch should not be superseded through his fault. "I was sorry I had unwittingly let it all out," Rosenfeld apologized.

That night Bohr wrote to his wife: "I was immediately frightened, as I had promised Frisch I would wait until Hahn's note appeared and his own was sent off." Bohr cabled Copenhagen, urging Frisch to act rapidly on the paper and to perform the experiments that would substantiate it.

While he waited for an answer, Bohr paced the courtyard outside the physics building, and he and Rosenfeld went furiously to work to set the record as straight as they could. They prepared a note for *Nature*, setting forth the confirmation their shipboard studies had provided. By January 20—in an unheard of two days—they had a six-hundred word article ready, "Disintegration of Heavy Nuclei."

Bohr explained that "through the kindness" of Meitner and Frisch, he had learned about the "remarkable findings" of Hahn and Strassmann and of the Meitner-Frisch explanation that the Berlin experiments indicated a new type of disintegration of heavy nuclei into two parts, with a tremendous release of energy. "Due to the extreme importance of the discovery I should like to add a few comments on the mechanism of the fission process from the point of view of general ideas . . . to account for the main features of nuclear reactions," said Bohr.

Bohr explained: the entry of the neutron would excite the

particles of the nucleus (create an intermediate stage in the formation of a compound nucleus) Ordinarily the added energy would be concentrated on one particle, which like one of the balls in the basin illustration would be shoved out of the nucleus. But if the added energy were converted into a "special mode of vibration" involving a considerable deformation of the nuclear surface, the nucleus, like a drop of water, might well split in two.

Still no answer had come to Bohr's cable asking for details from the institute. He and Rosenfeld labored over the wording of a second cable, and one of the Princeton men drove them to New York to file it at the central cable office. Both actions were a measure of Bohr's acute worry about the disclosure, and of the importance he attached to it. Still no reply came. Bohr learned later that no one at the institute could imagine what all the urgency was about. It was entirely foreign to the usual ways of Copenhagen.

On Thursday, January 26, the Fifth Washington Conference on Theoretical Physics was to be held at George Washington University under the sponsorship of the university and the Carnegie Institution of Washington.

Some years earlier, in 1931, on a visit to Copenhagen and other European university centers, Cloyd H. Marvin, the president of the university, had been impressed with the growing importance of theoretical physics. He talked at length to Bohr about it and met George Gamow and Edward Teller, two of the young physicists then at the institute.

Marvin snapped a famous photograph of the two young physicists on the snow-covered grounds of the institute. Both were "hamming it up" for the photographer and for two of Bohr's sons, Aage and Ernest, then respectively nine and seven years old. Gamow had leaned forward over the handlebars of his motorcycle with all the intensity of a speed racer. Teller had donned skis, ski cap, and a long plaid wool muffler and portrayed a skier about to take off on a fifty foot jump.

When Marvin succeeded, with the cooperation of the Carnegie Institution, in setting up his own department of theoretical physics, Gamow and Teller came from Europe to head it. One of their first steps was to organize conferences on the general model of the annual meetings in Copenhagen. Physicists from all parts of the country and generally some from abroad were invited to discuss the newest developments in physics.

In addition to Bohr those invited to the fifth conference included the leading figures in American physics, and some of the Copenhagen alumni. This brought together Vannevar Bush, M. A. Tuve, and L. R. Hafsted of the Carnegie Institution; Fermi, G. E. Uhlenbeck, and Harold C. Urey of Columbia. There were twenty-four in all.

On the first day of the conference, Bohr quickly met Fermi, and all of Bohr's misgivings materialized. Fermi told him that he had heard about the Meitner-Frisch explanations from Rabi who had been at the Journal Club meeting in Princeton. Fermi, as the one who had launched the slow neutron bombardment of the atom, instantly understood the implications, and he indicated that he had started making experiments.

That morning Erik Bohr had received a letter from his brother Hans, in which Hans remarked that Frisch had completed some experiments and sent his paper off to London. But there were no details.

Bohr told Fermi that he was sure that Frisch had made the experiment, but in the lack of detailed information he could not bear down too heavily on the point.

"Fermi . . . had no idea before that Frisch had made the experiment," Bohr wrote to Margrethe. "I had no right to prevent others from experimentation, but I emphasized that Frisch had also spoken of an experiment in his notes. I said that it was all my fault that they all heard about Frisch and Meitner's explanation, and I earnestly asked them to wait until I received a copy of Frisch's note to *Nature*, which I hoped would be waiting for me at Princeton."

Despite the word in his son Hans's letter, Bohr did not feel that he could report to the conference and lay the whole situation before them. There was still no indication that Hahn and Strassmann's work had been published, and Bohr was not released from his promise.

With all of this erupting under the surface, the meeting went ahead on its first day with its set program. Only two reporters, Robert B. Potter of *Science Service* and Thomas R. Henry, of the *Washington Star*, were covering the meetings. As Potter started for the second day's sessions, Watson Davis, the director of Science Service, handed him a copy of *Die Naturwissenschaften*, the German scientific publication. It had just arrived from Berlin.

Davis read German well enough to notice an unusual article by Hahn and Strassmann. Hahn reported the results of their work and said that the experiments were "at variance with all previous experiences in nuclear physics."

Hahn later explained that as a chemist he was reluctant to announce a revolutionary discovery in physics. Nevertheless, he hinted that there might have been a "bursting" of uranium and that this might have produced the barium. At the same time he warned that it was conceivable there had been a unique series of accidents.

Davis sensed something remarkable and suggested to Potter that he show the Berlin journal to Bohr and see what he might know about it. One glance at the journal and Bohr leaped from his seat. He was free at last of all restraints. The article, combined with the letter from his son, at last made it possible for him to speak. The Hahn-Strassmann experiment was in the public domain. From that time on, the more widely known the findings, the more certainly Hahn, Strassmann, Meitner, and Frisch would be assured of the proper credit.

Bohr asked if he might make an announcement of the utmost importance. The fairly humdrum conference exploded in its turn.

As Bohr spoke, Tuve whispered to Hafsted that he had "better go out" and put a new filament in the new Carnegie atom accelerator.

No sooner had Bohr finished than some of the physicists rushed for the door. They were on the way to phone their laboratories. Long distance calls went out to universities all across the country.

Gamow excitedly began to scribble figures on the blackboard

and Teller argued with him. Fermi started to leave—he had decided to return at once to his laboratory in New York—but on the way out he was stopped by Potter: "What did it all mean?"

Fermi told him that Hahn's chemical test of the "trans-uranic" substance he (Fermi) had discovered was the critical point. It had shown that uranium actually had split into two parts, one of barium and the other probably of krypton.

Fermi went back a moment to his experiments in Rome. He recalled that he had tested the new substances to see if they could be any of the elements in the six or seven places below uranium in the table of elements. It had not occurred to him, he said, to test for an element as far down in the table as No. 56.

That afternoon Potter wrote for *Science Service* a story that went out on January 28 to newspapers and magazines:

> New hope for releasing the enormous stores of energy within the atom has arisen from German experiments that are now creating a sensation among eminent physicists gathered here for the Conference on Theoretical Physics.
>
> It is calculated that only 5,000,000 electron volts of energy can release from an atom's heart 200,000,000 electron volts of energy, forty times the amount shot into it by a neutron (neutral atomic particle.)
>
> World famous Niels Bohr of Copenhagen and Enrico Fermi of Rome, both Nobel Prize winners, are among those who acclaim this experiment as one of the most important of recent years. American scientists join them in this acclaim.

Henry wrote a guarded story for the *Washington Star*. The story in the Washington evening newspaper and the *Science Service* story, going to a long list of newspapers and magazines, could have produced front page stories. But it was not that way. News editors noted that it was atoms of which the scientists were talking and not anything they considered practical or imminently important. The American press thus gave very little attention to the first revelation of the story of the century.

Physicists knew better. Within hours they were setting up their experiments to test the phenomenal hypothesis. The Carnegie's accelerator was quickly supplied with the new filament Tuve had called for and was ready to go into operation. Tuve invited Bohr, Rosenfeld, and Teller to watch the history-making test that night. The guests arrived about midnight and were led through the deserted library and halls to the laboratory.

The control board of the apparatus was a maze of dials, windows, and buttons, but all eyes were turned to the window of the oscilloscope. As Bohr and Rosenfeld walked in they could see great green pulses shooting to its very top. The shining, glowing line of green recorded the energy given off each time a uranium nucleus divided. The atom was being split in their presence, before their very eyes. "The state of excitement challenged description," said Rosenfeld. "I remember it as though it were yesterday."

Tuve was on the telephone. "Here's another one!" he would exclaim as the line pulsed upward again. In the exuberant spirit of that night Rosenfeld remembers that Tuve joked: "in my report [to the Carnegie Institution] how can I justify spending all this money [for the accelerator] to discover fission!" Bohr stood there transfixed and anxious.

"I had to stand and look at the first experiment without knowing certainly if Frisch had done the same experiment and sent a note to *Nature*," Bohr wrote to Margrethe in a letter in which he tried to set down "all that happened, and to bring peace and order in our thoughts."

"Had it not been for the remarks in Hans' letter to Erik, I could not have gotten the right recognition for what Meitner and Frisch did."

All of them lingered in the laboratory. They could not tear themselves away from the jabbing green pulse that was ending one era and opening another. The Washington sky was brightening with the dawn when Bohr and Rosenfeld left.

The next morning, January 27, was a revelation in itself of the speed with which science can move. The conference learned that the atom had not only been split at the Carnegie Institution the night before, or rather early that morning, but also at Baltimore and Berkeley. A little later word came in from Columbia that fission actually had been achieved in New York one night before, on January 25.

Just before Fermi was to leave for the Washington meeting, Rabi had returned from Princeton and told him about Rosenfeld's disclosures there. Fermi hurried in to have a talk with John R. Dunning, associate professor of physics at Columbia. If uranium splits into barium and krypton, he told Dunning, large amounts of energy will be released. It would be simple to test it, he argued. With all of his quick decisiveness, Fermi urged that the experiments should be done at once. Then he had to leave for Washington.

Dunning thought over the discussion of that afternoon of January 25. He saw how the experiment could be made, and after an early dinner went to the laboratory in a deep cavern basement in the Pupin Physics Laboratory. It was about 7 P.M. Two of his associates met him there.

They set up a holder, an ion chamber amplifier for the uranium to be bombarded, and an oscilloscope to show the energy given off. Dunning then was ready to insert a thin plate covered with uranium oxide. Suddenly brilliant green lines shot up in the circle of the oscilloscope. They leaped so high the stunned scientist almost expected them to jump from the screen.

"Tremendous kicks," Dunning noted in his laboratory record. "Believe we have a new phenomenon of far-reaching consequence." With Fermi away, Dunning decided to check and recheck the experiment, and to say nothing about it for the moment.

When Fermi rushed back from the conference and learned the details of the experiment, he let the Washington meeting know what had been accomplished. Any public announcement was withheld until some further checking could be done and until the university could make the proper arrangements for releasing the news to the press.

Confronted with the knowledge that the atom had been broken and its power experimentally released in at least four American universities, as well as at Copenhagen, the conference was pervaded with a sense of history and its concomitant, worry. For centuries the scientists had looked upon the power of the atom as forever, or distantly, inaccessible. Now that power had been brought within reach in a few short hours, it was almost too much for the mind to grasp.

Rutherford long ago had joked: "Some fool in a laboratory might blow up the universe unawares," and Bohr in his Maxwell speech of 1930 had called attention to the tremendous energies of the atom. Bohr's close associate, Kramers, in his book *The Atom and the Bohr Theory of Its Structure*, published in 1923, had written:

"One interested in speculating what would happen if it were possible to bring about artificially a transformation of elements propagating itself from atom to atom with the liberation of energy, would find food for serious thought in the fact that the quantities of energy that would be liberated in this way would be many, many times greater than those which we now know in connection with chemical processes. There is then offered the possibility of explosives more extensive and more violent than any the mind can conceive . . . But this is, of course, mere fanciful conjecture."

Non-scientific writers too had played dramatically with the idea and its possible consequences. Only seven years earlier Harold Nicholson had published a novel *Public Faces* that dealt extensively with atom bombs and rockets. At the moment the atom was split, J. B. Priestley's novel *Doomsday Men* was current and being widely read. A play *Wings Over Europe* had sounded the same dire note of world destruction, and the scientific fantasies of H. G. Wells haunted and intrigued the minds of others.

In those first dazzling and appalling hours some of the scientists felt that the public must be reassured. Potter conveyed the urgent concern in a story written for *Science Service* on January 27: "Scientists who have now confirmed the discovery of how to release enormous stores of energy bound up within the uranium atom, are fearful lest the public become worried about a 'revolution' in civilization." The scientists emphasized, Potter reported, that the transmutation of the elements had not led to the artificial production of gold and the wrecking of the money marts of the world, as some had feared. No more, they said, would atomic energy produce an immediate revolution. As far as the men at the conference could see, it would take more energy to release the atom's power than could be produced. Nothing was imminent; there was no immediate danger. The explosion was large but on a microscopic scale, and the 200,000,000 electron volts given off would not be sufficient to light a household lamp.

Some of the scientists, however, could not resist speculating on what atomic power could do. Potter in his story again reflected this ambivalent hope and fear: "They [the scientists at the conference] fear that there may be forecasts of the near possibility of running great ocean liners on the energy of the atoms contained in a glass of water . . . or of using 'atom smashers' as energy sources instead of steam and electric plants . . . or that atomic energy may be used as some super-explosive, or as a military weapon."

On that first morning of the public dawn of the atomic age, the atom's basic possibilities and problems clearly emerged—atomic power, atomic explosives, and the fearfulness with which civilized men would have to regard the phenomenal new conquest of nature.

Bohr in addition had to continue to worry about how the story would be written for history. Now more than ever he had to make certain that credit for one of the great discoveries went to Meitner and Frisch. Hahn and Strassmann had been fully protected by the publication of their work in the January 15 issue of *Naturwissenschaften*. It was most unlikely though that the Meitner-Frisch paper had been published. Bohr recognized that priorities might go to the first to make scientific publication of the experiments performed so hastily on the 25th and during the night of the 26th. He felt doubly responsible.

When he returned to Princeton on Sunday there was still nothing from the Copenhagen institute, but family letters confirmed that Frisch had succeeded "in verifying the presence of the high energy splitters." Bohr quickly phoned Tuve to tell him, and to urge him not to publish anything in a scientific journal until the actual text of the Meitner-Frisch note was at hand.

To Bohr's mounting alarm, Tuve said that things had already

gone too far to stop. He said Columbia University was preparing to release the report of their experiment to the New York papers. Bohr knew nothing about the material sent out by *Science Service* and the small story in the *Washington Star*, but he knew well that publication in the New York newspapers would expose the results and full story to the whole scientific world.

Bohr again cabled Copenhagen, insisting on confirmation of the Frisch experiments and some details. Then he phoned Professor George P. Pegram, dean of the Columbia Graduate Faculties, and urged that any statement to the newspapers should contain a reference to the Meitner-Frisch proof of "this wonderful phenomenon."

Pegram, who was often described as a "gentleman of the old school," was sympathetically understanding. The newspapers had the story, but he would phone at once and try to have an insertion made. It was, however, too late. The early editions were on the newsstands.

"Vast Energy Freed By Uranium Atom," read the headline on the first page of the second section of *The New York Times*. The news had been judged a science story and not of front page importance.

"The splitting of the uranium atom into two parts, each consisting of a gigantic 'cannon ball' of tremendous energy of 100,000,000 electron volts, the greatest amount of atomic energy so far liberated on earth, was announced here yesterday by the Columbia University Department of Physics," the story began.

The New York Times carefully explained Hahn's work and told how Meitner and Frisch "were the first to realize that what was happening was the actual splitting of the uranium atom, weight 238, into two lighter atoms, barium of atomic weight 137, and possibly krypton of atomic weight 82." It also explained that Bohr had brought the word to the United States. There was, however, no indication that Frisch and Meitner had performed the verifying experiment.

The New York *Herald Tribune* account, also well back in the paper, read "The largest conversion of mass to energy ever attained by man, the creation of 100,000,000 electron volts from a shattered atom, has been accomplished by Columbia University." There was no mention of Meitner and Frisch. The experiment was attributed solely to Columbia.

Bohr was well aware that the wire services would pick up such a story and he had nightmarish visions of the Copenhagen and Stockholm newspapers printing it, without having had a report of the work of Meitner and Frisch. He dreaded the thought of the two originators reading that an American university was announcing what was actually their work. Bohr was absolutely certain that no announcement of the Meitner-Frisch report would have been sent to the Copenhagen press. The institute never made first announcements of scientific findings to the press. Publication in a scientific journal or a report at a scientific meeting always came first.

Then word came that Fermi had made a speech on the radio without mentioning Meitner and Frisch. Bohr's gloom deepened. Justice was at stake, as well as the priorities that would determine scientific reputations and in all probability places in history. Bohr was highly conscious too that Frisch and Meitner had already suffered grievous blows in the political upheavals of Europe, and he was doubly determined that they should not suffer another for which he was partly responsible. He would continue to fight to see that the record was set straight.

At long last the next morning, on February 1, the specific information he needed came from Cophenhagen. It was a cable from Frisch. Bohr immediately sent off a letter to Fermi giving him the text of the cable. He expressed the strongest hope that steps would be taken to make certain that public recognition went to Meitner and Frisch.

Bohr also told Fermi that he, Bohr, would write an article for the American scientific publication *Science* setting forth the whole astounding development and giving proper credit to everyone. He said furthermore that he would try to have it ready by the end of the week, and that he looked forward to discussing it and everything else with Fermi when he came to New York on the next Saturday, February 4. Bohr was going to New York to attend a meeting of a Scandinavian-American group on that day.

Bohr also wrote to his wife. All the weariness and discouragement of the week, and yet his unquenchable joy in the progress that had been made, came forth in the letter: "This morning I received the first word from the institute. We long had waited for it. But the experiment is beautiful. It is a wonderful discovery. I have written a long letter to Frisch, and sent a telegram of congratulations to Dr. Meitner. As you will understand, say nothing more.

"Rosenfeld and I had a hard week, and it continued through today. The institute neglected to follow up my urgent cables and to send me letters and cables about the experiment.

"It is too bad," he lamented, "that something so nice should cause so much trouble."

Nothing in Bohr's background had prepared him for the kind of struggle in which he was so deeply involved. The race for deadlines and proper credits was completely foreign to Copenhagen. The struggles there were of a different kind, the struggles of the mind, and lately against unalloyed evil, the overt evil of Naziism. But Bohr did not relent, however painful the fight might be.

On Saturday, as he planned, he and Rosenfeld went to New York, and to see Fermi. Rosenfeld waited outside while the two giants of science talked. When they emerged both looked pale and strained. "Fermi didn't understand why Bohr took it so tragically," said Rosenfeld. "If it had been Bohr himself, he would have ignored it. But since it was Prof. Meitner and Dr. Frisch who were involved, Bohr took it entirely to heart."

Bohr wrote the article he planned, and lost no other opportunity to make certain that all physicists understood the true sequence of events and knew of the work of Hahn and Strassmann, Meitner and Frisch. Bohr's valiant fight on grounds so unfamiliar and difficult for him, and the full indisputable record succeeded. Newspapers and magazines, and of course the scientific publications, began to give full credit to all.

And when Columbia University celebrated the 25th anniversary of the experiment Dr. Dunning performed in the basement laboratory on January 25, 1939, the university in another press release carefully said: "Only one other person had conducted a similar experiment before—the physicist O. R. Frisch, about 10 days earlier, in Copenhagen. Neither Dunning nor other scientists in America knew of Frisch's work then. The Columbia experiment was the first of its kind in the New World."

In due time the Nobel Prize and the other honors of the world went to Hahn and Strassmann. In the histories of man's first conquest of the atom, Meitner and Frisch also receive full credit. Bohr thus won the struggle he had inadvertently caused and full justice was done.

13 /

The Essential Chain

IN EARLY FEBRUARY, 1939, fresh snow covered the Princeton quadrangles and outlined buildings and branches with delicate traceries of white. Standing in the windows of the Nassau Club, Bohr looked out appreciatively on the tranquil scene. In the few days since his return from New York he had also begun to feel that his fight for the proper recognition of the work of Meitner and Frisch would be won. The feeling of well-being held even when he picked up the newspapers on the way in to breakfast. They were faintly reassuring. England was talking about a five-year peace plan and a commercial treaty with Germany. Bohr had few illusions that the "peace" of Munich would hold, but for the moment at least Hitler was not in the headlines.

Bohr went into the dining room, looking eagerly to see if Placzek had arrived, as expected, from Copenhagen. The institute's always stimulating Bohemian was there, sitting with Rosenfeld, having arrived the night before. After all messages and greetings had been delivered and the group settled down for a second cup of coffee Placzek reported on some interesting news. Frisch and Meitner had proved that most of the other "trans-uranic elements" also were fission products.

Lise Meitner had suggested this possibility almost immediately after she realized that the mysterious substances produced by the bombardment of uranium were barium and krypton, and not "trans-uranic" elements. A few weeks later she went to Copenhagen and, working with Frisch, set out to test the other "trans-uranic elements" that had turned up in many laboratories.

The two used a technique that Lise Meitner had first employed about thirty years earlier. The aunt and nephew, Placzek said, bombarded uranium with neutrons and collected the fission products on water contained in a shallow trough of paraffin. The water was then evaporated and the precipitate tested. The materials they found were lower in mass than uranium, and they could only have reached the water by the disruption of the nuclei being tested. Thus Meitner and Frisch concluded that the trans-uranic elements of which they all talked for more than four years "originate by fission of the uranium nucleus." Even Placzek was satisfied.

Bohr, who had been listening with the closest attention, looked up with a big smile and commented: "For one good thing, we're free of trans-uranic elements."

Placzek, the reluctant and skeptical, conceded "Yes," and then added, "but now you're in a worse mess." How, Placzek asked, was Bohr going to explain the inconsistencies of fission? Why should slow rather than fast neutrons generally cause uranium to fission? And why should slow neutrons induce a modest amount of fissioning in uranium? In thorium slow neutrons were captured.

Bohr's face suddenly sent blank. He stopped almost in midsentence, as he did on the day at the institute when he conceived the idea of the compound nucleus. He pushed his chair back from the table and said to Rosenfeld: "Come with me, please." Placzek was forgotten.

Bohr, with Rosenfeld at his side, set off across the campus for Fine Hall where he had been assigned an office. For once Bohr paid no attention to the snowy scene that ordinarily would have delighted him. They climbed the stairs in silence and went down the hall to the office. Still without saying a word, Bohr went to the blackboard and began to put down figures and symbols. He made some rough sketches, working rapidly.

In about ten minutes he stopped, and turned to Rosenfeld with another of the broad, quizzical smiles for which he was known.

He had the answer to Placzek's questions, and to the major, the essential problem, posed by the fissioning of the nucleus.

Natural uranium, Bohr reminded Rosenfeld, actually is made up of several different isotopes, the ordinary isotope U-238 which constitutes about 99 percent of it, the rare isotope U-235, and traces of two others.

The U-238, Bohr said, merely captures slow neutrons; it does not undergo fission when a neutron penetrates into its nucleus. It is U-235, the rare isotope, that fissions with the intake of a slow neutron, Bohr declared.

As Bohr pointed it out on the blackboard, they all—Wheeler and Placzek had come in—saw it clearly. Another and critical insight into the uranium nucleus had been gained; indeed the distinction Bohr was making would make possible the whole future development of the atom. When fission occurred in the experiments everyone was undertaking, it was only because a slow neutron had found and hit one of the few nuclei of U-235.

Bohr's analysis was not based on an inspired guess, but on his understanding of the basic structure of the nucleus and its forces. U-235, as the number itself reveals, has an uneven number of neutrons and protons. Add one more neutron (235 + 1) and a uranium 236 would be formed. A nucleus with an even number of particles would be more tightly held together than one with an odd number. Vibrations set up by the arrival of the neutron would be violent enough to offset the surface forces holding the nucleus together—to undo the fastenings, as Lucretius said—and to cause the nucleus to break in two with a large release of energy.

U-238, on the other hand, would become U-239 (238 + 1) with the invasion of a neutron. As an odd numbered nucleus it would be less likely to fission; the neutron would simply be captured and retained.

The little group gathered around Bohr realized the implications. If U-235 could be separated from the abundant U-238 it would be highly fissile to slow neutrons. Fantastic amounts of power, or a tremendous explosion, could result if a reaction started. Only Placzek objected.

"A wonderful person, a man of the highest integrity, he was often a thoroughgoing skeptic about new ideas," said Wheeler. "He was skeptical at this time about the idea that U-235 is responsible for the fissility of natural uranium."

The big room with its leaded casements looking out on a Princeton courtyard was a scene of elation. In the high spirits of the moment, Wheeler, one of the most amiable of men, offered to bet Placzek 1,846 to one, $18.46 to a penny that Bohr was right. Wheeler and Placzek thus were wagering a proton against an electron.

More than a year later, on April 16, 1940, when it was verified experimentally that U-235 alone is responsible for low-energy fission, Wheeler received a money order telegram from Placzek for $.01. There was a one word message "Congratulations."

Bohr had learned his lesson of early publication. In two days, a one-thousand-word note was prepared and mailed, on February 7, to the editor of *The Physical Review*. The note was titled: "Resonance in Uranium and Thorium Disintegration and the Phenomenon of Nuclear Fission."

Bohr began the note with a few sentences of review, emphasizing who had done what—the fission credit battle was not yet wholly won: "The study of the nuclear transmutation by neutron bombardment in uranium and thorium initiated by Fermi and his colleagues, and followed up by Meitner, Hahn, and Strassmann, and by Curie and Savitch, has brought forth a number of most interesting phenomena."

Bohr also reviewed the stages in a nuclear reaction (his own work): (1) the formation of a compound nucleus in which the energy brought in by the impinging neutron is stored; and (2) the release of this energy in the form of radiation, or its conversion into a form suited to produce disintegration of the compound nucleus.

All who had worked on the interior world of the nucleus had found that the nucleus was most likely to get rid of the extra energy that was throwing it into a state of instability by radiation. Radiation thus was the most common way for the nucleus to return to stability.

Bohr went on to say that uranium's peculiar reaction to slow neutrons—its breaking apart with their impact—could not be explained if it were attributed to the formation of a compound nucleus of mass number 239 $(238 + 1)$.

Coming to the all-important point, he said: "We have the possibility of attributing the effect concerned to a fission of the excited nucleus of a mass 236, formed by the impact of the neutron on the rare isotope of mass 235 $(235 + 1)$." The tight binding of the even numbered 236, Bohr repeated, would increase the excitation and the probability of fission.

"Even for excitations produced by impact of slow neutrons we may therefore expect that the probability of fission of the nucleus 236 will be larger than that of radiative capture."

Bohr said that in a forthcoming paper with Professor Wheeler "a closer discussion" would be provided "of the fission mechanism and of the stability of heavy nuclei in their normal and excited states."

The initial reaction of many physicists to this fingering of U-235 as the fission material was as doubtful as that of Placzek. At best they thought that the attribution of fission to U-235 was a thesis to test. On a visit to New York Bohr explained his theory of U-235 to Dean Pegram. Pegram later suggested to Fermi and Dunning that Columbia undertake some tests of it. Fermi however was one of the non-believers.

Bohr later told Rosenfeld that he was "outraged" at the doubts. It was not that Bohr objected to a challenge of his ideas; on the contrary he always welcomed and sought disagreement, but in this case he was convinced: "It has to be so. Nobody can doubt the logic of this argument," he said to Rosenfeld. And so it proved to be. Once again a Bohr insight had set the whole direction of physics, and the nuclear age.

The attribution of fission to U-235 left large questions unanswered. Why are heavy nuclei stable? Why are there no nuclei in nature with a weight above the 238 of uranium? What specifically initiates fission?

One of the first questions with which the physicists struggled was; "what is the critical energy required to trigger the splitting of the nucleus?" How could one small chargeless neutron produce such a tremendous effect? It was like asking how a mosquito bite could bring down a hulking human? The effect was all out of proportion to the cause.

Before the nucleus upset everything, Bohr and Rosenfeld had expected to use their time in Princeton to work on a paper on electromagnetism. Now this expectation had to be set aside again as Bohr and Wheeler plunged into the detailed studies of the nucleus and disintegration.

"Many issues arose in the analysis," said Wheeler. "They illustrated Bohr's deep appreciation of the complexity of the fission process. An ideal background was supplied by the views of nuclear constitution that had come out of the previous three years."

The work was done principally in Fine Hall, an English manor house built into a Princeton quadrangle of varied architectural periods. Holly trees grew on either side of the carved stone entry and ivy climbed the walls. Carved oak stairs and corridors patterned in black and white tile diamonds led into Einstein's large and generous office, which Bohr was using. It not only boasted a fireplace and handsomely carved shelves and cabinets, but those academic rarities, draperies at the leaded windows and an Oriental rug on the floor. It had previously been assigned to Einstein, who was then working in Princeton, but Einstein had preferred a small cubbyhole of an office and left it vacant.

The corridors of Fine Hall were connected at the upper levels with the corridors of the Palmer Physical Laboratories. Here Georgian elegance gave way to plain plaster walls and echoing floors. Wheeler's office was just around the corridor from Bohr's in Fine, and as they worked there was a constant running back and forth.

Another advantage of their offices lay in the presence of the physics library directly overhead. Bohr and Wheeler were working on the surface energy of the nucleus before fission. Bohr remembered well from his early work in the Copenhagen gold medal competition about Lord Rayleigh's calculations on the surface energy of spherical droplets and on the energy required to break them up. The problem at hand held so many similarities that Bohr and Wheeler hurried up to the library where they soon found the six volumes of the *Collected Papers of Lord Rayleigh* on a lower shelf in one of the library alcoves. Bohr read off the Rayleigh formulas while Wheeler jotted them down. The figures proved highly useful. Bohr and Wheeler had only to extend them for their new purposes.

They continued to need the Rayleigh material. Bohr would bound up the steps, two at a time, in his best Copenhagen manner and be back with the books—presented by the Class of 1880—before Wheeler, a younger man, could stop him and go for the books himself.

In this work they established that uranium is stable against small changes in shape, a conclusion well backed by experience. If this were not the case the nucleus would not hold together over long periods. At this point they had a new answer to the old question of why nuclei heavier than uranium (238) do not exist in nature. In larger nuclear aggregations the surface tension over a period of time would not offset the interior forces of repulsion. More than 238 particles then would not stay together; thus, the natural elements of the universe ended with uranium.

As this intensive work went on, the pipe ashes and chalk often were strewn around Room 208, Bohr's borrowed office. Wheeler came in late one afternoon to find the janitor haranging Bohr about the mess that had been created in Fine Hall's finest office.

"The janitor stood in no awe of anyone," said Wheeler. "He was scolding Bohr, and Bohr being such a modest man did not dare to speak back. From then on when he left the office, Bohr carefully lifted up the rug and . . ."

Bohr and Wheeler went on to the important question of what happens when a deformation occurs in the nucleus and it breaks in two. "In heavy nuclei," they said, "to produce a critical deformation will require only a comparatively small energy and by the subsequent division of the nucleus a very large amount of energy will be set free."

The pieces set free, the two noted, would give off beta radiation, but, they continued in an observation of the first significance, "in addition it has been found that the fission process is accompanied by an emission of neutrons, some of which seem to be directly associated with the fission."

Bohr and Wheeler did not attempt to go into detail about what would happen to the neutrons released in fission. However, their theory predicted that the neutrons given off in the fission of natural uranium would likely be absorbed by the U-238. Since U-238 only rarely would fission at the impact of a neutron, the reaction would fizzle out instead of continuing like an exploding string of firecrackers.

Only if the neutron given off should hit another U-235 nucleus would there be another fissioning. The composition of natural uranium indicated that there would be very little chance of the neutrons finding those rare—one in 139—nuclei of U-235. Long before they came to one of them they would be absorbed by one of the numerous U-238 nuclei. If there was to be fission the neutrons would literally not only have to find the needle in the haystack, but they would have to avoid being absorbed by the straws of the stack.

The Bohr-Wheeler theory did not augur well for a chain reaction, and it was only in a chain reaction that the fabled production of power or explosion from uranium could become a practical matter. Without a chain reaction the cost of tapping nuclear power would be prohibitive and the power treasures locked in the nucleus might never be usable.

The Bohr-Wheeler analysis of the mechanism of nuclear fission opened another question. If U-238 does not fission when it captures a neutron, what does it do? "The method of analysis we were using made it possible to predict what would happen," said Wheeler.

The addition of one neutron to U-238 would bring its atomic weight to 239 (238 + 1 = 239) and 239 with its odd number of nuclear particles would be an unstable system. In seeking to return to stability, as all unstable systems do, an electron would be emitted. If the loss of the negative electron—subtracting a depressive influence—turned one of the neutrons into a positively charged proton, the nucleus would have 93 protons. An extra proton would mean the binding of an additional electron in the outer reaches of the atom and the creation of a new element, an element one step beyond uranium in the table of elements, and an element nonexistent in nature.

The Bohr-Wheeler predictions could be carried further. The proposed new element No. 93—later to be called neptunium—would be unstable. It would soon emit an electron and powerful X rays. Again by the loss of one unit of negative charge a neutron would be converted into a proton, and the addition of one more proton to the 93 already present would produce another new element, No. 94 (93 + 1)—later to be named plutonium. If the new, even numbered, relatively stable element 94 should absorb a bombarding neutron, increasing its atomic weight from 239 to 240, it might well undergo fission, much as U-235, farther down the elementary line. Another possibility of fission was opening, though at this stage it was only in a theory created out of Bohr's insight that U-235 is the fission material.

The neutrons given off in fission, and the chain reaction they might make possible, were foremost in the minds of many scientists in the spring of 1939.

Bohr indirectly involved Fermi in testing the whole thesis experimentally. On a visit to New York, he stopped at the Columbia physics building to see Fermi, but found him out. In his absence he met a young graduate student, Herbert L. Anderson, and they fell to talking in the exploratory, open, Bohr way. Bohr soon had Anderson excited about the possibilities of the nucleus.

The moment Fermi returned, Anderson rushed in and suggested that he would be happy to assist with some experiments with the Columbia cyclotron. Anderson had done much of the work of designing it, and thought that it would be just the machine to produce the neutrons they would need. Fermi was delighted with the idea, but he hesitated. He had been in the department for only a few months, and Dunning and Pegram were in charge of the cyclotron. Anderson, in his enthusiasm, was not to be stopped. He rounded up the two department chiefs to meet with Fermi and himself and a research program was soon planned. As Mrs. Fermi said, "Enrico emerged from the meeting again an experimental physicist." Others also were drawn into the project, particularly Leo Szilard and Walter H. Zinn, a young Canadian.

Szilard, born a Hungarian, had been at Berlin in the early thirties when student "brown shirts" began their attacks on all Jews. He had seen Einstein, who had helped to interest him in physics, hissed off the platform. Szilard left early in the Nazi regime, going first to England and then to the United States. He had no appointment at Columbia, but was permitted to work with Fermi and use the Columbia laboratories as a guest research scientist.

The Columbia group was confronted at the outset with the question, "which uranium fissions?" Dunning agreed with Bohr that it was the rare U-235, while Fermi continued to favor the abundant U-238. They agreed to what Bohr would have called a complementary investigation. Dunning would try to obtain a small sample of concentrated U-235, and Fermi would measure the number of neutrons produced in each fission. By the middle of March, Fermi was quite sure that it was two, much as he had predicted initially.

Fermi and Szilard put a neutron source in the center of a large tank of water and counted the number of slow neutrons produced in the water. The same experiment was tried without water. It was the second time that Fermi had used water to slow down neutrons, but this time he was working with a prosaic tank rather than with a fountain and pool in a romantic Italian garden, fragrant with lilies. The measurements convinced the scientists that a chain reaction would be possible if two conditions were met: if the neutrons were slowed by some material of low atomic weight, and if uranium was used rather than a lighter element.

The Columbia group completed its report saying in effect that a power millions of times greater than any previously known to man now lay within reach. They sent it to the publisher on March 16, 1939.

At 6 A.M. on the day before, March 15, German troops poured into what Hitler called the "rump Czech state," and the world knew as Czechoslovakia. After carving away the Sudentenland, Hitler had fomented separatist national movements in Slovakia and

Ruthenia. The Czechoslovak government had to halt the separatist movements or permit the destruction of the Czech democracy. On the other side of the cruel dilemma, they knew that any action would be seized upon by Hitler as a pretext for invasion to "protect" the nationalists. The aged president Hacha summoned his courage and decreed the dismissal of the separatist governments.

It was, as they feared, the excuse Hitler was waiting for. The German dictator ordered into effect long-laid plans for what was called the "liberation" of the two provinces. Hitler's secret orders captured at the end of World War Two read: "to the outside world it must clearly appear as a peaceful action and not a warlike undertaking."

The move across the border was delayed briefly when the Czech president asked for an audience with Hitler. At 1:15 A.M. he was finally admitted to Hitler's presence, only to be told that the order for an invasion by German troops had already been given and that Czechoslovakia was even then being incorporated in the German Reich.

Hacha, one member of his party reported, sat as though turned to stone. If the German troops met resistance, Hitler threatened that Czechoslovakia would be broken with brute force. Hacha fainted, but was revived enough to sign a proclamation that he "confidently placed the fate of the Czech people and country in the hands of the Fuehrer of the German Reich."

As the German troops made their entry into Czechoslovakia on the morning of the 15th, Hitler announced with great flourishes that he was forced to the conquest to put an end to "wild excesses and terror." Czechoslovakia "has ceased to exist," he shouted over the German radio.

That night Hitler slept in Hradschin Castle, the ancient seat of the kings of Bohemia and the seat of the first democracy established in Central Europe. The next day, March 16, he proclaimed the Protectorate of Bohemia and Moravia and acknowledged the separate status of Slovakia.

On this same March 16 two other events unproclaimed and

unknown except to a few, registered an ultimately more enduring effect on history. Fermi went to Washington to inform the American government that a new mammoth explosive had become a possibility, and Szilard journeyed to Princeton to consult Bohr and the Princeton group about the course of science in the unprecedented situation of 1939.

It was exactly two months to the day since Bohr had stepped from the *Drottningholm* with the first word in America that an atom of a chemical then used principally for illuminating the faces of clocks, might be broken into two pieces.

Troubled discussions had been going on at Columbia. Szilard watched the neutron experiments with a consuming sense of horror. Suppose the same experiments were going on in Germany and an atom bomb should be placed in Hitler's hands! With many misgivings, for nothing was sure, a decision was reached that the findings must somehow be brought to the attention of the American government.

Fermi was going to Washington on the 16th to speak before the Philosophical Society. Pegram asked him to call on Admiral S. C. Hooper, chief of naval operations, and present the findings to him.

Pegram addressed a letter of introduction to the admiral: "Experiments in the physics laboratories of Columbia University reveal that conditions may be found under which the chemical element uranium may be able to liberate its large excess of atomic energy and that this might mean the possibility that uranium might be used as an explosive that would liberate a million times as much energy per pound as any known explosive.

"My own feeling is that the probabilities are against this, but my colleagues and I think the bare possibility should not be disregarded and I therefore telephoned this morning chiefly to arrange a channel through which the results of our experiments might, if the occasion arises, be transmitted to the proper authorities in the United States Navy."

The dean explained that Fermi was a Nobel Prize winner and

highly competent in nuclear physics. He added that Fermi would stay permanently in the United States and become a citizen as soon as possible.

Fermi called as proposed, and talked to the admiral as well as representatives of the army's Bureau of Ordnance and several civilian scientists who had been invited by the admiral to be present. The Italian scientist spent more than an hour acquainting them with what had been learned and its potentialities. At the end of the meeting he was told that the armed services were anxious to be informed of the Columbia experiments and would have a representative call at the university.

The conference did not in itself bring the government and science together in an unheard-of combination for action on an esoteric critical discovery. It was, however, a continental divide in the scientists' recognition that the work of their laboratories held a direct national concern and that this had to be acknowledged. This had never happened before so directly.

Szilard on this same day of crossing of frontiers of varied kinds assembled a key group in Princeton. It was virtually a Copenhagen meeting—Bohr, Rosenfeld, Wheeler, Wigner, and Teller, all of whom had been fellows at Bohr's institute. Teller had come up from Washington. With the exception of Wheeler, all of them were Europeans and all knew directly and personally the desperation of the European situation.

Bohr had read the newspapers that day with anger and gloom. And the afternoon editions carried Chamberlain's statement to the House of Commons—the Slovak declaration of independence had put an end by internal disruption to the state whose frontiers Great Britain had guaranteed, and His Majesty's government "therefore could not hold themselves bound by this obligation." Britain and France were turning aside. Abandonment was added to assault.

The grim events of the day sounded a solemn obbligato to Szilard's warnings. If a chain reaction proved possible, and Szilard thought that it would, Hitler would have the final weapon needed for world conquest. Szilard went to the blackboard and wrote down all the data on the latest Columbia findings on the release of neutrons in fission. They indicated that at least two neutrons would be produced at each breakdown of a uranium nucleus. "It came out more than two," Rosenfeld remembered, and the "two" sounded like the knell of doom.

But here Bohr intervened. He argued again that a chain reaction could not be set up unless U-235 could be separated from U-238, and he thought that this would be an impossible, a prohibitive undertaking. It would take time and resources, Bohr said, that certainly were unavailable to science and probably to governments already strained under the burdens of the day.

Not a man there, however, doubted that the Germans would know as much as they did about the fantastic possibilities of fission, or questioned that Hitler would try to produce an atom bomb if it were possible. Four of the six had been educated in Germany and the other two, Bohr and Wheeler, had worked in the closest collaboration with the German scientists for many years and knew their competence.

Though Hitler had driven out some of Germany's highest scientific talents, an impressive group remained. Who in that room could doubt that Hahn, Heisenberg, Von Weizacker, and others were capable of the most difficult of scientific undertakings? Their authority was beyond question. Bohr cited Hahn. When Hahn had said only four months before—four months that seemed like years —"This is barium" no one questioned that he was right. If the statement had come from anyone else it would have been doubted and probably rejected.

By this time it was after midnight and Princeton was as silent as the scientists' thoughts were clangorous. Szilard was urging three steps. The scientists, he said, must contact the President of the United States and the armed services (at the time he did not know about Fermi's meeting with representatives of the services). If the additional experiments confirmed that the neutrons "come out larger than two" then Szilard pleaded "we must start a campaign," a campaign to push the work forward with every resource and dispatch. Szilard's third proposal was silence. "We must induce all physicists to stop all publicity about fission," he declared.

The very nature of physics had been openness. In large part this unusual method of procedure which had reached its epitome in physics, although its roots were deep, was the fashioning of Bohr. It was he who above all others had brought the physicists of the world together into what was actually a boundaryless single scientific community.

Bohr was the center, the exponent, the very symbol of the unique system that had brought physics so far in so short a time. The proposal of a voluntary censorship ran contrary to all of their instincts and particularly to his. Could physics function with the suppression of findings? And was censorship even possible, when any worker in any country might come upon the truths of nature at any time? The laws of nature could not be hidden.

Bohr also questioned whether as a practical matter censorship was possible. He had already had word that the Joliot-Curies were working on the neutron problem. Szilard proposed that they cable them the next day asking them to withhold any reports. Bohr also pointed out that *Nature* had already published, for all to read, the Frisch-Meitner papers. The newspapers had carried reports on fission.

Whether or not Bohr knew it, the newsstand at the Nassau Club had on sale that day the March 13, 1939, issue of *Time* with an article entitled "Big Game." It began: "Six weeks ago a report reached the United States about an atomic explosion which took place in a Berlin laboratory—the most violent ever accomplished by human agency.

"This news, known to only a few insiders, streaked over the physical world like a meteor. By last week half a dozen scientific journals were popping with reports confirming, extending, or interpreting the original report."

The work of Hahn and Strassmann, Frisch and Meitner, the Joliots and Savitch was cited. The same magazine on February 6 had reported Hahn's "release of 200,000,000 electron volts . . . an explosion on a microcosmic scale . . . with no flash, no crash, and in actuality not enough to knock a fly off the wall." Though Hahn had not detected the explosion, the magazine was correct in assuming in the light of later knowledge that it had been there. The March 13 issue also reported that the nucleus had been compared to a droplet and concluded: "The flood of reports made it appear that atomic physicists are off on the biggest big game hunt since the discovery of artificial radioactivity in 1934."

Calling a halt to the spread of information might well be an impossibility. "But that night no one knew what would come," said Rosenfeld. "The calculations could be mistaken."

The next day, while he was still in Princeton, Szilard cabled the Joliot-Curies, asking their cooperation in the voluntary halting of all publications about fission. They answered that it was too late. A paper by the Joliots with von Halban and Kowarski of their laboratories, had already been submitted to *Nature*. It reported, mistakenly as it turned out, that 3.5 neutrons might be produced with each fission of a uranium nucleus. The paper had been sent to London because French journals would have taken longer to get it into print.

The first attempt at voluntary censorship had failed. In the last six months of 1939 *Nature* published some twenty notes and articles bearing on fission, and a burst of articles was coming out in American publications. The correctly, soberly written scientific articles betrayed not a hint of the intense excitement of the authors or of the race in which they were engaged, but in all probability never before had so much appeared so rapidly about any discovery in physics. By the end of the year the score in the United States was an even one hundred articles. They were summarized and brought together in the January 1940 issue of *Reviews of Modern Physics*.

In April, 1939, Bohr's three months in Princeton were coming to a close. He had given several lectures to the students; now he gave another. The small dinners at which he met friends and colleagues multiplied. Everyone wanted to say farewell, but in addition in a university with its attention directed to world affairs many also wanted to know what Bohr thought about the increasingly ominous moves of Hitler and the deteriorating European scene. After his first statement to the House Chamberlain's attitude had stiffened. In a speech a few days later he asked: "Is this the last attack on a small state or is it to be followed by another? Is this in fact a step in the direction of an attempt to dominate the world by force?" The same questions echoed in the little dinners given for Bohr; the questions were put to him repeatedly.

Bohr had no real answer, but he could see little hope. With almost no pause after the seizure of Czechoslovakia Hitler began to apply the pressure to Poland. He was demanding Danzig and a highway and railroad across the corridor dividing East Prussia and the Reich. Bohr could feel and see the fire drawing ever closer to Denmark and the rest of Northern Europe. And Mussolini added to the turmoil by invading Albania.

More than once Bohr was strongly urged not to return to Copenhagen, but to bring his family to the United States. Almost any place that he might want in an American university would be open to him. Bohr refused; he would not be tempted. He felt that he had to return not only to maintain the institute and all those associated with it, but to keep a door open for the scientists fleeing from the widening Nazi zone of oppression. The institute by this time remained one of the few ways out. Bohr did not waver; he was going back.

Copenhagen on the surface seemed unchanged in that fateful summer of 1939. Crowds filled the narrow winding shopping street and flocked to Tivoli where the bands and the fountains played with all their accustomed gaiety and the firefly lights sparkled in the trees. Flowers bloomed in the clustered planters along the streets and the swans and ducks, trailed by their broods, sailed serenely across the quiet lakes where the old walls of the city once had stood. Visibly Copenhagen was unchanged. The institute too went its way, though the number of students from abroad was drastically reduced.

But beneath the quiet, clean order of Copenhagen anxiety was intense. Bohr and most Danes began to listen to every news broadcast on the radio. New faces would suddenly appear at the institute, not the young, keen faces of the physics graduate students, but faces ravaged by horrors just escaped and losses suffered.

The Bohrs did everything possible to lighten their load. If a refugee was coming in by train and they had advance notice, a member of the institute staff met him at the station. Many were taken into the Bohr's own home, the House of Honor, as guests. Margrethe spared no effort to give them surcease, and often this was not easy. Some were distracted and ill from worry and strain. And safety had to be sought beyond Denmark. Bohr made a trip to Sweden soon after his return from the United States to arrange for the reception of refugees there and to find as many places for them as might be possible.

In July while Britain, France, and Russia edgily talked of an alliance against Hitler, Bohr himself found some relief in the arrival of the proofs for the article on the mechanism of fission that he and Wheeler had completed during his American visit. "I read it with a great deal of pleasure," Bohr wrote to Wheeler, and then characteristically, "I still feel a few small alterations are advisable."

Worry gripped Copenhagen, situated as it was on the very edge of danger, but it did not exceed the anxiety of Szilard in the United States. Szilard lived with an ever present sense of doomsday. By July he became convinced that the government agencies which had been informed of fission would not or could not give it the full support needed if the work was to be done in time—in time to withstand Hitler. Restrictions on government contracts made it difficult for the navy to provide funds for the Columbia work.

Szilard's own research had offered evidence that if graphite were used to slow down the neutrons emitted in fission, a chain reaction might be established. He then regarded a bomb as a distinct possibility, and fervently thought that the time had come to keep as much uranium ore as possible out of the hands of the Nazis. He knew that Einstein knew the royal family of Belgium, and since the Belgian Congo had the largest uranium reserves outside of Europe, he decided that Einstein would be the right one to warn the Belgians of what the ore could mean. Szilard and Wigner went to see Einstein, and found him willing to get in touch with Belgian officials, if not with the king.

But somehow the American government also had to be pushed

into effective action. A friend of Szilard's directed him to Alexander Sachs, an economist with a New York financial firm, and occasionally a consultant of President Roosevelt on economic matters. What mattered was that Sachs had access to the president. Szilard found Sachs unexpectedly understanding. He had become interested in physics some years before upon attending a lecture of Lord Rutherford's and recently had seen Bohr's article in *Nature* on fission. Sachs suggested, however, that if the president were to be convinced of the unprecedented importance of fission, it would take the weightiest of testimony, say Einstein's.

Szilard and Wigner sought Einstein out at the cottage on Long Island where he was spending the summer. While they sat around a table on a screened porch they worked out the basis of a letter to President Roosevelt. "I feel it my duty," said Einstein, "to bring to your attention the following facts and recommendations . . . in the course of the last four months, it has been made probable that it may become possible to set up a nuclear chain reaction in a large mass of uranium, by which vast amounts of power and large quantities of new radium-like elements would be generated. Now it appears almost certain that this could be achieved in the immediate future.

"The new phenomenon would also lead to the construction of bombs and it is conceivable though less certain that extremely powerful bombs of a new type may thus be constructed. A single bomb of this type might very well destroy this whole port, together with some of the surrounding territory."

The letter urged the president to keep in touch with the physicists and to designate a proper government agency to make recommendations to him for what should be done. Einstein closed with a congealing warning: Germany was interested in uranium. And capping this, the Reich had just stopped the sale of uranium from Czechoslovak mines. As Einstein signed the letter he murmured, "For the first time in history men will use energy that does not come from the sun." Thus powerfully armed, Sachs sought an appointment with the president.

In England, at virtually the same time Professor George P.

Thomson, the son of J.J., and Professor W. L. Bragg of the Cavendish had decided that there was a chance that uranium might yield vast power and heat.

Like Szilard, Thomson realized that if this were the case, the Congo uranium ore and a Belgian ore stock pile must be kept away from the Nazis. Also, like Szilard, he did not know quite how to reach the right authorities in the government but he did reach General Ismay, secretary of the Committee of Imperial Defence. The General quickly consulted Sir Henry Tizard, chairman of the Committee On The Scientific Survey of Air Defence. Within four days, on May 10, 1939, the British were in contact with M. Sengier, president of the Union Minière Co., the holder of the Belgian uranium stocks. Some of the Belgian stockpile, about 100 tons, had recently been transferred to the United States, and only a small tonnage remained in Belgium. Sengier agreed to inform the British immediately if any abnormal demand arose for it.

It was also agreed in Great Britain that research must go forward. Nevertheless skepticism ran high. No one voiced it as tellingly as Winston Churchill, then a member of Parliament, prodding the government on the dangers of Hitler. After talking to Dr. Lindemann, Churchill wrote to the Secretary of State for War on August 5, 1939:

> Some weeks ago one of the Sunday papers splashed the story of the immense amount of energy which might be released by the recently discovered chain of processes which take place when this particular type of atom is split by neutrons.
> In view of this it is essential to realise that there is no danger that this discovery, however great its scientific interest and perhaps ultimate scientific importance, will lead to results capable of being put into operation on a large scale for several years.
> There are indications that tales will be deliberately circulated when international tension becomes acute about the adaptation of this process to produce some terrible new secret explosives, capable of wiping out London. Attempts will no doubt be made to induce us by means of this threat to accept another surrender. For this reason it is imperative to state the true position.
> First, the best authorities hold that only a minor constituent of uranium is effective in these processes, and that it will

be necessary to extract this before large scale results are possible. This will be a matter of many years.

[Lindemann had been reading Bohr, and had coached Churchill on the Bohr findings, though Churchill omitted the Bohr qualifications.]

Secondly the chain process can take place only if uranium is concentrated in a large mass. As soon as the energy develops it will explode with a mild detonation before any really violent effects can be produced.

[When Churchill reprinted this remarkable letter in his book *The Gathering Storm* he inserted a footnote on this point: "This difficulty was, of course, overcome later, but only be very elaborate methods after several years of research."]

It might be as good as our present day explosives, but it is unlikely to produce anything very much more dangerous.

Thirdly, these experiments cannot be carried out on a small scale. If they had been successfully done on a large scale (*i.e.* with the results with which we shall be threatened unless we submit to blackmail) it would be impossible to keep them secret.

Fourthly, only a comparatively small amount of uranium in the territories of what used to be Czechoslovakia is under the control of Berlin.

For all of these reasons the fear that this new discovery has provided the Nazis with some sinister new secret explosive with which to destroy their enemies is clearly without foundation. Dark hints will no doubt be dropped and terrifying whispers will be assiduously circulated, but it is to be hoped that nobody will be taken in by them.

CHURCHILL

Churchill spoke in rolling certainties. Bohr, upon whose analyses Churchill was relying at second hand, did not speak with such positiveness and in such absolutes.

And yet, in certain ways, as Churchill himself observed in *The Gathering Storm*, it was remarkable how accurate this forecast was. It was also a remarkable forecast in another sense—a first act involving the three men, Churchill, Bohr, and Roosevelt, in one of the most crucial dramas ever to be enacted on the world stage.

On August 23, 1939, another shock, as unexpected and shattering as an earthquake, stunned Europe and the world generally.

Hitler and Stalin, the reputed arch enemies, announced the signing of a nonaggression pact. In the midst of negotiations with the British and the French, Stalin cynically "bought time" by coming to the agreement with Hitler. Secretly, as the postwar records revealed, the dictators also agreed to the division of Poland and to a free hand for Russia in the east Baltic.

This time Britain stood firm. The government announced that the pact would "in no way affect" their obligation to Poland. Prime Minister Chamberlain was explicit: "That there should be no tragic misunderstanding . . . if the case should arise [His Majesty's government] are resolved and prepared to employ without delay all the forces at their command, and it is impossible to foresee the end of hostilities once engaged."

Hitler, shaken in his turn, halted the scheduled August 24 invasion of Poland. Mussolini wavered. But frantic and futile efforts to stop the inexorable march to war failed. At 4:45 A.M. on September 1, 1939, German troops invaded Poland, German planes roared overhead and bombs fell on Warsaw, the first but not the last to rain terror from the skies on undefended civilian populations.

In a final solemn ultimatum, the British government notified Hitler that unless his invasion halted by 11 A.M., September 3, a "state of war will exist between the two countries as from that hour." The French followed suit, and thus war came to Europe.

To the Bohrs in Copenhagen the shock was profound, though they had expected it and knew the alternative of Nazi domination was intolerable.

On May 31, exactly three months earlier Hitler had "presented" Denmark and the other Scandinavian states with "nonaggression" pacts. The Danes alone accepted one. It bound the Kingdom of Denmark and the German Reich "not to go to war or use armed forces of any kind against each other." The Danes, an eminently practical people, argued that it would have made no difference whether they said yes or no. Why, they asked, make a madman madder? They had no illusions whatsoever about the worth of the paper they signed.

The German armies rolled on across Poland despite the courageous, though disorganized, resistance of the Poles. Two and a half weeks after the German attack began, the Russians swarmed across the eastern frontier. Within a month all was over and a nation of 35,000,000 lay helpless and enslaved. Poland had been crushed by an overwhelming exhibition of what came to be known as Blitzkrieg, a seemingly irresistible thrust of armor through country prepared in advance by saboteurs and spies.

For all of the alarm and terror at the outbreak of the war, Denmark did the only thing she could do: her people tended their farms and businesses and the Institute for Theoretical Physics went on with its studies and experiments. Life does not stop for catastrophe. "In Europe life is surely most upset at present," Bohr wrote to Wheeler, "but in this small neutral country we try notwithstanding the increasing difficulties to go on with the work."

On September 1, the day war engulfed Europe, the Bohr-Wheeler article on the mechanism of fission was published. Almost from the moment of its appearance it was recognized as a classic. When the official American history of atomic energy *The New World* was written more than twenty years later, it called the paper the prime achievement of the period. The similar British book *Britain and Atomic Energy* referred to it as "fundamentally important." And so it was appraised by history. Bohr did not receive a copy of the journal for another month.

"The *Physical Review* of September 1 has just arrived with our long article which I was very happy to see in print," Bohr wrote to Wheeler on October 4. I admired when reading it again, perhaps still more, all the work you put into it and the many instructive drawings you composed. Our collaboration has given me the greatest pleasure indeed. As well Rosenfeld as Erik and I speak very often of the happy times we, in spite of the anxieties, spent in Princeton, where we also so much enjoyed the truly human and scientific spirit of the whole little community."

Bohr said that the work of the institute was going on "quite well" and that he was making some progress with the investigations in which he personally was engaged.

The Wheelers, shuddering at the reports of the bombs

dropped on Poland and at every mile the German armies advanced, wrote the Bohrs that they would be happy to have any one of the four Bohr sons come to Princeton to remain with them as long as there was danger.

"We are aware that a catastrophe might come any day," Bohr answered, "and I need not say how thankful my wife and I are for your and Mrs. Wheeler's kind thought of offering to have one of the boys with you for a time. This would of course under certain circumstances be extremely helpful for us and at the same time a source of the greatest pleasure for our boys. By hearing about it, Erik was indeed more than ready to return to America at once."

The Bohrs, however, did not think it best at the moment for the boys to leave them and their schools in Denmark. They wanted the family to remain together as long as it was at all possible. But Bohr knew that a different day might come and he expressed his especial appreciation for the Wheeler's added word that their invitation would hold for any time that the Bohrs might need it. Bohr said that they might want to accept it "if any great change for the worse should take place." The Bohrs knew that they were living on the sharp cliff of danger, but they were living and they would try to maintain their life together, precarious though the future might be.

The war was already disrupting the lines of communication, and Bohr, having learned his lesson in World War One, sent one copy of his letter by air and one by surface in the hope that at least one would arrive.

The Danish common sense about the war was not blindness, nor did it belie the alarm and fear that Hitler would move north. The moment war broke, Professor Chievitz insisted that all Danish hospitals be equipped to treat gas poisoning. He thought such an outrage might well be perpetrated by the Nazis. In preparation he developed a "nose catheter" for supplying a stream of oxygen to any victim of such an attack. No factory, in the instant press of orders that war brought, could undertake to make the tubes. Bohr said the institute would do the job; they knew after all how to make apparatus. All of the staff, from Bohr on down, donned overalls and manned the production lines they organized. Within a week 6,000 tubes and accessories were packed in cardboard boxes and sent to Denmark's hospitals. Fortunately they were never needed.

In the United States the European war superseded all else. President Roosevelt began a struggle with Congress for the repeal of the American arms embargo. In these hectic days, Sachs, waiting anxiously to present the Einstein letter, was unable to obtain an appointment until October 11. When Sachs was ushered into the oval office of the chief executive, Roosevelt, went straight to the point: "Alex, what are you up to?" Sachs read him the Einstein letter, and an accompanying document Szilard had prepared.

Sachs then told the president the story about a young American inventor who went to Napoleon and offered to build him a fleet of steamships that would permit him to sail against England in any kind of weather. Ships without sails? Napoleon scoffed at the idea. The inventor was Robert Fulton.

As Sachs finished his story Roosevelt scribbled a note and sent it out. An aide returned in a few minutes with a carefully wrapped package. It was a bottle of Napoleon brandy the Roosevelt family had kept for many years. The President filled two glasses. He nodded and raised his to Sachs. As Sachs started to leave, the President said to him: "Alex, what you are after is to see that the Nazis don't blow us up." That was it exactly.

The government had no machinery for dealing with the scientific community, for there had been few bridges between them. Roosevelt at the moment did not try to devise any, but took the easier course of handling the strange new problem through his own administrative set-up. He named an Advisory Committee on Uranium to investigate the momentous problem Sachs and Einstein had raised.

While the stillness of death and oppression settled over eastern Europe and western Europe stood locked in an unmoving struggle that Chamberlain called the "Twilight War," the one man with the potential power to develop the greatest of nature's powers had been reached and had taken the first necessary action to develop atomic energy. On November 1 the new committee reported to the President that a chain reaction was a possibility, though still unproved.

Bohr continued unconvinced that the practical problems of separating the mite of U-235 from the abundant U-238 could be overcome. Nothing he saw in the scores of papers coming from laboratories in the western countries changed his mind. The problem seemed impossibly formidable. By the end of the year the question still had not been settled experimentally.

The year 1939 had carried them to the peaks and plunged them into the abysses. The contrasts of joy and tragedy had been overwhelming to the little band called physicists. A few months had placed in their hands the lever that could move the earth. But would it?

There was only uncertainty, and perhaps no one realized that every catastrophe Hitler visited upon Europe was matched by a dawning of a new, different, and greater power. It was one of history's most eerie counterpoints, an unprecedented and incredible coupling of death and life.

14 /

The Occupation—Night Comes

Oɴ ᴛʜᴇ ᴄʜɪʟʟ blustery night of April 8, 1940, King Haakon cordially welcomed Niels Bohr to the Royal Palace in Oslo. There was, nevertheless, no mistaking the tension in the gaunt face of the king or the uneasiness of several leaders of the government who also had been invited to dine with the sovereign.

The government knew that the British might mine the Norwegian "Leads," the passage taken by the ships carrying Swedish iron ore from Narvik in the north to Germany. They also were acutely aware that the iron supplied the German war machine and the German economy. If the passage were closed German reprisals would be certain to follow. Disquieting reports also had come in of Nazi troop concentrations at northern German ports. The king, however, entertained the distinguished Danish scientist on the last night of his visit to Oslo.

Bohr had come to Norway to give several lectures on the transformation of the atomic nucleus. He was able to report to the Norwegian scientists about the momentous establishment of fission, but he also had further scientific news of importance to tell them— Element 93 had just been discovered in the United States and U-235 had been experimentally identified as the isotope of uranium which fissions at the impact of slow neutrons. Both developments had been predicted in the Bohr-Wheeler paper of September 1, 1939.

In March, 1940, Alfred O. Nier, 27-year-old physicist of the University of Minnesota had succeeded in separating a few micrograms of U-235. When it was measured at Columbia University there was easy proof that it was indeed, as Bohr had said, the part of natural uranium that fissions. The experiments left little doubt that a chain reaction could take place, if enough U-235 could be separated.

Bohr was doubtful that the proviso could be fulfilled on a scale necessary to produce an extraordinary explosion. The Americans themselves estimated that the largest plant capable of performing such a job would require ten days, working twenty-four hours a day to produce a thousand-millionth of a gram of U-235. At that rate it would take 26,445 years to turn out one gram. This was why Bohr, who knew how infinitesimally small the mills would grind, had said initially that U-235 was not practical for weapons. He still was skeptical about the possibility.

Bohr also had heard through an early prepublication report about the discovery of Element 93. As soon as the news of fission had reached Berkeley—and that was by a phone call on the day Bohr announced the splitting of the nucleus at the Washington meeting—Edwin M. McMillan, a young assistant professor of physics, began trying to measure the energies of the fragments thrown off as the nucleus split. Among the main fragments he found a strange radioactive substance with a half-life of 2.3 days. McMillan thought that it might be an isotope produced by U-238's capture of a neutron and the neutron's conversion into a proton. If this were the case one proton would have been added to Uranium's 92 and a substance with 93 protons would have been produced. This would mean that it was the new element Bohr had predicted.

McMillan and Philip H. Abelson promptly confirmed this analysis with positive chemical identification. A new element had been created, an element that did not exist in nature. McMillan appropriately named it neptunium, for the next planet beyond Uranus. With this breakthrough the search was pushed forward for Element 94, which Bohr and Wheeler had predicted might be formed from Element 93, if 93 in its turn should pick up a neutron. Whole new possibilities for fission were opening.

In between lectures in which he discussed these principles and findings, Bohr also quietly conferred with leading Norwegian officials. Bohr was fearful that the war would spread to Scandinavia and wanted to talk to the Norwegians about the dangers they all faced. At dinner at the palace that night the talk also turned inevitably, though discreetly, to the subject uppermost in all their minds—the war and its menace to all of the world and particularly to their part of the world.

Immediately after the dinner Bohr boarded the night train for Copenhagen. The sleeping cars were shifted to a ferry for the crossing of the Kattegat to allow the passengers to make the trip undisturbed. As the train left the ferry the next morning and pulled into Elsinor, in the very shadow of Hamlet's castle, Bohr was awakened by loud pounding and shouting. It was the Danish police. "Invasion," they shouted with all the urgency of horror and shock. "The Nazis are invading Denmark! Norway is being attacked!"

At that moment planes bearing the black crosses of the Germans were roaring ominously overhead. Bohr, for all of his fears that this would happen, was stunned at the reality. He made the police repeat their words.

Everything confirmed the dread news—the Nazi planes, the blaring radios, the turmoil, the milling people, the woman crying silently—everything confirmed the credible and yet incredible invasion. If the ferry passengers had been awake during the night they might well have seen some of the Nazi ships moving northward, stealthily, without lights in the darkness. For several days the Danes had seen them on their way through their waters, but had assumed that they had some other object than the conquest of peaceful, unmenacing Denmark and Norway.

Bohr's train was permitted to go on to Copenhagen. After making certain of the safety of his family, Bohr rushed to the institute. Then there was nothing that could be done, nothing except to watch the Nazi planes zooming over the city, to hang despairingly over every news bulletin, to talk to friends who knew no more than one did; to pick up one of the manifestos the German planes were dropping and to laugh bitterly over their misspelled Danish and their patent absurdity—"The German troops do not set foot on Danish soil as enemies. . . . German military operations aim exclusively at protecting the north against the proposed occupation of bases by the Anglo-French forces. . . . The German Government has no intention of infringing . . . on the territorial integrity and political independence of the Kingdom of Denmark now or in the future. . . . The Reich Government therefore expects that the Danish people will . . . offer no resistance. Any resistance will have to be broken by all possible means . . . and would therefore lead only to an absolutely useless bloodshed."

At 4:20 A.M., an hour before dawn, Denmark gradually learned, the German envoy had routed the Danish foreign minister out of bed and presented him with an ultimatum. It demanded that the Danes "accept the protection of the Reich" on that instant and without resistance.

An hour later German troopships appeared off Copenhagen and other Danish harbors. At the capital the Nazi ships passed between the harbor forts without a shot being fired and tied up at the Langelinie (the Little Mermaid) Pier in the very heart of the city, only a short distance from the headquarters of the Danish army and the Royal Palace. The Little Mermaid herself sat wistfully as usual on her shining brown rocks.

A small Nazi force took over the army building and the palace, where a few shots rang out. Upstairs the king and his government were meeting in extraordinary session. The ministers recommended nonresistance. As the debate went on, the German general became impatient. He phoned for additional bombers to zoom over Copenhagen and "force the Danes to accept."

The Danes yielded. Unlike the Norwegians, they did not fight initially or deliberately. All resistance was ordered halted. A little skirmishing at several airfields ended. By the time the morning was over and the big clock in the townhall was striking the noon hour the Nazis had completed the occupation of the ancient land of Denmark and had entrapped its 4,500,000 people. In a few hours, Denmark's thousand-year-old tradition of independence was at an end. Denmark had been overrun, even as Czechoslovakia and Poland.

As Norway fought valiantly, and her king and government fled to the mountains to continue the fight, the Danes, including the Bohrs, endlessly discussed Denmark's surrender. Denmark could not have been defended. The flat lands and well-tended roads of Jutland lay open to the Nazi Panzers. There were no mountains to which to retreat, and the Danish islands that made the coast a lacework of land and sea were as accessible as the mainland. No hope was held for help from abroad. A heroic suicide would have accomplished nothing.

When Palle Lauring wrote the history of this fateful day, some twenty-five years later he said:

> The reaction was not fully understood abroad. The Danes hardly understood it themselves that sunny spring morning when the peaceful, well organized little country suddenly discovered that she had suffered the same fate as Czechoslovakia. . . . It was an eventuality to which every Dane had given careful thought . . . but never seriously believed possible. The reaction that morning was very Danish.
>
> Just as Christianity was introduced in Denmark by a decision at a Thing-meeting, just as the Reformation was brought about more peacefully than anywhere else in Europe, just as absolute monarchy was introduced by an announcement in 1660 and discarded again in 1848 at the request of a delegation of honest citizens clad in top hats and frock coats, so in similar fashion did Denmark react on April 9, 1940.

Everything could be explained and justified, the Danes told themselves again and again, and yet as the Norwegians continued the fight there was bitterness and recrimination about Denmark's sensible, if unheroic, course. The assurances that nothing else could have been done did not wholly assuage the Danish conscience then or later.

Bohr, like his countrymen, was torn. The circles deepened under his eyes and the already strong bones of his massive face stood forth with new solidity. Bohr listened to the news bulletins with his head bowed to catch every word, but he was not long frozen into inaction. Even as the Nazis moved into Denmark and while the planes still were making the city hideous with their noise, Bohr began the destruction of all the institute records on the scientists the refugee committee had brought out of Germany and other occupied countries. It was vital that this data should not fall into Nazi hands.

Bohr also hurried off to see the chancellor of the University of Copenhagen to ask his assistance in protecting Polish-born members of the institute staff. Within the next few days he also went to see members of the government to solicit their help if the Germans attempted to put their racial laws into effect in Denmark. Bohr urged them all to resolute resistance against racial persecution. Bohr argued that Denmark did not have to yield to the Nazi racial hysteria and brutality, and could, with determination, withstand it.

The rest of the world was as shocked by the new Nazi transgressions as the Danes and the Norwegians themselves. The day after the occupation, cables began to arrive from American friends and universities inviting, and pleading, with Bohr and his family to come to the United States. There was grave worry about the dangers the Bohrs would face in a Nazi-controlled Denmark. The American embassy also was asked to do everything possible to insure Bohr's safety. American embassy officials called to say that they were certain they could arrange for a journey to the United States if Bohr cared to accept one of the American invitations.

Bohr's answer was the same he had given in Princeton. He must stay to do everything in his power to protect the staff of the institute and the institute itself. And officials of the Danish Foreign Office to whom Bohr talked told him flatly that his departure would weaken the morale of the entire country and come as a particular blow to the Jews of Denmark. Bohr's mother had been a member of the Adler family, one of the outstanding Jewish families of the country. Though Bohr had not been reared in the Jewish religion or tradition and personally would not have been subject to the Nazi racial laws, his work for the Jewish scientists was well known and he stood as a symbol of strength and support in those desperate days to the Jews of Denmark and the refugees who had found sanctuary there. Niels, Harald, and Margrethe in long sessions while walking in the gardens and arcades of Carlsberg resolved again to stay in Denmark as long as it was possible for them to do so.

A second decision was made to keep the institute in as normal and full operation as possible. It could no longer serve as the training ground for the physicists of the world and as the world center of physics. Nearly all of the foreign fellows were already gone and the foreign physicists had not been able to come freely for almost two years. The last foreign student of 1939–40, a young Swede, left when the academic year ended. Frisch, who had gone to England to carry on some studies and work there, would not return.

Nevertheless there were Danish students to educate in physics, one of them Bohr's own son Aage. Advanced training in physics, Bohr pointed out, had become more important then ever before for the Danes; it was one of the best ways in which they could maintain their own culture and potentialities and prepare for whatever the future might hold.

The day after the occupation, Bohr had cabled Frisch, urging him to remain in England and assuring him that all were unharmed. The message concluded: "Tell Cockcroft and Maud Ray Kent." Frisch quickly delivered the message to Cockcroft, but what did Maud Ray Kent mean? Was Bohr trying to convey some concealed message about uranium? Could Maud Ray be an anagram for radium? Could it mean that the institute's supply of radium had been seized? None of the English group could interpret the obscure words, turn them around, decode them, and consider them as they would. Maud Ray Kent quickly became a well-known conundrum in British physics circles, and shortly, with a bow to Bohr, the name MAUD (M.A.U.D.) was bestowed on the new committee the British were setting up to conduct their excursion into atomic energy and the possible development of an atomic bomb.

It was only several years later that they all learned that Maud Ray was a governess who lived in Kent. She had spent several years with the Bohrs teaching English to the children and Bohr wanted her to know that they were safe.

Just about a year before, Professor Thomson at Cambridge had begun a study of fission and uranium. When war broke out and the talk began that an atom bomb might be Hitler's vaunted "secret weapon," Chadwick, the famed discoverer of the neutron was asked to appraise the possibilities of uranium. At first Chadwick very much doubted that an explosion could be achieved. Before he fully made up his mind, however, he went back and studied the Bohr-Wheeler paper and some of Bohr's earlier work on the compound nucleus. Upon this reconsideration he reported to the government: "I think one can say that this explosion is almost certain to occur if a chain reaction would be possible." The report nevertheless was most cautiously put and Lord Hankey, a minister in the War Cabinet, remarked after reading it: "I gather we may sleep fairly comfortably in our beds." At about the same time Thomson concluded from his studies and from experiments he was making that a war project for atomic energy would have very little chance of yielding any results. Any war would be over, Thomson thought, before enough material could be obtained for an effective bomb or for power.

The idea of an atom bomb was about to be dismissed as a "wild goose chase" so far as the war was concerned when the visitor to England, Frisch, startled British scientists and the British government with a very different kind of report. Frisch was working with Peierls at the University of Birmingham.

Frisch also had at first been skeptical, but as Margaret Gowing reports in the official history of the British atomic energy effort, *Britain and Atomic Energy 1939–45*, when Frisch and Peierls made a new study of the Bohr-Wheeler paper on U-235, they changed their minds:

> Bohr has put forward strong arguments for the suggestion that the fission observed with slow neutrons is to be ascribed to the rare isotope 235-U and that this isotope has on the whole a much greater fission probability than the common isotope 238-U. Effective methods for the separation of isotopes have been developed recently, of which the method of thermal diffusion is simple enough to permit separation on a fairly large scale.
> This permits, in principle, the use of nearly pure 235-U in such a bomb, a possibility which apparently has not so far been

seriously considered. We have discussed the possibility and come to the conclusion that a moderate amount of 235-U would indeed constitute an extremely efficient explosion.

Frisch and Peierls held, on the basis of their calculations, that every collision with fast or slow neutrons would produce fission in U-235, and they estimated that a sphere of 235-U weighing about one kilogram (2.2 pounds) would be a "suitable size for a bomb." A five-kilogram (11 pounds) bomb would be equivalent to several thousand tons of dynamite, the scientists said. Such a bomb, they continued, would have to be made in two hemispheres, for the instant the uranium came together in a lump of suitable size an explosion would occur.

Frisch and Peierls went into the possibility of producing U-235 and concluded that it could be done by a process of gaseous diffusion. By turning the uranium into a gas, the light isotope, U-235, could be collected near a hot surface. With about 100,000 tubes for the collecting, the two scientists estimated that 100 grams a day of 90 percent pure U-235 might be produced. A plant not much larger than some existing munitions plants would be sufficient, they thought.

"In addition to the destructive effect of the explosion itself," the scientists said in their memorandum, "the whole material of the bomb would be transformed into a highly radioactive state . . . the radiation would be fatal to living beings even a long time after the explosion. . . . Most of the [radiation] will probably be blown into the air and carried away by the wind. If it rained the danger would be even worse because active material would be carried down to the ground. . . . Effective protection is hardly possible. . . . The irradiation is not felt until hours later. . . . It would be very important to have an organization to determine the exact extent of the danger area . . . so that people could be warned against entering it."

In three typed foolscap pages Frisch and Peierls had set forth virtually all the facts about the bomb—size, how it could be detonated, how the U-235 might be produced, and what the effects of an atomic explosion would be on humans and the earth. Years of work and the expenditure of hundreds of millions of dollars later went into determining the same facts and the fine details required to manufacture the new weapon Frisch and Peierls had sketched in their amazingly perceptive and accurate memorandum. Here too were most of the "secrets" the United States was later to guard so zealously. By returning to the theory of fission which Bohr had published they "had drawn a startlingly simple conclusion," said Margaret Gowing.

Word also came that the French had succeeded by fast action in acquiring the whole stock of heavy water manufactured by the Norwegian Hydro-Electric Co., and had brought all the twenty-six cans of it to Paris. The Germans also had been trying to acquire the world's only supply of heavy water. The French physicists wanted it to slow down neutrons for a chain reaction in uranium. Who could doubt that the Germans were after it for the same reason?

At this point British officials and scientists did a complete and abrupt reversal. Fully alerted by the Frisch-Peierls memorandum and by the news of the heavy water race in Norway, they realized that the decisive, the final, weapon had to be considered as a possibility for the war they were waging and that in Hitler's hands it could mean the end. A small but scientifically high powered subcommittee of the Committee for Scientific Survey of Air Warfare was quickly organized. In April, when the Nazis invaded Denmark and Norway, the subcommittee was given independent status within the Ministry of Aircraft Production. A new, innocuous, nonrevealing name was needed. What better than MAUD, a simple, disarming name, a mystery as baffling as the one the committee would tackle, and a recognition of the man to whose guidance they owed it all. And besides it would be a delectable private joke to cheer them through dark hours. MAUD was thus named.

And so Bohr's findings and Bohr's men opened the way in both Great Britain and the United States to the atom bomb, atomic power, and undoubtedly, the future. Bohr's explanation of the compound nucleus, his singling out of U-235 as the part of natural uranium that fissioned to slow neutrons, and his prediction of two new elements, one of them fissionable, were the essentials. One man, a Dane, then watching with heartbreak the subjugation of his own beloved country, had almost single handedly and for the second time, shaped and changed the direction of his time.

Would it have happened without Bohr's insights? Without the men Bohr trained and guided and inspired? Without Bohr? The question is an unanswerable one, hidden forever in the intricate evolution of the past. The edge in every country was against the possibility of a war project in atomic energy until Bohr's findings and Bohr's men weighted it in the other direction. Bohr did not intend this. He sought only the understanding of the truths of nature and to bring the physicists of the world together in a cooperating, interacting whole in the search for truth. But so it came to be.

In Copenhagen the man whose insights were about to launch one of the world's largest undertakings knew almost nothing of it. Denmark was shut off from much of Europe, and the outlook was as black as the blackout imposed on Copenhagen—though the lights of Malmo across the strait in Sweden could always be seen sparkling in the distance.

The British inflicted severe defeats on Nazi shipping in Norwegian waters, but the Norwegians and the British who brought aid were driven under the pounding of the Nazi dive bombers farther and farther to the north until they finally held only a limited area not far below the Arctic circle.

On May 10 the early stalemate in the west came to a frightful end, as the Nazis, without so much as an ultimatum, overran Belgium and Holland, both states whose neutrality they had repeatedly promised to respect. Four days later a hundred-mile long column of Nazi armor massively broke through the French lines and rolled almost unopposed across northern France, behind the French, British, and Belgian armies drawn up in Belgium. On the fifth day, after parachute landings in the rear and the wanton bombing of the heart of Rotterdam, the Dutch surrendered.

By May 20, only ten days after the breakthrough the German tanks reached Abbéville on the sea. They had completely crossed northern France. At that moment the Allied forces in Belgium were cut off and pinned against the sea. On May 28 the king of the

Belgians gave himself up, announcing that the Allied cause was lost.

Then came the "miracle of Dunkirk." While the Royal Air Force drove the Luftwaffe away from the narrow strip into which the Allied armies had been driven, the British navy and the small boats of England, converging from every river and harbor, snatched 338,226 British and French troops from the trap and carried them safely to England. British spirits soared, but Churchill, who had become prime minister as the attack broke, warned the House of Commons "wars are not won by evacuations." The British had lost their armament, and the islands stood almost defenseless before the German hordes drawn up on the French and Belgian channel coasts. The British navy indisputably controlled the seas, but its vulnerability to landbased aircraft had been shown in Norway. In this grim juncture Churchill spoke: "Even though large tracts of Europe and many old and famous states have fallen or may fall into the grip of the Gestapo and all the odious apparatus of Nazi rule, we shall not flag or fail. We shall go on to the end, we shall fight in France, we shall fight in the seas and oceans . . . we shall defend our island whatever the cost may be, we shall fight on the beaches."

The Bohrs listening to the BBC and living in one of the states that had fallen were stirred and heartened to the depths of their beings by these brave, unyielding words. On June 5 the Nazis broke through on the Somme and on June 14 were in Paris, flying the swastika from the Eiffel Tower. In this moment of collapse and degradation Mussolini, after making certain that the French troops could not attack, entered the war.

Disaster piled upon disaster until on June 22, 1940, only one month and twelve days after the Germans had attacked Belgium and Holland, the French capitulated. An armistice was signed, at the revengeful inspiration of Hitler, at Compiègne where in 1918 Germany had surrendered to the Allies.

It was one of the darkest hours in the life of Niels Bohr. His own country lay under the heel of the Nazi and England, which he loved almost as his own land, stood alone, menaced by all the indomitable evil of the Nazis. Western civilization itself was threatened with extinction. Bohr's heart was filled with foreboding, but even at this nadir he did not give up. This was a man of courage and tenacity as well as intellect.

Three courses lay open to Bohr—to resist the Nazis with every possible means at his command, to press on with his own work, and to keep the institute intact as a resource of Denmark and an influence the Nazis could not destroy.

Initially in Denmark the Nazis were stiffly correct, but rapacious reality soon began to reveal itself. The Danish cooperatives and factories were offered high prices for their bountiful produce, but the bills were made payable by the banks of Denmark. In effect the Nazis made the Danes pay for their own products which were then shipped off to Germany. The Nazis also demanded the surrender of Danish shipping; restrictions grew one by one.

Bohr refused all cooperation whatsoever with the Nazis and with those who tolerated them. He served on no committees tinged with collaboration and boycotted scientific meetings sponsored directly or indirectly by the Nazis. On the other hand, Bohr joined in the celebration of the king's seventieth birthday and in the other national historical observances at which the Danes poured forth all their pent-up feelings. Bohr was very likely to be seated in the front row at such Danish celebrations. His resolute, kindly face and tweed-clad figure became something of a symbol, as were the king's daily horseback rides through the capital. The tall, thin monarch rode unattended through the streets of the city. The Germans, unable to understand such simplicity, asked, "Who looks after him?" The answer was, "We all do." It was true. The Danes gave their king, who calmly ignored the Germans, a fierce devotion the Germans knew could never be theirs. A measure of it also went to Bohr, another undiminished, unyielding strength in the darkness that surrounded them.

During the early part of the occupation Bohr threw himself into his work, in itself a defiance of all that the Nazi stood for. A highly important unsettled question related to the fragments produced in fission and their course as they left the bursting nucleus. Danish physicists at the institute, at Bohr's suggestion, began studies of the tracks left by the fragments as they hurtled through a cloud chamber. Their passage left a trail reminiscent of the trails made by jet planes flying in the upper strata of the air.

The tracks, the experimenters found, were surprising. They were quite unlike those left by nuclear particles. They burst across the cloud chamber in big looping curves from which many smaller tracks branched forth. In contrast alpha particles made initially straight tracks that curved only when slowed down toward the end of flight. And the alpha tracks did not branch.

Bohr was soon able to clear up the mystery of the curving tracks. He proposed an explanation: The fragments on their headlong flight sometimes collide violently with the nuclei they encounter in the chamber. Each collision sends both fragment and the nucleus it hits careening off in an altered direction, and a forked branch appears in the tracks. Smaller near-collisions produce smaller deviations in the flightpath, many of them too small to be seen as forks, but effective enough to produce the curving of the tracks.

"Branches due to close nuclear collisions are the rule rather than the exception," Bohr said, "and the scattering and stopping effect of less violent collisions is clearly shown in the irregular gradual bending of the tracks."

Bohr wanted this interesting finding to reach those who were working on fission and all its problems. Communications with England by all regular means had been completely shut down, but the United States was still a neutral and mails could get through, subject only to the delays of war. This note and two others on the same subject which Bohr completed during the devastating summer of 1940 were sent to the *American Physical Review*. The first arrived on July 12 and was published on November 15. Bohr said in a concluding word that the calculations would be presented soon in greater detail in the Communications of the Copenhagen Academy. He was unduly optimistic about the time. The volume that ultimately carried the full paper did not appear until 1948 and it bore the remarkable date 1940–48.

During this summer while all Europe crumbled, Bohr further sorted out the roles of U-235 and U-238 in the fission of uranium.

He had predicted at Princeton that only U-235 would fission at the impact of a slow neutron. The question remained what would happen if either U-235 or U-238 were hit by a fast neutron.

Bohr's studies showed first that in natural uranium fast neutrons would have a negligible effect on the rare U-235. Bohr then analyzed what would happen if a fast neutron hit the abundant U-238. An Italian team had demonstrated experimentally that there was a probability of U-238's fissioning if the common isotope were hit by a neutron with an energy above 10,000,000 volts. Hit by between 1,000,000 and 10,000,000 volts, the chances were that the bombarding neutron would escape and the nucleus would not split. In an above 10,000,000 volt impact, Bohr argued, a neutron might be knocked out of the nucleus and the excitation might continue so high that the residual nucleus then would fission.

Bohr offered a subtle explanation of what would transpire in the nucleus in such a case. The impinging neutron would increase the weight of the nucleus to 239 (238 + 1). However if a neutron were knocked out, the nucleus would again be reduced to 238, but an agitated, excited 238. Great agitation in a nucleus of even mass and charge would very likely produce fission, exactly as it does in U-236 (U-235 plus the 1 of the bombarding neutron). In essence Bohr was showing that the march up the hill and back down again would leave a very excited U-238, and a U-238 very likely to split violently in two.

Ordinary U-238, Bohr pointed out, is not excited. When it absorbs a slow- or intermediate-speed neutron, and its mass is increased to 239 it becomes, like other odd numbered nuclei, very unlikely to fission. The difference would be in the violent reaction produced by the high speed "bullet."

Bohr thus provided the explanation of why U-238 does not fission to slow- or intermediate-speed neutrons, and predicted why and how it could fission if struck by high-speed neutrons. "A study of these phenomena offers a means by enlarging greatly the number of different nuclei in which the fission process may be investigated," Bohr said in concluding the paper he wrote. This paper also was sent to the United States, to the American Physical Review for publication. It was received on August 12 and published on November 15.

Bohr also could combat the Nazis by building up the equipment of the institute. It was a satisfying way to defy them, and to compensate for the isolation their war and occupation had imposed. Bohr and the staff undertook to build a larger and more powerful accelerator. Though the metal they needed was generally unobtainable, Bohr had only to drop a word to the institute's great and good friend, the Carlsberg breweries. The brewery quietly diverted some of the metal it was allotted to the institute.

The long sparks that would spring from the new generator with its more than 2,000,000 volt energies would have required a larger room or the raising of the ceiling and the moving of the walls of the room they had been using. As such changes would have been impossible, the institute placed the generator in a tank with air held below seven atmospheres in pressure. The length of the spark was reduced and the new generator fitted into the space available for it. At first, however, there seemed no way to build the compressed air container. When the Helsingor shipyards learned of the institute's need, they supplied the metal and made the tank despite all the Nazi restrictions on coal and metals. An isotope separator also was added during the grim years of the Nazi occupation.

"The isolation and the nervous strain were hard to bear," said Rozental, "but from the standpoint of work the period was not a bad one." That it was not, constituted a triumph of the spirit and brain over the Nazi repressions and depredations.

But the Nazi night grew ever blacker. All through the latter part of this summer of defeat, Hitler was assembling barges and boats or an invasion of England. In August, 1940, Goering and the Luftwaffe began the air attack planned to drive the RAF from the skies and to "bring Britain to her knees," weakened and ready for conquest. The attack began on British shipping and the south ports. Despite the grave damage inflicted, the RAF was not lured out, and came back day after day to pound the German invasion equipment drawn up in the French and Belgian channel ports.

The Luftwaffe shifted to an attack on the RAF landing fields.

Some of the German planes in these attacks came from bases established in Denmark; the Danes unhappily watched them take off and rejoiced when many of them failed to return. Again the damage was great and the rain of bombs terrible to bear, but the British shot down more planes than they lost, and far from surrendering to panic as Hitler had confidently predicted, only grew more resolute, and fought and worked the harder.

On August 23, despite the damage they had suffered, the British dropped their first bombs on Berlin, and continued the bombing night after night. In retaliation Hitler abandoned the crucial and crippling attacks on the RAF fields and ordered an overwhelming assault on London. On September 7, some 625 bombers, protected by 648 fighters came up the Thames in a first wave. Wave after wave of about 200 planes each followed until the dawn. The great city shook and the flames leaped so high that it scarcely seemed that London could survive, but survive it did. Emboldened, Goering undertook a daylight assault on September 15. This time the Luftwaffe losses were so heavy that the day attacks were not renewed. This was the crisis. Though the brutal bombing of the British Isles continued for another 57 consecutive nights, Hitler had effectively been stopped. The invasion, except as a pretense, was finally called off in September and Britain stood magnificent, resolute, and secure in its island. The Battle of Britain had been won in Britain's "finest hour," as Churchill unforgettably said.

The lesson, the fight against nearly insuperable odds and adversity, was not lost on the Danes. Hitler's power on the continent still was unbroken, but the Danes resolved that if the Third Reich should last for a thousand years, as Hitler boasted, the spirit of Denmark would endure for even longer. The Danish Society, which Bohr had helped to found, planned a definitive book on Danish culture. It would preserve the history, the tradition, the very soul of Denmark as long as the written word should survive. Bohr was asked to write the introduction.

What did it mean to preserve a nation's soul? What makes a nation—or a man? In times of trouble, said Wheeler, the answer is clear. "Mankind is formless clay; his spirit derives from his heroes, his traditions from well accepted thoughts and ways of life, from storied legends and firmly held standards of value." These were the answers to which Bohr came as he walked the peristyle at Carlsberg and debated the questions with Rozental who was assisting him in the extraordinary work. Bohr drew upon all the resources and thought of a lifetime:

> If we ask ourselves in which sense we can speak of an especially Danish culture then the answer will depend entirely upon the point of view on which it is based.
> On the one hand, common human features will become the more apparent, the deeper we delve into the foundations of our existence. . . . On the other hand, the differences from other communities will be felt all the stronger the more we strive to maintain an image of all the facets of our existence.
> As any knowledge of how human life can develop under differing circumstances must be the basis of any assessment of a community's cultural stage, an investigation of our place among other communities must be a natural starting point for any attempt to clarify what is peculiar in us.
> If in spite of the limits of my knowledge, I have carried out the editor's request to write a few words as an introduction to this great work, it is because by working together here in Denmark and abroad with scientists of many different nations I have had many opportunities of concerning myself with new relationships to the outside world, and of meditating on the traditions which give the Danish attitude to life its special characteristics.

As Bohr examined the genius of Danish culture, he unconsciously explained his own genius for drawing the science of the world into a coordinate and strengthened, rather than a competitive and diminishing, whole, and of making it an interplay, enhanced and enriched by the contributing viewpoints.

Bohr continued: "The reminder of how much we had to learn has been too emphatic to let us be misled by the glamor of the past . . . or into forgetting our humble place among the nations. On the other hand there arose among the Scandinavian countries a feeling of cosmopolitanism, a feeling which was more harmonious in its expression than in many larger countries, where the temptation to consider their own cultures as independent organic unities was so much stronger."

Bohr quoted Hans Christian Andersen whose words had become Denmark itself: "In Denmark I was born and there my home is . . . from there my world begins." Bohr dwelt on the word "from," "*from* there my world begins," from the soft colored land jutting out into the northern seas, from there the Danish culture began; it did not end. It did not draw in to itself, contained and limited. On the contrary from this point it looked out to "participation in that intellectual world shared by all humanity."

Although what we understand by Danish culture offers so many aspects which are indissolubly involved with one another, the view of fellowship between nations, which our history has developed, can be regarded as culturally our most significant feature. We can feel proud of the way we have utilized our circumstances for our own development and for . . . working together for the advance of human culture.

What fate has in store for us, as for others, is hidden from our gaze, but however far reaching may be the consequences for all spheres of human life of this crisis in which the world finds itself at present, we have a right to hope that our nation in the future too can with honor serve the cause of mankind, provided that we retain our freedom to develop this attitude which has its roots so deep within us.

Bohr spent two months writing the nine-page article. The society knew it was publishing the book at its peril, for the very concept and every word were denials of the tenets of Nazism, but there was no question about doing it. The editor clamored for Bohr's manuscript. When it was finally finished Bohr had to have seven proofs before he could relinquish it. The soul of a country is not lightly summoned or put into final words.

Danish Culture when it appeared in 1941 was not alone a splendid record of all that Denmark most distinctively was, but a source of sustenance and hope and pride at a time when the three were seriously needed.

The book helped to bring a turn in the spirit of the Danes. The Danes now adopted Churchill's famous V for Victory sign. It was used to greet friends and upon every occasion. Small, crudely printed news sheets (*Free Denmark* was one title) began to come from small presses hidden away in cellars and attics. They were the first of more than five-hundred clandestinely printed sheets. In the dark, and in the full light of day when the Nazis were not looking, the papers were slipped under doormats, or left in birdhouses, or rolled into the paper that wrapped the fish.

As 1941 dragged into 1942 sabotage increased. Bombs began to go off in plants supplying the Nazis. German trains carrying supplies to the south or troops to the north began to be wrecked by dynamite planted on the tracks. A strong underground resistance movement was growing throughout the country.

From the beginning Bohr and many of Denmark's men of learning aligned themselves with the resistance movement. Bohr always knew what was going on and could quickly reach the underground forces in times of emergency.

As an open non-collaborator and a suspected advisor of the underground, Bohr was kept under surveillance by the Gestapo. Ironically the Gestapo occupied the building immediately to the north of the institute. It had been built by the Freemasons. Tall Doric columns formed the entrance, but the flanking stucco walls were unbroken by windows, and the building had the appearance of a fortress. Wires that tapped the telephones of the institute ran directly into the Gestapo building.

The Gestapo made a number of attempts to trap Bohr into an act that would justify his arrest. One day a young man slipped into the institute. He appealed to Bohr for help. He told a plausible tale; he had been fighting the Germans and if he were captured he would be tortured and killed. Even before he arrived, however, Bohr had been warned that a trap was to be laid and skillfully sidestepped it. When help was really needed, it was given. Even after the occupation Bohr continued to assist refugees. He made the arrangements to slip uncounted numbers of them across the strait into Sweden.

Harald Bohr was as much under suspicion as Niels. He had early incurred the enmity of the Nazi by disputing some of the claims of their mathematicians. The Nazis, however, did not quite dare to take direct action against the Bohrs. An attempt still was being made to keep Denmark as quiet as possible. Danish production was important to the Germans and Hitler could ill afford the additional troops who would be needed if the Danish populace was further aroused, as it undoubtedly would have been by the arrest of the country's most famous scientists and its most honored citizens. A Danish Nazi party which had been formed before the occupation emerged briefly when the Germans marched in, but it was so despised and so miserably led that the Germans were unable to rely upon it for civil control.

In the summer of 1941 Hitler indeed stood in greater need of troops than anyone realized. When the Nazis were held at bay by Great Britain, Hitler turned to the east and south. He swept through the Balkans and rescued Mussolini in his badly bogged down attacks on Greece and Albania. Rommel was sent to North Africa and with daring, bold tactics drove the British back across the northern desert to Egypt.

By June the megalomaniac dictator was ready for bigger prey. On June 22, 1941, Hitler turned with the usual surprise on his recent ally and sometime friend Stalin. Along hundreds of miles the German guns blasted an end to the misalliance. By autumn of 1941 the German panzers had pushed 400 miles into Russia, and Moscow lay only 200 miles to the east along the old highroad Napoleon had taken to the capital and ultimate defeat. The Nazis also were approaching Leningrad.

In this tumultuous autumn Bohr, in Copenhagen, had a surprise visitor. Heisenberg came to see him: Heisenberg the brilliant young German Bohr had plucked from a lecture hall at Göttingen and brought to Copenhagen for three years of work. It was this work and the insights growing out of it that had won Heisenberg the Nobel Prize. Few of the institute fellows had been closer to Bohr.

Now all was changed. Heisenberg had come to Copenhagen ostensibly to lecture at a scientific meeting. It was a meeting that the institute was boycotting as collaborationist. Even before this

Bohr had been hearing disturbing reports from Germany. It was said that Heisenberg had defended the German invasion of Poland. This might have been overlooked as a mistaken report, but Heisenberg had indubitably taken over the direction of the Kaiser Wilhelm Institute of Physics.

Following Lise Meitner's necessary flight, the prestigious directorship had gone to the Dutch physicist, Peter Debye, who had been working in Germany since 1909. The Nazis, however, had insisted that he either become a German citizen or publish a book extolling National Socialism. Debye rejected the demands with scorn. He was soon afterward invited to lecture in the United States, went, and did not return to Germany. The post then was given to Heisenberg. His acceptance under the circumstances signified at least a willingness to work with the Hitler government. Non-Nazi physicists were outraged. All of this was known to Bohr when Heisenberg appeared at his house at Carlsberg one late autumn night.

Both knew that Bohr was under surveillance, and decided to a walk through the nearly dark streets around the House of Honor and the big buildings and parks of the brewery. But Heisenberg sensed that the old ease and confidence were gone. The German scientist too was on edge and fearful that if reports of his talk with Bohr should get back to Germany his life would be in danger.

"I tried to conduct the talk in such a way as to preclude putting my life into immediate danger," Heisenberg said in a letter written some twenty years later in response to a request by Robert Jungk for his first-hand report on what was said in the famous conversation. (Jungk was gathering material for his book *Brighter Than a Thousand Suns*.) "The talk probably started with my question as to whether or not it was right for physicists to devote themselves in wartime to the uranium problem—as there was the possibility that progress in this sphere could lead to grave consequences in the techniques of war," said Heisenberg.

> Bohr understood the meaning of this question immediately, as I realized from his slightly frightened reaction. He replied as far as I can remember with a counter question: "Do you really

think that uranium fission could be utilized for the construction of weapons?"

I may have replied "I know that this is in principle possible, but it would require a terrific technical effort, which one can only hope cannot be realized in this war." Bohr was shocked by my reply, obviously assuming that I had intended to convey to him that Germany had made great progress in the direction of manufacturing atomic weapons.

Although I tried subsequently to correct the false impression, I probably did not succeed in winning Bohr's complete trust, especially as I only dared to speak guardedly (which was definitely a mistake on my part), being afraid that some phrase or other could later be held against me. I was very unhappy about the result of this conversation.

Heisenberg maintained that the real reason for his visit to Copenhagen was to enlist Bohr and through him other physicists in an effort to convince the governments of the world that atomic bombs could not be produced for use in the war. This would block attempts at their production.

On September 26, 1939, only a little more than two weeks after war engulfed Europe, nine German physicists had formed a "Uranium Society." Four weeks later Heisenberg and Von Weizsacker, another former associate of the Copenhagen institute, joined the committee. Experiments were planned, but Jungk maintains that the scientists soon found that all the Czechoslovakian uranium ore had been preempted by an army department which wanted to use it as an alloy in armor-piercing shells.

Heisenberg and Von Weizsacker, nevertheless, proceeded with studies, which indicated to them that power might be obtained from U-238 and that new substances, Elements 93 or 94, might yield an explosive. But the scientists later insisted they could see no practical method of producing an atom bomb with the resources available to Germany.

Another German scientist, who had fled to Russia with the rise of the Nazis, also was involved in the uranium studies. Fritz Houtermans had fallen into the hands of the Russian secret police and had undergone days of torture before he was released, following the Hitler-Stalin pact. The Russians, however, returned him to Germany. Houtermans succeeded in finding work with a German scientist and by July, 1941, had convinced himself that an atom bomb could be produced if enough of the new Element 94 could be obtained. He did not try to publish his ideas, but he did confide them to Heisenberg and Von Weizsacker. The three reasoned, Heisenberg said, that if Germany could not produce a bomb and the United States and Great Britain were made aware of this possibility, the other two countries might halt any efforts to produce one on their own. Production of the extraordinary explosive thus could be averted. It was to convey this to Bohr, Heisenberg said, that he went to Copenhagen.

As Heisenberg sensed, Bohr was agitated by their conversation as they walked the dark streets. Bohr immediately told members of his family and Rozental that he was convinced that Heisenberg was trying to find out what he might know about fission. He also had the impression that the Germans were working very hard on the uranium problem, and that Heisenberg thought it might decide the outcome of the war, if the war were prolonged. Nothing said later in the least changed Bohr's mind on this score.

The two accounts agree on the point that Bohr was extremely cautious and revealed nothing to Heisenberg. And Bohr, like Heisenberg, was most unhappy about the conversation.

Bohr was more convinced than ever that the Germans were pushing uranium research. At least one scientific paper had been published in Germany in 1939 showing that the Germans understood the scientific base of fission. The Danes also knew that the Germans were making every effort to build up production of heavy water in Norway, and felt positive that it was to be used in slowing down neutrons in uranium experiments. "There was no certainty that Germany could not produce a bomb," said Rozental. "No one could guess what Hitler would do."

At the end of the war the German records on the whole uranium project were captured and the German atomic physicists, including Heisenberg, Von Weizsacker, and Hahn, the discoverer of fission, were taken to England where they were closely questioned about what was done in Germany. In England when the United States dropped the atomic bomb at Hiroshima, they were dumbfounded at the news and at first believed that it was a bluff.

In the first shock of Hiroshima, the Germans issued a statement from the English manor house, "Farm Hall," where they were being held in custody: "At the beginning of the war a group of research workers was formed with instructions to investigate the practical application of these energies. Towards the end of 1941, the preliminary scientific work had shown that it would be possible to use the nuclear energies for the production of heat and thereby to drive machinery. On the other hand it did not appear feasible at the time to produce a bomb with the technical facilities available in Germany. Therefore the subsequent work was concentrated on the problems of the engine for which, apart from uranium, heavy water is necessary."

The Germans' own words and their private conversations show, said General Leslie R. Groves in his book *Now It Can Be Told*, that the Germans had not thought of using the bomb designs developed in the United States. "They thought," said Groves, "that they would have to drop a whole reactor, and to achieve a reasonable weight they would need an enormous amount of U-235 (two tons)."

Heisenberg, in a private conversation at Farm Hall, summed up the German dilemma. The scientists, he said, did not have the courage to recommend to Hitler that he channel a large share of German resources into the development of a fission weapon. If they had been wrong and it had not worked out, the retribution would have been swift and certain. They did not take the chance.

At last, in the final months of 1941, the tide of war began to change. The Japanese struck at Pearl Harbor, and the United States, with its vast resources entered the war, both in the East and the West. Hitler had made the latter step inevitable by declaring war on the United States in support of the Japanese. Through 1942 and on into 1943, disaster, defeat, and rout began to come to the Nazi forces. They were turned back at Stalingrad and El Alamein.

The Nazi setbacks, combined with increasing repression in

Denmark, further stirred the Danish forces of resistance. All through 1942 the movement spread and moved into a new phase. Close communications were established with England.

On moonless nights RAF planes swooped low over remote fields to drop five-foot-long metal canisters filled with arms and materials for sabotage. The rumble of explosions could increasingly be heard in Copenhagen, and German rail service was constantly disrupted by the blowing up of trains and tracks. Some of the main Copenhagen tracks ran alongside the House of Honor. Although the tracks were well screened by plantings, the Bohrs could see wrecked equipment being hauled away, and they watched emergency crews being rushed into action to reestablish service. It was a heartening sight.

15 /

Escape!

PETER RECEIVED A "most important" secret message from "Jarlen." It told the Danish underground worker that "Justitsraaden," another member of the underground, would soon receive a bunch of keys containing a very important message from the British Government to Professor Niels Bohr. Peter was asked to see that Professor Bohr received the keys and to explain to him how to find the message.

The "most important message" contained a diagram. It showed that a tiny hole had been drilled in the handles of the two keys. "Very, very small" duplicate microfilm messages had been inserted in the two holes and the holes completely sealed. When all of this preparation was finished, the two-inch-long keys looked as old and rust-flecked as any that had ever opened antique bric-a-brac cabinets. Peter was instructed to tell Professor Bohr that the message could be found by gently filing the keys at the points indicated in the diagram. When the holes appeared, the professor was to "syringe or float out" the letter onto a microslide. The message added that the microfilm should be handled very delicately.

Soon afterward in the early winter of 1943 Captain V. Gyth of the Danish army came to call at Carlsberg. After tea he and Bohr stepped out into the House of Honor's Winter Garden, a glass-roofed, glass-walled room filled with a profusion of tropical plants. The glossy green leaves and bright flowers stood in captivating contrast to the swirling snow and early dusk of the Copenhagen winter. Besides, if microphones were hidden in the house, the greenhouse probably was a safe place.

The captain, or "Peter," not only told Bohr about the message but offered to dig it out of the keys and make a full-size copy that Bohr could read. Bohr gratefully accepted. Thus the bits of microfilm, each no longer than the head of a common pin, were removed and typed in the same building occupied by the German command. The emptied keys were then carefully buried in the garden at Carlsberg.

Bohr and Margrethe, feeling like the principals in a spy melodrama, as indeed they were, read a warm, personal, and yet undoubtedly official letter from James Chadwick, the British Nobel Prize winner who discovered the neutron: "I have heard in a roundabout way that you have considered coming to this country if the opportunity should offer. I need not tell you how delighted I myself would be to see you again. There is no other scientist in the world who would be looked upon with more favour both by the university people and the general public."

Chadwick assured Bohr that he would be able to work on scientific problems in which he was highly interested. Not a word was written even in this highly secret letter about uranium, but Chadwick said that he had in mind a particular project in which Bohr's assistance would be of great help.

"But I do not want to influence your decision, for only you can weigh all the various circumstances, and I have the most complete faith in your decision whatever it may be. All I want to do is to assure you that if you decide to come you will have a very warm welcome and an opportunity of service to the common cause."

When British Intelligence had received word from Denmark that Bohr was in danger of being deported to Germany, the report was taken to Sir John Anderson, the Lord President of the Council. Anderson knew that Bohr's services would be invaluable to the British. On the other hand every pressure would be put on Bohr to aid the Germans. Niels Bohr, Anderson ordered, must "be got out of Denmark at once." The Chadwick letter opened the way. The captain in his polite conversation told Bohr that all the necessary arrangements could be made to help him escape, if that were his decision.

The Bohrs again faced a rending decision. Had the time come to go? Was Bohr needed in England? Conditions were becoming increasingly grim and strained in Denmark. Would Bohr soon face arrest? He did not delude himself about the risk; he knew it was real, and probably imminent. But there were always the overriding considerations of loyalty to one's country in a time of pressure and peril, leaving those who could not escape, and possibly exposing his associates and family to Nazi reprisals. Bohr again answered that he could not go as long as it was possible for him to stay. How long that would be he did not guess.

Bohr responded with gratitude that his friends had not forgotten him. He ardently wanted to contribute to the fight for freedom and human dignity. He wrote with deep feeling:

> I feel it my duty in our desperate situation to help in resisting the threats to our free institutions and in protecting those scientists in exile who have sought refuge here. But neither such duties nor even the dangers of retaliation against my colleagues and family would have sufficient weight to hold me here, if I felt that I could be of real help in any other way, though this is hardly probable.
> I feel convinced, in spite of any future prospects, that any immediate use of the latest marvelous discoveries of atomic physics is impractical. However, there may, and perhaps in the near future, come a moment when things look different and where I, if not in other ways, might be able modestly to assist in the restoration of international collaboration in human progress. At that moment I shall gladly make an effort to join my friends.

Bohr had clearly understood Chadwick's guarded reference to the "problem" with which he could help. If it were a matter of the life and death of civilization, as it would be if Hitler could develop a nuclear weapon, Bohr without question would put aside all other considerations and go. Even in 1943, however, Bohr did not believe that the materials for an atomic weapon could be produced. He knew nothing of Frisch's memorandum or of the pace of atomic development in the United States. The channels to Denmark had all but closed, as the elaborate cloak and dagger expedient of the key message dramatically demonstrated.

Bohr sent his reply by the same concealed route through which the Chadwick letter had come. Then he placed Chadwick's letter and his own reply in a small metal cylinder and buried it in a carefully remembered spot in the gardens at the rear of the House of Honor.

At this time the Russians at Stalingrad were visiting defeat, total and retributive, on the invading German forces. As the news of the destruction of the German armies came through to Copenhagen, the forces of resistance took new heart. Sabotage increased and took on a new effectiveness.

The Germans constantly demanded that the government halt the bombing and wrecking of any factory or supply that served the Germans. Members of the underground were hunted down with every device of electronics and brutality, and when even this did not yield the saboteurs, hostages were seized and shot in reprisal. The Danes' resolve was only hardened, and they began to call strikes. In Copenhagen, in Odense, the home of Hans Christian Andersen, and in other cities whole populations came to a standstill. Not a factory moved, not a piece of bacon was shipped. The Nazis began to shoot, and civilians were gunned down and beaten in the streets. More than five hundred Danes were arrested following one wave of strikes.

Then crisis came. On August 28, 1943, the Germans handed the still-existant government an ultimatum demanding the death penalty for saboteurs and other measures so stringent and odious that the government knew that it did not dare to accept them. It rejected the demands, and at last resigned.

The Germans struck back the next day, the 29th, with a declaration of martial law. At dawn small Danish garrisons were attacked, and the Germans attempted to seize the Danish navy. This time the Danes had learned. Orders had been given the day before, and even as the Germans approached the docks, blasts shook the ships. They were scuttled in their berths.

Not knowing that the all but inevitable final break would come at this moment, the Bohrs had gone to Tisvilde for a few days of respite. In the peace of the country and walking in the forest they could momentarily forget the constant strain and uncertainty, or brutal certainty, of life in an occupied country. Then on a sunny Sunday morning when only the chatter of the birds on the terrace and in the forest broke the quiet, the radio suddenly blared forth the news—the Danish government had resigned. The Bohrs knew that it would portend the end, the end of even the pretense that Denmark could live under the Nazis. What new repressions would be ordered they did not even want to think. Grimly Bohr jumped on his bicycle and set off for Copenhagen. It was nearly a thirty-mile trip through the forest, a small cathedral town, the countryside, and the city itself. For once Bohr, eager to get to the institute, had no eyes for the country rolling down to the sea. He did not know what retaliatory measures might be taken, and he pedaled hard.

Mrs. Bohr remained in the country until the next morning. When she went to the station to take the train into the city, the stationmaster told her all service had been discontinued, and advised her to take any means of transportation into the city if she had to be there. Mrs. Bohr found a driver and made it in to Copenhagen just before the roads were barricaded. As she drove up to the institute Bohr was standing out in front, anxiously waiting for her. Hundreds of hostages were being seized, including people the Bohrs knew well, and no one was safe. But still no moves were made against Bohr or members of his family. How long would they escape, they asked again? And again they did not know.

At the same time, in the moment of crisis and showdown, the Bohrs and Denmark also felt a sense of relief, and almost of jubilation. Denmark at last had declared herself. The sham of independent government was gone, and in actuality the people of Denmark were now at war with the Nazis. More grievous times certainly lay ahead, but for all of their forebodings they rejoiced.

But time was running out; the signs were unmistakable. Arrests increased; professors, workers, shopkeepers. "The phone rang constantly," Mrs. Bohr recalled. "Friends and relatives were calling and their anxious first question always was 'Are you there?' "

The British, who well knew the dangers, sent a second message to Bohr, this time by word of mouth—"We still are waiting for you."

Within two weeks Bohr heard from two of his highly reliable sources that the Germans were planning to intern all "undesirable aliens." This meant all Jewish refugees who had found haven in Denmark. Bohr quickly called Rozental into his office and told him that the time had come for him to try to get across to Sweden.

"I was put in touch with a group who would help me with the 'journey' " said Rozental, who had been born a Pole. "A few days were taken up in arranging things and then I went to Carlsberg to say goodbye. On the table in the little room behind the Pompeian court (the familiar workroom with its big table and blackboard) lay some money Bohr had taken out of the bank in case I might need it. While he kept walking around the table, he made a little speech. The war would not last much longer (Mussolini had fallen in July). Perhaps it would be no more than another six months. It would not be long until we would see one another again."

Bohr later acknowledged that he was talking to keep up Rozental's courage. Far from expecting that they would soon meet again, Bohr believed that he would be arrested and imprisoned in Germany unless he received word in time to escape. However, he did not want to add to Rozental's worries.

Many years later at the Nuremberg trials it was revealed that the Nazis had intended to arrest Bohr on the day martial law was declared. There was, however, a dispute about it, and the decision was made to put off his arrest and deportation to Germany until the Nazis began their roundup of the Jews. In the general excitement, they thought the arrest of Bohr would attract less attention and the furor would be less.

As Bohr said goodbye to his old friend and colleague he handed Rozental the final copy of the paper on the passage of charged particles, and asked him to take it to Sweden. He wanted to have it in a safe place until it could be sent to some other country for publication.

A few days later Rozental phoned in a prearranged code to tell the Bohrs that he would leave the next day. Underground workers obtained a rowboat from one of the city's lakes and carted it to a beach near Copenhagen. Rozental and two other refugees, a woman and an escaping French soldier, waited on the beach until dark. In the hurried confusion of getting into the boat, the attaché case in which Rozental had placed Bohr's paper somehow was left behind.

Rozental did not miss it at first; he had more urgent problems to face. With the aid of an underground worker the refugees expected to row across the Kattegat in four hours, but they were not very far out when a storm broke. There could be no turning back. The little park rowboat was tossed about on the high choppy waves and the water poured in over the gunwales. They nearly capsized. The "journey" that should have taken four hours took nine. Finally they landed safely, though in a state of exhaustion, at Landskrona.

The Swedish police were waiting. The refugees were helped ashore and assured of protection. It was then, however, that Rozental missed the attaché case with the paper. Arrangements were immediately made through the underground to have a search made for it on the beach near Copenhagen. Nothing more could be done, and Rozental took the train to Stockholm. In the arched old station, Oskar Klein, one of the early fellows of the institute and a leading Swedish physicist, was waiting. Never had the concerned, reassuring face of a friend or his impulsive embrace been more welcome.

Back in Copenhagen, the personal crisis the Bohrs had known must come came swiftly. On September 28 a diplomat dropped by for tea at Carlsberg. He remarked with pointed emphasis that many people were leaving Denmark, "even professors." "He did not say so directly," said Mrs. Bohr, "but we knew that the time had come when we must go."

All papers that should not fall into German hands had already been destroyed. Now Bohr acted to protect the gold Nobel medals Franck and Van Laue had left in his safe keeping. Both medals were dissolved in acid and put in a seemingly innocuous bottle on a shelf filled with unimportant bottles. The medals could be recast after the war. Bohr's own Nobel medal had earlier been donated to a Finnish war aid drive.[1]

The next morning Margrethe's brother-in-law hurried in. He brought word from an unimpeachable source, an anti-Nazi in the German diplomatic service, that orders had been issued in Berlin for the arrest of Niels and Harald Bohr and their removal to Germany. Absolutely no time could be lost.

"We had to get away the same day," Mrs. Bohr said. "And the boys would have to follow later. But many were helping. Friends arranged for a boat, and we were told that we could take one small bag."

In the crushing finality and heartbreak of having to flee their own country, of leaving behind family, friends, their home, and all that they had spent a lifetime building, the Bohrs and those helping them had to plan a secret departure. They had to get away without arousing the suspicions of the Gestapo agents who were keeping a close watch on them. Escape was no peaceful journey, but a hazardous risking of one's life.

A tight curfew had been clamped on Copenhagen. Anyone moving in the streets and disobeying the regulations was shot on sight. The Bohrs had to reach the beach from which they would depart before the curfew hour.

Following instructions, they walked late in the afternoon along a crowded Copenhagen street. As they reached a specified corner a scientist whom Bohr knew passed them and nodded in greeting. This was the signal. All was well. The Bohrs made their way to one of the outlying fields where Copenhageners maintain small summer gardens. Most of the gardeners build tiny shacks, just big enough to hold their tools and maybe a chair or two, and the Bohrs took refuge in one of them. Then they had to wait.

At last darkness came. At a set time they walked the short

[1] Later another donation was made in place of Bohr's, and the gold symbol of Bohr's Nobel Prize was returned to Denmark for permanent exhibition in a Danish museum.

distance to the nearby beach. But at this hour the moon was rising across the strait. It was far from a welcome sight, for its light turned the water to silver and cast any boat into sharp silhouette. Suddenly the stillness was broken by the sound of a motor. A small boat putt-putted in to the beach; the Bohrs were quickly motioned aboard. The boat moved out over the low surf toward the lights of Sweden sparkling in the distance, but the little boat was only to take them to a larger one.

A boat appeared and voices carried out across the water. Was it the boat they were expecting, or were those German voices? "Get down, keep down," came the sibilantly whispered orders. But there had been no slipup; it was the awaited fishing boat. In an hour and a half the Bohrs landed safely at Limhamn, a small harbor near Malmo. As they looked back to Denmark, their land, it lay low and massive against the still faintly glowing sky. No light burned there.

Mrs. Bohr remained in Scania to await the arrival of their sons and other members of the family. Bohr went immediately on to Stockholm, for he had very urgent business there.

When Bohr learned of the plans to arrest him, he also was told that the Nazis were about to begin their long feared roundup of the Danish Jews. Throughout the occupation the small Jewish population of Denmark had been relatively unmolested. There were only about 7,000 Jews in Denmark and they were well integrated into the life of the community.

The Danish underground organizations had just united into a "Freedom Council" and, as soon as word came of the impending move to arrest the Jews and send them off to German concentration camps, the underground made plans to save their fellow citizens from the unthinkable fate that would await them in Germany. The alert was sounded throughout Denmark. Plans were laid for a mass exodus by boat to Sweden. The dangerous rowboats were to be barred, and a fleet of small boats capable of navigating the often stormy waters of the Kattegat organized. The way also had to be prepared in Sweden for the reception of the refugees. This was Bohr's special task.

Through a phone call made by the chancellor of Lund University, an appointment was made for Bohr with the Secretary of State for Foreign Affairs, Boheman, on the day of Bohr's arrival in Stockholm, September 30. Two Swedish army officers attached to the intelligence service called for him to take him to the foreign office. Until all of Bohr's family could escape and the Nazis knew he was gone, it was considered necessary to conceal his presence in Sweden. In the best cloak and dagger tradition, the two officers drove Bohr up to a building in one car, led him through a maze of rooms, and put him into still another car to take him to his appointment. Stockholm swarmed with Nazi agents; the officers wanted to lose any followers.

Bohr urged the Swedish government to take all possible steps to dissuade the Nazis from making their proposed attack on the Jews of Denmark. Boheman, who later became ambassador to the United States, told Bohr that the Swedish ambassador to Berlin had already tried to intervene. When the Swedes learned of the impending move they informed the Nazis such a step would be looked upon with great concern by the people of Sweden. The German foreign minister told the Swedish ambassador that the report was only a rumor spread by the Gestapo to scare the rebellious Danes. Both Bohr and Boheman gave a bitter little laugh at this. They knew exactly how much reliance could be placed upon such a statement by the Nazis.

On October 2, the "rumor" became reality. The Nazis started their arrests of the Jews of Denmark. Scores were seized in their homes and thrown on board a ship anchored in the Copenhagen harbor. It was to take them to a German concentration camp.

Bohr again appealed to Boheman and the Swedish Foreign Minister, Gunther. In this crisis, Bohr proposed that the Swedish government ask the Nazis to reroute the ships to Sweden and that Sweden offer to intern the refugees—a formality—until the end of the war. Gunther said that Sweden already had notified the Germans that she stood ready to receive all refugees. Gunther, however, thought that the proposal for the rerouting of the ships was a good one. The Germans had their hands full trying to rescue Mussolini and combat both the Allied advance in Italy and the Russian attack in the east. They might be more willing to listen than in the past.

But Bohr's proposal involved state policy, Gunther said, and he would want to take it to the king. Again there was no time to be lost; the ships were partly loaded with their human victims. Bohr immediately called on Princess Ingeborg, a member of the Swedish royal family, and the sister of the Danish king, and expressed a desire to be received by the king. The audience was granted that same afternoon. Bohr and the Danish ambassador, still shepherded by the secret agents, hurried to the Royal Palace to be greeted by King Gustaf V.

The king pointed out that Sweden had made a somewhat similar approach, offered asylum, to the Germans when the Nazis began deporting the Jews of Norway. The humanitarian effort was abruptly rejected as unjustifiable interference in the affairs of Germany, and the Swedes were told that they were showing a regrettable lack of understanding of the principles of National Socialism.

Bohr again pressed the point that the Allied victories made the situation very different in the fall of 1943 than it had been in the summer of 1940. The king promised that he would immediately discuss the proposal with his government, but he emphasized the difficulty of succeeding in such an unusual undertaking.

Bohr knew that he was fighting for the life of many people; in this desperation he went even further. If all else should fail, he begged the king to try a personal appeal to Hitler. He also urged Sweden to make a public announcement of her willingness to accept responsibility for the Jewish refugees. A few hours later Bohr received a message from the palace that an announcement would be broadcast that night. Bohr, Rozental, and Klein listened at Klein's home to the formal announcement that Sweden would offer asylum, and invite the Nazis to reroute their prison ships to Swedish ports. A great nation had acted again for humanity.

The Nazis, however, turned down the pleas almost as arrogantly as they had the earlier ones. Only one recourse was left; the rescue would have to be made by boat. Nevertheless the formal

Swedish offer greatly facilitated the full-scale rescue operation the Danish underground was already putting into effect. Swedish ships went out to the limit of the territorial waters to embark refugees from the Danish boats. Every night a fleet of small boats shuttled back and forth, bent on their dangerous mission.

Within a few weeks almost 6,000 reached safety and sanctuary in Sweden. About 300 refugees were captured or lost as the Nazis tried to halt the great rescue. Altogether 472 Jews, including some who were unable to attempt the hazardous escape, were sent to the dread Nazi concentration camps. Forty-three died there. The toll would have been vastly greater if the people of one enslaved state and one free state had not taken grave risks to halt further Nazi inhumanity.

Bohr's absence from Copenhagen was rapidly discovered by the Gestapo, through it was not announced publicly for several months. As the Swedes had feared, it was soon clear that Bohr was again under Nazi surveillance. Bohr's friends, Danish and Swedish, feared that an attempt would be made on his life. An armed guard was posted at Klein's home where Bohr first stayed and later at the home of the Danish diplomatic official to which Bohr moved. It was also suggested that Bohr used a "cover name" to conceal his movements. Bohr cooperated fully, but Rozental said that when he was called to the phone he kept forgetting and would automatically answer, "Bohr speaking."

Aage Bohr narrowly escaped capture when he made the run from Denmark to Sweden, but he arrived safely as did the other three sons and Harald. Erik's baby daughter was carried to Sweden in the market basket of the wife of a Swedish embassy official. When all the family were together they went to Stockholm where they stayed for the remainder of the war.

Rozental was invited to work with Klein at the Institute for Mathematical Physics at the Technical University of Stockholm. Bohr and Rozental were soon reunited in Stockholm. They had no sooner greeted one another with heartfelt relief, and even managed a laugh about their reunion only a week after what both secretly thought might be a final parting, than Bohr asked about his paper on the particles. Rozental had to tell him the story of its loss. Underground members who had gone to the beach to obliterate all traces of the escape, had found it and sent it to Sweden by another group of refugees. There it disappeared again. Every effort to trace it failed. In later years the story spread that the lost manuscript contained all the secrets of the atom bomb—quite an enlargement for a treatise on the passing of charged particles through other substances. A copy was later obtained from Copenhagen.

The story also spread and flourished that when Bohr left Copenhagen he intended to take with him the institute's supply of heavy water. According to the most popular account, Bohr had concealed the heavy water in an ordinary beer bottle. The story continued that Bohr in the haste of his escape picked up the wrong bottle and thus carried an ordinary bottle of beer all the way to Sweden. Later, it was said, British agents, in a daring foray, rescued the real bottle of heavy water from an institute refrigerator. The prosaic truth was quite different. Heavy water was used in the institute's accelerator. When the supply had to be replenished, someone remembered that Bohr had once had a bottle the Norwegian manufacturer had presented to him. The institute men were unable to find it and sent an underground query to Rozental in Stockholm. Rozental sent back full instructions about where to find it and pointed out that the bottle was unlabeled. Out of such materials are legends born.

Not legend, but more fantastic than invention, were the realities of the next few weeks. Soon after Bohr's arrival in Stockholm a telegram came from England, not from Bohr's friend Chadwick, but from Lord Cherwell—formerly Dr. Lindemann—Churchill's close personal advisor on scientific matters. It was virtually an official invitation to come to England. Bohr gladly accepted, and said that he would like his son Aage, then 21 and a physicist, to accompany him as his personal assistant. Bohr knew it would be impossible for Mrs. Bohr or the rest of the family to go with him.

On October 6, 1943, a British Mosquito bomber landed at the Stockholm airport, completely unarmed in order not to violate Swedish neutrality. At the request of the secret service and Tube

Alloys, as the British atomic energy project was then called, the unarmed plane flew over hostile seas and enemy airfields in Norway to bring to Britain a man whose brain and knowledge were regarded as a priceless asset to a nation at war.

There was only one place in the Mosquito for the special passenger; the empty bomb bay had been prepared for him. Bohr was put into a flying suit, a parachute was strapped to his back, and he was handed a helmet equipped with earphones so that he might communicate with the pilots. As Bohr climbed, with a skier's ease, into the bare compartment, several flares were placed at his hand. If the plane should be attacked and it should be impossible to escape the enemy planes, the bomb bay doors would be released. Bohr was to parachute into the sea and light the flares. Many pilots had thus been fished out of the North Sea, he was assured.

The Mosquito took off. To avoid enemy attack it flew extremely high. The pilots instructed Bohr to turn on his oxygen supply. Bohr, however, did not hear the instructions. The aviator's helmet rode so high on Bohr's massive, high domed head that the earphones did not cover his ears. The escaping scientist soon fainted for lack of oxygen, and so, unconscious, he made the trip to Britain.

When the pilots could get no response from the famous professor they had snatched at such risk from the clutches of the Nazis they were afraid that he was dead. As soon as they had passed the Luftwaffe airfields in Norway, however, they came down to lower altitudes. By the time they landed in Scotland, Bohr had regained consciousness, and was then flown on to London.

Chadwick and other British scientists were waiting at the airfield to give Bohr the warmest and most appreciative of welcomes. Not only were they personally happy to see him, but also felt that they had scored perhaps decisively for their country and for the cause of freedom.

A revelation awaited Bohr. Since the end of 1940 he had known little or nothing about what the British and Americans were doing on atomic research. A double barrier shut him off from all the information that in normal times would have flowed to Copenhagen. Early in 1941 the scientists of the United States and Great Britain had simultaneously and spontaneously done what they could not accomplish in 1940 when the subject was brought to Bohr at Princeton: they had completely halted all publication on uranium research.

The French constituted no problem this time. When France fell Joliot decided to remain in his own country, but von Halban and Kowarski, his colleagues, fled to England with the twenty-six cans containing the 185 kilograms of heavy water France had obtained from Norway at the outbreak of war. It was the French, not Bohr, who escaped with heavy water. At first the containers were lodged in Wormwood Scrubs prison, but later they were shifted to the opposite extreme of the social scale, to Windsor Castle where they were placed in the care of the librarian. Von Halban and Kowarski continued their work in Britain.

In addition war shut off the visitors who would have brought personal reports to Bohr. The sudden silence in itself was revealing to Bohr, as it was to others, for anyone with any knowledge of science and the prewar burst of uranium experimentation knew that it had not halted at one set moment. Nevertheless Bohr was not prepared for what he heard.

An atomic bomb and atomic power were no longer academic hypotheses. The United States and Great Britain were committed to producing them, and were putting into the tremendous mission whatever resources might be required. Both believed that the fate of the world rested upon their developing a bomb before Hitler did so. They could not gamble on annihilation.

Cherwell had put the issue for Churchill. In a "minute" to the prime minister, in the earlier stages of research when there remained much doubt about the possibilities of a fission weapon, he wrote: "The chances are two to one against a bomb being produced within two years . . . but I am quite clear we must go forward. It would be unforgivable if we let the Germans develop a process ahead of us by means of which they could defeat us in war or reverse the verdict after they have been defeated."

Chadwick laid the whole development before Bohr as a preliminary to his meeting all the top scientists in the work and visiting the sites where the research was underway. Bohr was amused when he heard about the Maud Committee and how it received its name. He promptly set the British physicists straight about Maud, and the story was told and retold with glee whenever two physicists got together.

In eighteen months four small teams at four universities working separately and yet in full cooperation, had achieved amazing scientific results. Peierls and Frisch at Birmingham worked out the actual size of a U-235 bomb, and Francis Simon outlined a gaseous diffusion plant, estimating the manpower and money needed. At Cambridge the French physicists von Halban and Kowarski demonstrated the possibility of a chain reaction with uranium and heavy water, while others established the fission possibilities of Element 94. At Liverpool Chadwick's own team studied the behavior of the isotopes U-235 and U-238 and found that it accorded exactly with Bohr's predictions. Birmingham concentrated on the problems of producing uranium metal.

While part of this work was being done, the sufficiently destructive bombs of World War Two were raining down on Britain and each day was consumed with the problem of surviving until the next. Nevertheless the uranium work went on in the hands of this small, dedicated group of scientists, with almost no resources at their command and at virtually no expenditure of government money. They were working in the old Rutherford tradition of accomplishing much with very little.

By July of 1941 the MAUD scientists were able to report to their government that a bomb was feasible and "likely to lead to decisive results in the war." The committee named for the nurse who had cared for the Bohr children recommended that the work be continued on the highest priority to obtain the weapon in the shortest possible time. "There was," said Margaret Gowing in her official history, "no other choice." The project was reorganized and given official status under the unrevealing name of Tube Alloys, and Sir John Anderson, a member of the war cabinet, was placed in charge.

Bohr was also informed of the astounding events taking place in the United States. Up to mid-1941 research in the United States had lagged behind Britain's, and the British were urgently invited over to contribute their findings. Even late in 1941 no chain reaction had actually been achieved. For all of the progress, the bomb remained a bomb in theory only.

Then on December 7 the Japanese attacked Pearl Harbor. The surprise crippled a formidable fleet and produced war instantly. President Roosevelt did not hesitate but authorized construction of an atom bomb forthwith. Expenditures of over half a billion were approved and all of the power of one of the world's great powers was directed into the unprecedented undertaking. Nevertheless there first had to be proof absolute of the chain reaction on which the bomb was predicated. To speed the experiment all work was concentrated at the University of Chicago.

Bohr heard the story of the historic experiment conducted on December 2, 1942, in the squash court under the stands of the university's Stagg Field. At the command of Enrico Fermi, who was in charge, the last control rod, nicknamed "Zip," was slowly withdrawn from a pile made up of lumps of uranium embedded in graphite. The counters clicked madly and the line on the graph paper began to rise. Neutrons were giving rise to neutrons and the line rose exponentially before the control rods were slammed back in and the experiment halted short of a disastrous explosion. For the first time in history man had initiated a self-sustaining chain reaction.

"Physicists like to think in figures," said Samuel Allison who was present. "One atom of coal yields two or three electron volts. One atom of uranium 235 in fission yields 200,000,000 electron volts. Two and 200,000,000! A transcendental source of energy had been tapped."

Bohr listened with growing amazement. Only a little more than three years before he had brought word to the United States of the splitting of a few atoms of uranium, but now the full might of one of the great nations was to be arrayed to produce a bomb based on that same fissioning. In believing that a bomb would never be produced, Bohr had not conceived of an effort on the scope and scale the United States was about to undertake.

Within a few days Bohr met Sir John Anderson, who was the chancellor of the exchequer as well as the cabinet director of Tube Alloys. He was a man of sensitivity as well as acuity. Though he was engrossed in Britain's urgent problems of the moment he also took the long view. He and Bohr immediately understood one another. Anderson had described Bohr as one of the few living scientists "worthy to rank in every respect with a Newton or a Rutherford," and he did not conceal that he wanted Bohr on the British rather than the American atomic energy team.

Roosevelt and Churchill had decided early that the large atomic energy plants should be built in the United States where they would be free from bombing, but that the British would contribute men and research and thus share in the undertaking as a partner. The ideal quickly broke down in the compartmentalization and secrecy with which the Manhattan project was organized. Soon the British were being barred and hampered in their share of the work.

It took the president and the prime minister to get matters straightened out again. At their meeting at Quebec in August, 1943, the two heads of state signed an agreement for full and effective collaboration. It provided for full interchange of information between members of a Combined Policy Committee and, at the working level, between groups working in the same sections.

Anderson had conducted all the negotiations that led up to the agreement. All seemed to be squared away, but Anderson knew that many snags would lie ahead and he counted on Bohr as a weighty enough contribution to even the tables and thus maintain the partners as equals rather than as principal and junior. General Leslie R. Groves, who had been placed in charge of the American project, wanted Bohr on the American side.

Anderson suggested that for the time being Bohr see what was under way in England—much of it in the hands of friends and former colleagues. Bohr and his son Aage, who had been brought to England a week later, settled down in a hotel in Westminster. Much of the city lay in ruins, but Bohr had lived so much in England that he felt thoroughly at home, and more than this, a part of England. There were, however, many changes in the life he formerly had known. As he walked through the lobby, he soon learned to recognize the secret service men assigned to guard him. Some of them even wore trench coats and snap brimmed hats. The Bohrs were rapidly becoming experts on surveillance in all its forms. "They were like characters out of a novel," said Aage.

The Battle of Britain had ended almost a year before, but occasionally the air raid sirens still wailed. Bohr liked to tell his English friends that England was a place of peace compared to the tension of an occupied country. He took the rationing and shortages without complaint, and skillfully darned his own and Aage's socks or sewed on buttons. Bohr had always been skillful with his hands.

Bohr was also assigned an office in Old Queen Street, near the London headquarters of Tube Alloys, which was convenient to Whitehall and all the government offices.

Most of the two months the Bohrs spent in England went into traveling around to the universities and research plants. It was a reunion and a homecoming, though some of Bohr's closest friends already had gone to America on the British team. Cambridge had lost some of its serenity, but the Cam still flowed quietly between the buildings that had always stirred Bohr's heart and that now, when so much had been destroyed, were all the more to be treasured. They had to endure.

Visiting the laboratories, sitting at a bench and talking to the scientists, Bohr had no doubt now that the bomb would be built. Only technical, not fundamental, problems stood in the way—though this had been true almost from the beginning. But the scale indicated that even the technical problems would be met.

Another question was arising in Bohr's mind. How would the presence of this transcendental weapon affect an all too imperfect world? Could it be turned to good, rather than destruction? Were there ways in which this power of a new dimension could insure peace? Bohr's concern was for the future, a future close upon them and a future altered forever by all that was then being done in the laboratories.

Bohr went back to see Anderson, who also recognized the problem. Bohr raised the question of what the possession of a definitive weapon would do to the relations of the possessor and other states. Would others, particularly the U.S.S.R., stand by or would the bomb start an arms race? And could this not lead to the destruction of civilization itself?

Bohr pointed out that cooperation between allies whose political systems differed as drastically as did those of the democracies and Russia was a tender, uncertain relationship at best. It was difficult even when they faced a common enemy. Bohr knew at first hand the frustrations of trying to deal with the Russians. He had participated in the rescue of several scientists from Russia and had attempted to promote scientific cooperation. And the passionate anti-Communism lying not far below the surface among many Americans, also did not augur well for cooperation. "It may be very hard," said Bohr, "to find a basis for cooperation between the East and the West." Anderson, acutely aware of the problem of maintaining a partnership even when allies are not basically antagonistic, agreed in full with Bohr.

Bohr did not stop at defining the obstacles that lay ahead. He was not concerned with warnings, but with discovering worthwhile opportunities for overcoming the difficulties themselves. Would not this sudden advance to a whole new level of scientific achievement offer an unprecedented opportunity to advance to an equally new level of international cooperation, he suggested. An extraordinary advance, Bohr argued in his discussions with Anderson, opens an extraordinary opportunity. A degree of cooperation unthinkable and unattainable in the past might now become possible, with atomic energy as the key. The alternative—a precarious existence dominated by the constant threat of obliteration—would strengthen the argument for cooperation. In talks at the office and at dinner, Bohr and Anderson debated the altered future Tube Alloys and the

Manhattan District were decreeing for a world still unaware of the atomic threat.

More immediate problems had to be settled too. Following the Quebec agreement the British were preparing to send all their other top physicists, headed by Chadwick, to the United States. Anderson wanted Bohr to go with them or to follow soon afterwards. Bohr pointed out that he had close ties with the United States, as well as with Great Britain and that he did not want to be designated exclusively a member of one team or the other. He hoped that he could serve the common cause. In the end a formal decision was not made on Bohr's status. Anderson said that he would try to persuade Groves to allow Bohr to have an overall look at the work under way in the United States and then decide in what capacity he could best serve.

Bohr promised, however, that he would not allow himself to become aligned exclusively with the Americans. It was agreed that he would assist the common effort in any way he could, and that he would give all possible help toward developing a genuine Anglo-American partnership with a full reciprocal sharing of scientific and technical knowledge. If this arrangement should prove impossible, Bohr said, he would return to work in Britain. Officially Bohr was appointed a "Consultant to the British Directorate of Tube Alloys." Aage was designated a "Junior Scientific Officer." Plans were made for both of the Bohrs to sail for America early in December.

The pall of secrecy had already settled down upon them. Mrs. Bohr, with Hans, Erik, and Ernest, had found an apartment in Stockholm, but Bohr could give her no idea of his address or of when or why he was going to America. The open explanation was that he was working on postwar scientific cooperation. Niels could tell Margrethe little more.

Whenever Bohr had been away he had written to Margrethe, telling her about all that was happening to him and his innermost reactions to events and people. Margrethe, with her perceptiveness, understood there must be a strong reason for the new vagueness of her husband's and son's letters. What might lie behind it she did not know. She could only guess that he was working on matters of the highest importance. "He thought we realized much more than we did," said Mrs. Bohr. His family and friends could write to him only in care of British Intelligence. There was no regular mail to Britain.

Just before Bohr was scheduled to depart for the United States, two of the young Danes who had continued working in the institute arrived in Stockholm with the news that the institute had been seized by the Germans. Early one morning a squad of German military police marched in and took possession. Jorgen Boggild, the physicist in charge, and the chief laboratory assistant, Holger Olsen, were taken off to jail, though they were not formally placed under arrest. Miss Schultz, Bohr's secretary who lived at the institute, was permitted to leave. Word was rushed to Bohr in England. Many years later at the Nuremberg trials, it came out that the seizure was recommended by a Gestapo member who hoped to make a reputation for himself by proposing the action. The records show that the Germans did not know what they wanted to do with the institute.

After the take-over there were rumors that a German physicist would be placed in charge and the insititute put back in operation. The staff responded to this threat by going underground. Rumors also spread that the cyclotron and other large pieces of apparatus would be dismantled and shipped to Germany. Both ideas were given up as impractical, for by the end of 1943 the Germans were confronted with more pressing matters.

An Austrian physicist and friend of Bohr's who passed through Copenhagen spread the word of the take-over among the German physicists, including Heisenberg who suggested a mission to Copenhagen to investigate. To allay suspicion of his personal interest in going, he arranged for a Nazi physicist to accompany him. The two made the inspection and, finding that no war work had been done at the institute, reported it of no value to the German war effort. The way thus was opened for an official return of the institute to the Danes. However no one in Copenhagen knew that Heisenberg was trying to come to the rescue. Seeing the seizure, the underground concluded that the institute would be converted to some highly important German war work, perhaps the making of a secret weapon. From this it was only a step to a decision that the institute should be blown up. Underground workers crept through the sewers and planted enough dynamite to wreck completely the institute buildings and everything in them.

Shortly before the demolition was to take place, Professor Chievitz, one of the leaders of the resistance movement, and a lifelong friend and childhood schoolmate of Bohr's, heard of it. He rushed into action, pleading with the underground not to light the fuse until they could get in touch with Bohr and find out if the demolition were really necessary. The query was quickly slipped through to Stockholm. Mrs. Bohr and Rozental in a frenzy of worry, went to the highest sources to get the question to Bohr with no delay. "And then we tensely waited for an answer," said Rozental. "But it did not come. We grew more and more worried that the resistance group would finally lose patience. Our worry became even greater when an engineer, who came legitimately from Copenhagen to Stockholm on a business trip, brought with him a report of where the mines were placed, just in case the members of the resistance who laid them should be arrested or killed in action."

Rozental and Mrs. Bohr did not know that Bohr had acted directly. He did not dare to delay to send the word "No" through Stockholm. That night a British plane dropped Bohr's message by parachute to underground contacts in Denmark. Bohr said the institute would be of no war use to the Nazis and urged in the strongest terms that it should not be destroyed. The institute thus survived by the narrowest of margins. Several months later after involved negotiations, the Gestapo marched out as abruptly as it had marched in.

Bohr and Aage sailed for the United States with a new sense of relief, leaving behind them, or so they thought, messages in keys, escapes in the dead of night, flights in bomb bays, secret service agents, and threats of bombings. It had been a tumultuous, incredible year, far more likely in a mystery thriller than in the life of a 58-year-old Nobel Prize winning professor. The gray haired, solidly built gentleman in the cap who took his morning constitutional around the deck did not look like the central figure in a spy thriller. And the voyage was uneventful.

16 /

The Clue

THE CONVOY DROPPED AWAY and the tugs eased Bohr's ship into dock in wartime New York. Army security agents at once boarded the ship and found the Bohrs. Before Bohr stepped again on American soil or entered the massive city of skyscrapers he and Aage were watching from the deck, they were enmeshed again in guards and all the fantastic toils of secrecy.

Under no circumstances was Bohr's presence in the United States to be publicly revealed. The mere fact of his arrival would tip off the enemy that the United States probably was working on an atomic bomb or nuclear project of some nature. The circumstances of Bohr's flight from Sweden, which were of course known to the Germans, indicated the high significance attached to his presence. At the time of Bohr's arrival American agents in Europe were making every effort to trace the whereabouts of the German nuclear physicists, certain that this would indicate whether Germany was producing a bomb. Find the few physicists capable of working on fission and most of the answer would be in hand, the intelligence services well knew.

The secret service was ready for Bohr. He was assigned the pseudonym of Nicholas Baker and Aage became James Baker (like the other top physicists Bohr retained his own initials in his "cover name"). Both the professor and Aage were presented with papers necessary to identify them by their new names. They also were told that a guard would be constantly with them. When they moved from one place to another the guard would change, but there always would be someone to protect their lives against possible attempts at assassination. Bohr cheerfully accepted the irksome precautions and adapted to them, but he and Aage always were amused when they later saw each guard who assumed responsibility sign a receipt for them.

The newly created Bakers went on to Washington to confer with British and American officials and scientists about the work Bohr would do. Bohr was also granted permission to visit the Danish legation under his own name. If he had not done so, there was a strong chance that some Danish official would meet him by chance and begin official inquiries. It was thought better for Bohr to go openly to the legation and explain that he was in the United States to prepare for scientific cooperation after the war. The Danish minister Kauffmann and Bohr rapidly became friends, and the Bohrs were invited to live at the legation whenever they were in Washington.

At the embassy nothing was known about the Bohrs' identities as Bakers, and the double system of names occasionally gave rise to complications. Bohr left his watch at a shop where he was buying a suit. He had, as required, given his name as Baker. His watch, however, was engraved "Niels Bohr." Aage went to call for it, explaining that he was Mr. Baker, secretary to Mr. Bohr. When the shop looked askance at this shift in names, Aage then suggested that they phone the Danish legation for confirmation.

Bohr, as it happened, was not in when the call came. A secretary who answered said that nothing was known there about any Bakers; Professor Bohr's secretary also was named Bohr. There was nothing then for Aage to do but to explain that they were using the name Baker while traveling in the United States. "The watch was handed over to me with a stern reprimand that in the U.S.A. it is against the law to use a false name," said Aage.

Before Bohr left England, Sir John Anderson had arranged for him to meet Lord Halifax, the British ambassador in Washington. Anderson wanted Bohr to lay his ideas before the ambassador and also to keep in touch with him through the diplomatic service. At the big red brick British embassy on Massachusetts Avenue Bohr talked at length to the immensely tall, bony, and thoughtful ambassador, and to Sir Ronald Campbell, the British minister. The diplomats knew only generally about the definitive weapon their own government and the United States were bent upon producing. Bohr explained its force and dimensions, and in a number of talks with them argued with all his persuasiveness and logic that it would alter the course of the postwar world.

Unless the basic information about nuclear power—not the bomb—were made available to all, Bohr argued, the world would be driven into a nuclear arms race. Others would not sit by and permit the United States and Great Britain to have the sole access to this nearly unlimited source of power and the final weapon to which it could give rise.

Other states, particularly Russia, would spare no efforts or resources to build their own nuclear weapons. Bohr cited their ability to do so. Nuclear physics and nuclear physicists were not limited to the United States and Great Britain. Bohr's earnestness and emphasis increased. On the other hand the very magnitude of the development offered a new opportunity for nations to work together on a new basis. Measures unthinkable in the past would become possible if the basic information were freely shared and access made available to all.

Halifax and Campbell at first found it difficult to listen to Bohr's low voice, and his step-by-step development of his argument, leading to conclusions that were not divulged until the whole case had been prepared for them. Bohr had once protested when his brother and a friend interrupted such a discourse: "Of course you cannot understand what I am trying to say now; how could you possibly understand until you have heard the story as a whole and know the end." The ambassador bent his long frame forward to hear the nearly inaudible words and waited for the end, for he understood the importance of what was being said. Out of his full diplomatic experience Halifax was soon convinced, Margaret Gowing said, "of the sense of Bohr's main proposals." Both Halifax and Campbell could see, said the official historian, that "as long as no competition in atomic weapons had begun—and it might begin immediately after the surrender of Germany—America and Britain would have in their hands a card which they could use in their negotiations with others for the improvement of the world situation. "Halifax and Campbell thought there was hope in this and that in any case it called for urgent and deep consideration by the Prime Minister and Sir John Anderson."

The ambassador emphasized that the President of the United States would have to take a prime role in the decisions. Halifax invited Bohr to keep in the closest touch with the embassy.

Bohr fully understood that the American interest would have to be primary: development of the bomb was taking place in America and the United States was paying the formidable bills. Great Britain might be a partner, but the United States was the prime mover. Bohr racked his brains about how to reach the president and bring to his attention the necessity for early postwar planning of atomic policy. The Danish ambassador introduced him to Supreme Court Justice Felix Frankfurter. The former dean of the Harvard Law School was a frequent advisor to the president and had furnished him with a number of the brilliant young lawyers who staffed and ran many of the New Deal and war agencies.

But how could Bohr discuss atomic problems with the justice? If Frankfurter did not know that the United States and Great Britain were developing an atomic bomb, Bohr could not tell him. He could not even ask if Frankfurter knew. His lips were sealed by the intense secrecy requirements of the Manhattan District. At tea at the Danish legation, Bohr and the justice had a highly interesting discussion on the problems of Europe and the war generally. The justice then invited Bohr to lunch at his office in the Supreme Court building.

Bohr was quick to act upon the invitation and as the two men talked again, Frankfurter, who knew of Bohr's work on fission and the structure of the nucleus, had no trouble guessing why he was in the United States. He also sensed that Bohr had something urgent on his mind, and obliquely made a remark about "X" in such a way that he told Bohr he knew about the atomic bomb project. As it happened, a troubled young American physicist had come to Frankfurter about the bomb not long before. He was worried about the difficulties in which the project was then embroiled, and not only appealed to Frankfurter but to Bernard Baruch and to Mrs. Roosevelt.

The young physicist was a member of the theoretical group working under Eugene P. Wigner at the University of Chicago. Early in 1943 the Dupont Company had been brought in to build the huge Hanford plant at which plutonium was to be produced for the bomb. In the first months, however, the work the theoretical group had done on this subject was ignored. Decisions were taken that the group had already ruled out as in error, and other decisions were made which the physicists feared would result in months of delay. As the young physicist saw it, the race to produce the bomb might be lost, and the Allies beaten to the decisive weapon by Hitler. In his desperation he wanted to reach the president with a warning in time for the work to be reorganized and saved. Thus he went to those who could convey the message to President Roosevelt, and Frankfurter learned about the bomb project, though he knew none of the technical details. In his own mind he called it "X," and this was the term he used to Bohr.

At the mention of "X" Bohr was suddenly free to talk "in an innocent, remote way," as he said, not about the technical aspects of the bomb, but about the postwar arms race that he feared unless early action was taken to avert it and establish a new, more promising order. Frankfurter quickly grasped the point of Bohr's argument and agreed with it. On his own initiative he offered the opinion that Roosevelt should be acquainted with the whole situation and with Bohr's views. Frankfurter thought that the president would not only want to know, but that the proposal of using the new power to build a stable peace would accord exactly with the president's own hopes and aims. Frankfurter said that he would talk to the president as soon as possible and let Bohr know what had transpired as soon as Bohr returned from a projected trip. Bohr walked out through the great columns of the court portico and, with hope high in his heart, stopped to gaze at the Capitol directly beyond, and on toward the White House. He felt that the way was opening. "Bohr was a man weighted down with a conscience and an almost overwhelming solicitude for the dangers of our people," said the justice, a man of remarkably keen insights and an instinct for the issues and manipulations that shaped events.

Bohr and Aage were scheduled to leave for Los Alamos, a place unknown to virtually the entire population of the United States and unmentionable even by those going to it. Bohr had been told, of course, that most of the research on the construction of the bomb was centered there. He also knew that so many of the physicists assembled there had been connected in one way or another with his institute that it would be almost like an institute reunion.

For most of the trip west the Bohrs were joined by General Leslie R. Groves, the army director of the entire Manhattan District project. The thick-set former army construction boss wanted to talk to Bohr before he joined the other physicists, whom Groves privately called his "crackpots." The general began by laying down the rules in his commanding voice. To obtain the secrecy that was his fetish, the production of the bomb had been compartmentalized. A worker in one section knew little or nothing of what was going on in another. Only a few at the top had an overall knowledge of the work, and Bohr, by necessity, came close to the top. Groves told him what could be talked about and what was barred. The general also was eager to learn anything Bohr possibly could tell him about atomic research in Germany and the scientists who would carry it on. He was hopeful that Bohr would know more than he actually did.

As the train moved west across the Mississippi and the plains of Texas, the balance of the conversation began to shift. Bohr did more and more of the talking and the general strained forward to hear him. The clicking of the wheels and the clanking of the cars almost drowned out Bohr's voice. The morning after their arrival at Los Alamos, Oppenheimer, the scientific director, met the general and noticed that he seemed stiff and limped a little. He asked what the trouble was. The general's laconic answer, which Oppenheimer understood very well was: "I've been listening to Bohr."

The Bohrs and the general had arrived at Los Alamos in the early evening. The general took his important new prizes directly to the Oppenheimers's for dinner and then tactfully disappeared, while old friends met again.

Not until the next morning did the startling, distant magnificence of the scene and its immediate chaos break upon Bohr's surprised vision. Los Alamos stood upon a high mesa, the weathered and flattened cone of an ancient and long extinct volcano. To the west loomed the green domes of the Jemez Mountains—the rim of the ancient crater—and to the east the snow-crowned peaks of the Sangre de Cristo range. The walls of the mesa, with their wondrous stripings of reds and oranges, dropped sharply into the dry desert of the valley below. Only where the valley sloped down to the Rio Grande River did green break into the sands, the mauves and the ever-changing lavenders of the desert. The stripes of green along the curves of the river and along the creeks cutting down through the canyon walls marked the sites of the ancient Indian pueblos and Spanish villages. On the mesa grew long-needled pines, turned a grayish green by the dust of construction and the sands that sometimes swirled up from the desert.

The mesa was a construction camp; the sound of hammering and pounding filled the air as new apartments and new laboratories were hastily slapped together. Trucks roared up and down the streets; wires looped everywhere—all was new and raw. Only the rustic buildings of the boys' school, which formerly occupied the mesa fitted into the setting, and its grounds were planted with grass and flowers.

The mesa had been chosen for the laboratory because of its isolation. The city of Sante Fe lay about twenty miles away and only a rutted, hair-pin curved road connected the two, linking the laboratory to trains and planes. Those on the mesa could work freely, completely separated from the outside world and all unwelcome attention. In addition the canyons formed by the rush of water down the steep sides of the mountains offered excellent sites for the testing of explosives.

Work on the wilderness laboratory started early in 1943. Before the shoddy boxes of the physical laboratory were finished, Los Alamos also was assigned the task of purifying the plutonium, the element 94 Bohr and Wheeler had predicted, which was to be produced at Hanford. The laboratory also was to engineer the final weapon. The town thus had to be expanded again, and it was still growing when Bohr first saw it. It was then a city, of unacknowledged existence, of some 6,000 persons. Correspondence was addressed only to Post Office Box 1663.

Bohr immediately received a badge reading "Nicholas Baker" and Aage one for "James Baker." But as Bohr walked down the muddy streets or stepped into a laboratory the shouts of recognition and delight would go up. There was only one Niels Bohr—even a half glimpse at a distance of the sturdy figure, the massive head, the posture with the forward flexed knees was enough to identify him for most physicists. Nicholas Baker! The laughs rang out but the greetings were fervent and emotional. Frisch, Weisskopf, Chadwick, and dozens of others crowded around. So much had happened since Bohr's last meetings with all of them that even the most phlegmatic were shaken at their coming together again on this far western mesa. But the discipline to which they had subjected themselves was stern and implacable. Niels Bohr even in this isolation, became "Uncle Nick" and Aage "Jim." "Uncle Nick" proved so right for Bohr, it so well reflected his relationship to the physicists—the benignity of his smile, his graying hair and the beginning dewlaps that emphasized rather than concealed the bones of the face—that "Uncle Nick" persisted long after Los Alamos was left behind.

One of the shouts of recognition came from ordinarily quiet and scholarly Rudolf Peierls. Peierls had originally gone from Germany to England and, as a member of the British team, had come to the United States and the west. His wife Genia liked nothing better than mothering friends separated from their families and the Bohrs were invited to their house for dinner whenever a shopping trip to Sante Fe had yielded some unrationed meat or some other find for the table.

The Fermis lived directly above the Peierls in one of the four-family apartment buildings, and Mrs. Fermi tells in her book how the Fermis always knew when the Bohrs were guests of the Peierls: "Through the floor of our living room characteristic sounds would reach our ears at characteristic intervals: loud and prolonged peals of laughter would alternate with perfect stillness. Uncle Nick's whispering voice as he told a joke to the Peierls did not carry upstairs, but a mere wooden floor could not dampen the sound Genia made when laughing. Bohr must have told many jokes and they must all have been funny."

Bohr was never free of his troubled worry about the fate of Europe and sometimes seemed lost in abstraction as he walked around Los Alamos, with "Jim" at his side to protect him from the speed of the construction trucks. It seemed to Mrs. Fermi, though, that his concern was less poignant than on that day in 1939 when she met him at the ship in New York and he was gripped with fears of Nazi conquest. Actuality, no matter how fearful, was less paralyzing than anticipation.

Bohr could be diverted at least temporarily from the burdens that weighed so heavily upon him by the wonders of the country in which he so unexpectedly found himself. One Sunday Bohr and the Fermis hiked down the Frijoles Canyon and followed its little stream to its confluence with the Rio Grande. The day was chilly and they chose a sheltered path at the bottom of the canyon. Suddenly a small black animal with white stripes down its back made its appearance along their trail. Before anyone else had noticed it, Bohr had squatted down close to it and was admiring its bright eyes and coquettish movements of its head. For a moment no one dared to speak or move. Very carefully and gently they persuaded Bohr to back away from the fascinating animal—a skunk. He had never seen one before, but somehow escaped making its full acquaintance then.

"Bohr surprised us with his agility," said Mrs. Fermi. "We had to cross the stream numberless times, and he never stopped to consider its width or the best place to cross it. He jumped it. And while he did so his body straightened and his eyes glowed with pleasure." As they climbed back up the steep walls of the canyon most of the party needed all their breath for the ascent. Not Bohr.

He started to talk about Europe, pausing only now and then for another flying leap across the curving creek. Most of his words though were lost in its noisy rush.

On another Sunday when snow covered the Jemez Hills, many of the Los Alamos staff went out to Sawyer Hill for skiing. Fermi had gone off on a cross-country trek, but Mrs. Fermi was running one of the lower slopes when Bohr appeared with some friends. Bohr looked so longingly at the fresh, powdery snow that one of the young physicists offered him the use of his skis. Unce Nick happily strapped them on and in a few minutes was on the slopes, bare headed and wearing only a light sweater.

"He gave himself to elegant curves and to expert snowplows, to dead stops at fast speeds and to stylish jumps that no one else on the slope could perform," recounted Mrs. Fermi. "He went on with no pause for rest, with no thought of the man who had taken his place at the bottom of the hill, ski-less. He quit only when the sun went down and darkness and a chill descended upon the snow."

In conferences with Groves and Oppenheimer it was decided that Bohr's main duty would be to review all phases of the work to make certain that nothing had been overlooked. With his understanding of the structure of the atom he was outstandingly equipped to appraise and detect any omissions. Bohr was assigned an office and he was there at eight o'clock every morning when the second whistle sounded. Oppie whistled first, as they all put it, at seven o'clock.

Bohr's first question as he saw the scope of the work at Los Alamos was, "is it big enough?" Oppenheimer, to whom the pointed question was directed, was willing to concede that it was not, but he demonstrated to Bohr that the project at Los Alamos, at Oak Ridge, and Hanford soon would be. Bohr continued to be acutely aware of the tremendous size of the plant essential to separate uranium and produce enough plutonium for the effective weapon which Bohr referred to in his Danish accent as "the bum."

In truth, as Bohr once told a friend, "They didn't need me to make the atom bomb." After U-235 had been identified as the fissionable part of uranium and the possibility of producing fission in a new element—plutonium 94—had been established, the production of the bomb was largely a technical and engineering matter.

Without planning or intent, Bohr came to perform quite another function at Los Alamos. The physicists knew then, despite still potent difficulties and frustrations, that the way they had found into the atom would produce a weapon of definitive power. They faced the fact that an atomic bomb could visit destruction of terrible dimensions on whole populations and that the radiation aftermath could be even deadlier than the gigantic explosion. What had they done? What Pandora's box had they opened? Were they precipitating a new warfare that would obliterate civilization and the very globe itself? These were the long thoughts, the consuming fears and doubts that came to the physicists on their remote mesa. Few men in history have been compelled to ask themselves more awful, more shattering questions.

As the Allied armies advanced up the Italian peninsula in the spring of 1944 and the United States began to win back territory in the Pacific the questions became all the more agonizing. Perhaps the bomb was unnecessary. Perhaps the war could be won without it. The fears and doubts did not halt the work that went on day and night, but they could be disruptive, even paralyzing, if they grew.

After dinner and in sessions that went on far into the night the physicists poured out their desperate worry and guilt to Bohr. He above all others understood; he spoke the same language. He shared the responsibility, in fact, a major portion of it. Bohr paced the thin creaky floors of Los Alamos and talked. He did not need to remind them of the degradation and illimitable evil wrought by Hitler on a whole world and civilization, nor of the absolute necessity of bringing it to a conclusive end. An atom bomb in Hitler's hands could save him from defeat and permanently enslave the world. They still did not dare to take that chance.

Bohr, though, carried the argument to another plane. The unprecedented power they were creating offered an unprecedented opportunity to establish a new and better civilization and peace than had ever been possible in the past. The construction of the bomb was not only a necessity for combating rampant evil, but an opportunity for a new and far better start.

"He made the enterprise, which often looked so macabre, seem hopeful," said Oppenheimer. "He spoke with contempt of Hitler who with a few hundred tanks and planes had hoped to enslave Europe. He said nothing like that would ever happen again; and his own high hopes that the outcome would be good, and that in this the role of objectivity, friendliness and cooperation incarnate in science would play a helpful part, all this was something we very much wanted to believe."

Others who listened to Bohr felt the same upsurge of hope and the same lessening of the doubts that so brutally assailed them. "Many of us were beginning to worry about what the future might hold for a humanity in possession of such a dreadful weapon," said Frisch, "and once again it was Bohr who taught us to think constructively and hopefully about that situation."

Weisskopf was still another who was moved by Bohr:

In Los Alamos we were working on something which is perhaps the most questionable, the most problematic thing a scientist can be faced with. At that time physics, our beloved science, was pushed into the most cruel part of reality and we had to live it through. We were, most of us at least, young and somewhat inexperienced in human affairs, I would say. But suddenly in the midst of it, Bohr appeared in Los Alamos.

It was the first time we became aware of the sense in all these terrible things, because Bohr right away participated not only in the work, but in our discussions. Every great and deep difficulty bears in itself its own solution, and therefore the greater the hardship, the greater would be the reward that would come out of it. This we learned from him.

. . . . .

It was through him that what one might call the political movement of the scientists started. It is of course much more than that, it is a new kind of thinking by a group of people . . . placed in very important positions in this world. And I feel very strongly that the fact that we still live here and that our children will live on is due largely to him and to his influence in those days.

Bohr had always been able to lift the minds of men. In the colloquia at the institute and in discussions he had drawn out the physicists. Out of this heady, enthralling intellectual advance and adventure came the great discoveries of modern physics, nearly all of them having their genesis in this unique stimulation Bohr provided.

Perhaps it was the distinctive attribute of the era to produce men with a unique ability to sway others and to lead them on—Hitler could mesmerize millions with his demonic rantings; Churchill could invoke a matchless courage and resolution with the majesty of his words and his unyieldingness; Roosevelt could reassure and enlist by the very sound of his voice; and Bohr, though he spoke only to a few, a select few, could stir these men of brilliant abstract minds, not with his soft indistinct voice and ordinary words, but with ideas and concepts that led the world into the scientific era.

To give Bohr a wider view of the work, and to permit him to extend his review, Groves asked him to make a trip to Oak Ridge, Tennessee. With the usual guard, Bohr and Aage took the crowded train from Santa Fe and traveled across the Mississippi to the new South. Bohr had known well how large a plant would be required to separate U-235 and to produce plutonium—this had been his reason for believing that it would never be done. None of this, however, prepared him for the actual plant that he saw. Huge bastions of concrete, suggesting the walls of a medieval castle, nearly filled the valley at Oak Ridge. A human approaching them was dwarfed; they matched the scale of the hills around them. They stretched down the valley as far as the eye could see, and spread over eighty square miles. "It was an installation of indescribable dimensions and at the same time was founded on the highest technical standards," said Aage Bohr.

The pile at the Clinton Laboratories, one of the sections of the Oak Ridge installation, was in full operation. It was a high stack of graphite interlaced with 1,200 channels into which uranium slugs were introduced. When Bohr arrived in the Tennessee Valley, plutonium by the gram was coming from the laboratories, where before scientists had never separated more than millionths of a gram. A gaseous diffusion plant and an electromagnetic plant produced U-235.

Bohr did not visit or was not permitted to visit the giant plutonium plant on the Columbia River in the state of Washington, but he had seen enough and learned enough to have the answer to his first question: the plant was big enough. As Bohr studied the huge plant he thought wonderingly of the day at Princeton when he had suddenly understood that it was U-235 which fissions and that a new element No. 94—later called plutonium—might be created by fission. The idea was then a few strokes of chalk.

Bohr was grave as he started to dictate a letter to Anderson: "The more I have learned and thought about this new field of science and technique, the more I am convinced that no kind of customary measures [of control] will suffice for the purpose and that no real safety can be achieved without a universal agreement based on mutual confidence."

While Bohr was at Los Alamos and Oak Ridge, Frankfurter, as he had promised, had talked to President Roosevelt. Roosevelt was at first taken aback that Frankfurter knew about the bomb. Without making any admissions, the justice remembers the president conceded that the whole postwar atomic problem "worried him to death." Frankfurter urged him to talk to Bohr about Bohr's extremely promising ideas. Bohr's proposals also had set the wheels in motion in England. Halifax had been so impressed by his conversations with Bohr that he sent Campbell to England to talk to Anderson. Anderson was already convinced and felt that no time could be lost if a disastrous postwar armaments race was to be averted. In consultation with Lord Cherwell he prepared a long "minute" (a memorandum) for Churchill.

He urged first that the time had come to inform the War Cabinet, the Service Ministers, and the Chiefs of Staff. He then went into the international effects of the bomb. The American-British effort, Anderson said, seemed certain to produce a bomb before the Germans could get one; in fact there was no evidence the Germans were near production. But, Anderson warned, it would be foolish to assume that the Russians would not make every effort to build an atomic bomb the moment they were free of the war. Anderson also predicted that as knowledge increased and processes were simplified the bomb would come within the capacity of a number of other states.

There are only two alternatives, said the cabinet member in charge of the British atomic energy project, "a vicious arms race in which the United States and Great Britain would at first enjoy a precarious and uneasy advantage," or international control. "It may well be that our thinking on these matters must now be on an entirely new plane," he continued. "I am myself convinced that we must work for effective international control."

Anderson, like Bohr and Halifax, recognized that the British could not act without the Americans. He emphasized this to Churchill, but he also told him that several persons close to Roosevelt were urging the president to consider the postwar atomic control problem. All, he said, believed that the president and the prime minister would have to take the lead. Since the question undoubtedly would arise, Anderson proposed that Churchill begin preparations for meeting it.

One of the pressing questions would be the Russians, Anderson predicted. Should they be told? If it were decided to work for international control, Anderson continued, accepting Bohr's thesis, there would be much to be said for communicating to Russia in the near future the bare fact that the Americans expected by a given date to have a devastating weapon, and for inviting them to collaborate in preparing a scheme for international control. If the Russians were told nothing, Anderson warned, they would learn sooner or later what was afoot and might then be less disposed to cooperate in international control.

Churchill read the note carefully. He "peppered it" with disapproving comments and at the end wrote with crushing finality: "I do not agree." Mrs. Gowing reports that the prime minister saw no urgency about international control and was adamantly opposed to any relaxation of secrecy. He would not even consent to Anderson's making a presentation to his own war cabinet.

None of this was known to Bohr. He and Aage returned to Washington anxious to see Frankfurter. The justice then told Bohr that President Roosevelt was deeply interested in the prospects that atomic energy would open. He was particularly eager to see it turned to the purposes of peace rather than destruction. Roosevelt made clear to Frankfurter that he recognized it as a problem that he and Churchill would have to confront. And then came the word that lifted Bohr's head and hopes. The President would welcome suggestions from Churchill about how they might best approach this momentous matter. Bohr was authorized to convey this message to the prime minister.

Bohr hastily sought out Lord Halifax. The ambassador was as excited as Bohr when he heard of the message. He instantly appreciated the opening it offered; it was the essential key. Bohr, Halifax, and Campbell conferred earnestly and at length. Halifax attached such prime importance to getting the message to Churchill that he asked Bohr to go to England to deliver it in person.

Bohr happily and eagerly assented. It was almost incredible that he, a scientist, might be moving the president and prime minister toward action that he felt would hold the fate of the world in balance. Bohr's innate modesty made him marvel that he should be called upon to participate in the highest affairs of state. He hoped that the science which had so profoundly altered the physical world could contribute some of its special competence to the solution of political problems. Politics was a most imperfect art; perhaps science could make a contribution to it.

In April the professor and his son took a military plane to Britain. Anderson warmly welcomed them to London. In a number of meetings Bohr gave Anderson a complete account of the trenchant Washington conversations, emphasizing the climax—the invitation to Churchill to open the question with the president. So important did Anderson believe this development that he decided not to tell Bohr about the rebuff he had already suffered from

Churchill. He did not want to imperil Bohr's success by dampening his enthusiasm. He agreed that Bohr should try to see the prime minister as soon as possible.

While steps were taken to get Bohr to Churchill, Bohr had a surprise. Somehow the word of his presence in London had gotten around, for he received a letter from the counsellor of the Soviet Embassy in London. The letter invited him to come to Russia! It came from his old friend of Cambridge days Peter Kapitza—Kapitza whom they had all struggled to get out of Russia when Stalin refused to let him return to Cambridge, Kapitza who had a crocodile carved over the door of his Cambridge institute in sly tribute to "That old crocodile" Rutherford; Kapitza to whom Rutherford sent all his apparatus when it became clear that leaving Russia was impossible. The letter was dated October 28, 1943, Moscow, and had originally been sent to Bohr in Sweden. The Russians were only slightly behind Great Britain and the United States in trying to attract Bohr to their laboratories and cause. In an era of science, a great mind was much prized.

After explaining that he had learned of Bohr's escape to Sweden, Kapitza went on to say that all Russian scientists were worried about Bohr's safety, and told him that if he wanted to come to the Soviet Union he would be very welcome. Everything would be done to provide for him and his family, and everything necessary for carrying on his scientific work would be made available. Bohr had only to let Kapitza know what he required; he was certain that all would be ready whenever it suited Bohr and his family.

Kapitza, who certainly had once had his own disagreements with the regime, spoke of the complete unity of the Russian people and their determination to free themselves from the savage invasion of Hitler. He said the scientists had done everything to put themselves and their service at the disposition of the military.

At the institute, he said, Bohr would find many friends, and he named the scientists, some of whom had studied at the institute in Copenhagen. The mere prospect that Bohr might come to Russia had already aroused the enthusiasm of the physicists there,

Kapitza continued. He added that the Russian scientists, like the English, were hard at work fighting the common enemy, Nazism.

Kapitza stressed that he considered Bohr not only a great scientist but also a friend of their country, and would be honored to assist Bohr and his family in every possible way. He said that in his own mind he always linked Bohr's name with that of Rutherford, and recalled the strong bonds of affection between all of them. Kapitza spoke of his own family and inquired about Mrs. Bohr and the Bohr sons. He attached a postscript saying that Bohr might answer through the same channels through which the letter came to him.

In many ways the letter, both in proposals and tone, was almost ludicrously like the letter Bohr had received from Chadwick. It was not dissimilar to the invitations Bohr had received to go to the United States.

The United States, Great Britain, and the Soviet Union were allies, but Bohr, far better than most, knew the tenuous character of the friendship. Bohr showed the letter to Anderson and the ever present British secret service. They advised him to answer it in the same friendly tone in which it was written, though of course giving no indication of the reason for his presence in England or the United States.

In his answer, dated April 29, 1944, Bohr followed these instructions explicitly:

Dear Kapitza,
    I do not know how to thank you for your letter of October 29, which through the counsellor of the Soviet Embassy in London, I received a few days ago on my return from a visit to America. I am deeply touched by your faithful friendship and most gratified for the extreme generosity and hospitality expressed in the invitation to my family and me to come to Moscow. You know the deep interest with which I have always followed the cultural endeavors within the Soviet Union, and I need not say what pleasure it would be for me for a time to participate with you and my other Russian friends in work on our common scientific interests.
    At the moment, however, my plans are quite unsettled. My

wife and three of our boys are still in Sweden, while I, with my fourth boy, who in later years has been a scientific assistant to me, came over to England. I came to England in October and I hope that it will be possible for my wife to join me here and that it will not be long before we both can come to Moscow to see you and your family again. Ever since our first visit to Russia we have constantly had you all in our thoughts and have treasured the memory of our stay at your beautiful institute.

It has also been a very great pleasure to me to learn what certainly was no surprise, how wonderfully you have succeeded in making the facilities which you with such liberal support have created there, fruitful to science and your country and thereby to the benefit of mankind.

On my travels in England and America it has been most encouraging to meet a greater enthusiasm than ever for international scientific cooperation in which you know I have always seen one of the brightest hopes for a true universal understanding. Just about this subject, I had a most pleasant and interesting talk with Mr. Zinchenko and we spoke especially of the new promises inspired by the mutual sympathy and respect among the United Nations arisen from the comradeship in the fight for the ideals of freedom and humanity. Indeed, it is hardly possible to describe the admiration and thankfulness which the almost unbelievable achievements of the Soviet Union in these years have evoked everywhere and not in the least in the countries which have experienced the cruel German suppression.

Notwithstanding my urgent desire in some modest way to try to help in the war efforts of the United Nations, I felt it my duty to stay in Denmark as long as I had any possibility of supporting the spiritual resistance against the invaders and assisting in the protection of the many refugee scientists who after 1933 had escaped to Denmark and found work there. When however, last September I learned that they all, besides a large number of Danes, like my brother and myself, were to be arrested and taken to Germany my family and I had the great luck of escaping at the last moment to Sweden, among the many others who due to the unity of the whole Danish population succeeded in counterfoiling the most elaborate measures of the Gestapo.

For many reasons indeed I am hoping that I shall soon be able to accept your most kind invitation and come to Russia for a longer or shorter visit, and as soon as I know a little more about my plans I shall write to you again. Today it is above all upon my heart to express my deepfelt thanks to you and my warmest wishes to you and your family and our common friends in Moscow.

Yours ever,
NIELS BOHR

Bohr asked the secret service to check the whole text of his letter. When he delivered it to the counsellor of the embassy, Zinchenko, he had the distinct impression that the Russians knew the Americans and British were at work on a nuclear weapon. He suspected, though he did not know, that the Russians wanted him to work on fission and probably on a bomb. Nevertheless Bohr responded only to the surface words. For all the graciousness of his reply it was a firm *no*.

Bohr was not making any progress in seeing the prime minister. Anderson was pressing Cherwell to help Bohr, but he did not feel that he personally could approach Churchill about the appointment. Nevertheless Anderson was so impressed anew with the paramount importance of acting to forestall a postwar arms race that he felt he himself should again try to convince Churchill.

A War Cabinet paper, including memoranda on postwar world organization, gave Anderson the opportunity he sought: "I cannot help feeling," he wrote, "that plans for world security which do not take account of Tube Alloys must be quite unreal. When the work on Tube Alloys comes to fruition the future of the world will in fact depend on whether it is used for the benefit or destruction of mankind."

Anderson again reminded the prime minister that Roosevelt was thinking about the problems, and would certainly not be averse to hearing the prime minister's views. He pleaded with Churchill to send Roosevelt a cable which would "break the ice but not be open to misinterpretation," and even prepared a draft in case the prime minister should be willing. Once again Churchill returned an abrupt reply: "I do not think any such telegram is necessary nor do I wish to widen the circle who are informed."

Bohr knew that Churchill was overwhelmingly busy—he was actually absorbed in plans for the invasion of France—but he could not understand why there was no sign that the prime minister would see him. He knew that Halifax had requested an appointment for him, and he hoped that Anderson was working at it too. But the days went by. Bohr felt that he could not give up, that he had to see the prime minister both to deliver Roosevelt's message and to appeal to a man whose vision he believed to be wide. As he had done in the United States, he sought someone else who could intercede in his behalf. He went first to Sir Henry Dale, the president of the Royal Society, who as a member of the Tube Alloys Consultative Council knew "the secret" and the issues at stake.

Bohr reviewed the dangers and possibilities with Dale in the most secure place either of them could think of—seated on an iron bench in the middle of a wide expanse of grass in Hyde Park. Dale also was worried about a postwar world dominated by an atom bomb and thought it essential to get Bohr to the prime minister.

He wrote to Churchill and asked Cherwell to see that his letter reached the hands of the prime minister himself: "I cannot," he said, "avoid the conviction that science is even now approaching the realization of a project which may bring either disaster or benefit, on a scale hitherto unimaginable, to the future of mankind." In urging the prime minister to see Bohr, Dale emphasized Bohr's special position: "I think it probable that a vote of the world's scientists would place him first among all the men of all countries who are now active in any department of science."

The weapon that would put the mastery of the world in the hands of those possessing it, Dale said, will have been created by the work of leading men of science of the U.S.A. and Britain, largely on the basis of Bohr's theoretical work and recently with his active participation. "Those men of science cannot concern themselves with its tremendous political implications," Dale wrote. "It is impossible, nevertheless, for a man of science who has been allowed to see what is happening and what may be involved to neglect any opportunity of furthering the timely consideration of these implications by the only two men in whose power it may yet be to take effective action, yourself and President Roosevelt.

"It is my serious belief that it may be in your power even in the next six months to take decisions which will determine the future course of human history. It is in that belief that I dare ask you, even now, to give Professor Bohr an opportunity of brief access to you."

Bohr also appealed for help to Field Marshal Smuts of South Africa though Smuts did not then know about the bomb and Bohr could not tell him why he wished to see the prime minister. Smuts, a distinguished scientist himself, had such a regard for Bohr that he was willing to help in anything Bohr considered paramount. When Lady Anderson invited the field marshal to lunch to meet Bohr he exclaimed: "This is tremendous, as though one were meeting Shakespeare or Napoleon—someone who is changing the history of the world."

Cherwell, too, transmitting Dale's letter to Churchill, was aroused to the need for action. He was further spurred into joining Bohr's cause by a talk with Mackenzie King, the prime minister of Canada. King told Cherwell that he thought international action would be necessary if atomic energy was not to master its masters. Canada of course was a participant, along with Great Britain, in the development of the bomb. Cherwell wrote a communication of his own to the prime minister: "I must confess," he said, "that I think plans and preparations for the postwar world are utterly illusory so long as this crucial factor is left out of account." Cherwell proposed that since the prime minister himself had no time to give to such subjects, that Field Marshal Smuts might be informed and asked to advise him.

Churchill readily consented to bringing Smuts in as an advisor, but he repeated that he would not stand for any lifting of secrecy. The prime minister feared that any opening of negotiations might upset the Quebec Agreement, and he wanted none of that. He indicated in a note to Cherwell that his principal concern about postwar control lay in the effect of the Quebec Agreement on Britain, and Britain's own access of nuclear power and materials. "Our association with the United States must be permanent and I have no fear that they will maltreat us or cheat us," he said.

April was turning into May. The parks and fields of England were green again and the sun of spring brought relief from a cold weary winter of war, but more than the spring stirred the spirit of England. Preparations were well advanced for an invasion of northern France and a final attack on the bastion of the foe. The prime minister had no desire to interrupt this heartening and all-demanding effort to discuss a distasteful subject with a Danish professor.

Bohr still knew none of this. He could only watch day after day slipping away. He could not stay indefinitely in England waiting for an interview with the prime minister. He had promised Groves that he would return in a short time. The waiting, the uncertainty, and the feeling that he was losing an opportunity that might never recur, an opportunity vital to humanity, were hard to bear. Bohr had need of all his fortitude. He waited and waited through one anxious day after another.

17 /

The Prime Minister, the President, and the Scientist

THE PRIME MINISTER GLOWERED. Cherwell, Smuts, and now Dale, all were urging him to see the Danish professor with his message from Roosevelt. The president had never before shown any inability to deliver his own messages. But Churchill took another look at Dale's letter: "there are only two men in whose power it may lie to take effective action, yourself and President Roosevelt. It is my serious belief that it may be in your power even in the next six months to take decisions which will determine the future course of human history. It is in that belief that I dare to ask you, even now, to give Professor Bohr the opportunity of brief access to you."

These were words not easily denied or dismissed. Churchill grumpily told Cherwell to have Bohr come on May 16. Bohr walked through the famous doorway of No. 10 Downing Street with the highest of hopes. The prime minister, on the other hand, met the famous scientist with the most meager of expectations. Bohr began to talk in his low voice and to build up his argument, point by point, as he had done with Halifax, Anderson, and others. But Churchill's patience did not last as long as theirs; he liked incisiveness. Why didn't Bohr come to the point?

Cherwell, who had been asked by the prime minister to be present, put in a word. His remark sounded to Churchill like a criticism of the Quebec Agreement, and the two were off into an argument of their own that consumed virtually all of the half hour allotted for the interview. Bohr did not even get to his argument that the new knowledge of the atom offered untold possibilities for a new level of relationships of nations, or, conversely, if it were held as a secret monopoly, untold potentialities for evil—an arms race and even ultimate destruction of civilization itself. He did not get to tell the prime minister that the president would welcome suggestions for control of atomic energy. The time was up. As he started out, Bohr asked if he might send a memorandum to the prime minister, developing the points he wanted to emphasize. "It will be an honor for me to receive a letter from you," answered the prime minister. And then he added tartly, "But not about politics." The prime minister was annoyed. He had consented to the interview against his better judgment and his time had been wasted.

Bohr, who had crossed the submarine-beset Atlantic and had waited weeks for the moment that might open the door to a literally new world, was sick with despair and discouragement. "We did not even speak the same language," he lamented.

The meeting had been an unmitigated disaster and both principals knew it. On such a clash of personalities and incompatibilities did history, and perhaps the ultimate fate of mankind, turn.

Up to the point of the Bohr-Churchill meeting the prospects could not have been more promising. The head of the British atomic energy project, the president of the Royal Society, the British ambassador to the United States, the American president and one of his chief advisers, and thus most of the men of power and decision in government and science who knew about the bomb were willing to consider a new approach as demanded by a new force in the world. They were convinced by Bohr that mortal danger could lie in the alternative, a nuclear arms race. Among the few who knew that a new order was being created and had the power to influence it, only one, only Churchill, rejected even an attempt at change.

From the day of the ill-fated interview the direction was set.

The struggle would continue and modifications would be made, but the British prime minister, through conviction and pique, had made it very unlikely that a political breakthrough comparable in magnitude to the breakthrough into the heart of the atom would be made.

Bohr, and those who heard of the sad outcome of the meeting asked repeatedly what went wrong. Anderson, who knew the prime minister well, had feared from the beginning that the meeting would fail or be inconclusive. He told Cherwell of his misgivings: "Bohr's mild, philosophical vagueness of expression and his inarticulate whisper may prevent his making the desperately preoccupied prime minister understand him." Anderson was all too right. Bohr and the prime minister were not only diametrically different, the differences were of a nature calculated to produce irritation and disagreement between the two. Incisiveness and discursiveness are not easily reconciled. Nor is a man whose voice has rallied nations apt to like, or be willing to abide, a whisper.

Though no record was made of the unhappy Bohr-Churchill meeting, publication of the Churchill papers disclosed that Churchill was fundamentally opposed to what Bohr sought before Bohr saw him. The prime minister's rejection of Anderson's minutes and his sharp comments on them demonstrated a disagreement that underlay any personal animosity toward Bohr. Bohr did not have a chance. Churchill wanted nothing to upset the status quo or the secrecy, or that would open any further approaches to Russia.

Bohr, who had listened to the ringing speeches of Churchill, had mistaken matchless courage and indomitable purpose for political imagination and willingness to progress. He had forgotten or he did not know that Churchill was no social or political innovator. The prime minister was a preserver of past glories, not a creator of new ones. Bohr also failed to realize that a prime minister with Churchill's sense of the fitness of things would not be overjoyed at dealing with an intermediary, and particularly an intermediary who was an outsider. No distinction in science could offset this handicap. Prime ministers properly spoke to presidents and presidents to prime ministers. The Bohr-Churchill meeting clearly was doomed from the outset. But it was crucial and irreparable and its consequences were long.

As stricken and defeated as he felt, Bohr did not surrender. He carefully drafted the letter Churchill had grudgingly said that he would accept. Bohr tried to put his statement on atomic energy into a form that he hoped would catch the imagination of the prime minister. And he at last conveyed the Roosevelt message that had brought him to England.

Bohr also met again with Anderson and Smuts. This time he could talk freely to the South African leader. Acting with the prime minister's permission, Cherwell and Anderson had informed Smuts of the development of the bomb and had asked him to begin a study of postwar control of the new force. Smuts's first step was to meet with Bohr. This time Bohr talked to a man willing to listen and to a scientist-statesman uniquely equipped to understand. By June Smuts wrote to the prime minister: "The discovery of the use of atomic energy is both for war and peace, for destruction and beneficial use, the most important ever made by science."

However, Bohr did not succeed in convincing Smuts that any immediate approach should be made to other nations—which meant particularly to Russia since the Soviet Union alone had the potential to become a major competitor. Smuts also felt that any initiative would have to come from the American president. Nevertheless, Smuts told the prime minister that atomic energy could not long remain a secret and that it could set off a destructive competition in armaments. At the first opportunity, Smuts urged, the prime minister and the president should consider the whole fateful question, and particularly whether or not the new knowledge should be shared with the Russians. Essentially this was all that Bohr was seeking as a first step. "If ever there was a matter for international control," Smuts wrote, "this is it. Immediately after the war steps should be taken to regularize production and use of atomic energy." Smuts also recommended the establishment of a small committee of physicists to advise the powers on the carrying out of atomic policy "in the interests of mankind."

On June 6 American and British troops stormed ashore on the

beaches of Normandy. The landings went well; the Germans were taken by surprise (they had expected the attack in another section) and were slow in counterattacking. Hitler had given orders that no troop movements were to be taken without his explicit consent, and thus caused delays that further handicapped the Germans.

As acute an observer as Bohr could not doubt that the invasion marked the beginning of the end. A few days later Bohr returned to the United States, buoyed up by the thought that the war was entering its final phase, but deeply discouraged by his failure to enlist Churchill in a new approach to peace. He held some hope, however, that Anderson and Smuts still might persuade the prime minister to open discussions with Roosevelt.

As soon as possible after his arrival in Washington Bohr saw Justice Frankfurter and told him the whole frustrating story of the meeting with Churchill and the encouraging one of the continuing work of Anderson and the recruitment of Smuts. The justice lost no time in reporting it all to the president. Roosevelt, as was his wont, threw back his head and laughed at anyone's tackling Winston in one of his belligerent moods. The president was intrigued and said that he wanted to talk to Bohr. To guard against the indirect speech that had so riled the prime minister, Bohr was asked as a preliminary to put his ideas into a memorandum.

Through the last days of June and the early days of July, Washington baked in one of its typical summer heat waves. By 10 o'clock in the morning the temperature was up to 90 degrees and it continued to rise until it passed 100 in the late afternoon. Heat shimmered over the softening asphalt of the streets and Washingtonians sought relief under the shade of the trees or in the newer air-cooled buildings. Through the hot days, Bohr, with the assistance of his son, worked on the memorandum. Bohr would pace the room dictating, only to change the paragraph on another circuit of the room. The subject was much too secret to bring in any secretarial help. Such a revelation of state secrets would have been unthinkable. Aage therefore took down his father's words and typed them. After one of Washington's hot nights—not conducive to sleep for one from a northern climate—Bohr would come in with new ideas of how a sentence or a section might be changed to encompass every fine shade of his meaning.

The pages Aage had so laboriously typed would be destroyed and Aage would bend over the typewriter again. While he pecked out the new version Bohr would sew on buttons and do any other required mending. And the next day they would start all over again; no effort was too great, no sentence ever sufficiently accurate or trenchant for all the memorandum had to accomplish. The failure with Churchill made it all the more imperative to find the words that would move the mind and heart of Franklin Delano Roosevelt. On July 3, the seven-page memorandum, went to the White House. It was marked on receipt *Top Secret*. Only excerpts of the memorandum itself can convey the powerfully knit argument and the feeling of Bohr's own involved style:

It certainly surpasses the imagination of anyone to survey the consequences of the project in years to come, where, in the long run, the enormous energy sources which will be available may be expected to revolutionize industry and transport. The fact of immediate preponderance is, however, that a weapon of an unparalleled power is being created which will completely change all future conditions of warfare.

Quite apart from the question of how soon the weapon will be ready for use and what role it may play in the present war, this situation raises a number of problems which call for most urgent attention. Unless, indeed, some agreement about the control of the use of the new active materials can be obtained in due time, any temporary advantage, however great, may be outweighed by a perpetual menace to human security.

. . . . .

Without impeding the immediate military objectives, an initiative, aiming at forestalling a fateful competition, should serve to uproot any cause of distrust between the powers on whose harmonious collaboration the fate of coming generations will depend. . . .

. . . . .

Many reasons, indeed, would seem to justify the conviction that an approach with the object of establishing common security from ominous menaces, without excluding any nation from participating in the promising industrial development which the accomplishment of the project entails, will be welcomed and will be responded to by a loyal cooperation on the enforcement of the necessary far-reaching control measures.

Soon afterward, Bohr asked permission to add a few paragraphs:

Just in such respects helpful support may perhaps be afforded by the world-wide scientific collaboration which for years has embodied such bright promises. . . . Personal connections between scientists of different nations might even offer means of establishing preliminary and unofficial contact.

It need hardly be added that any such remark or suggestion implies no underrating of the difficulty and delicacy of the steps to be taken by the statesmen in order to obtain an arrangement satisfactory to all concerned, but aims only at pointing to some aspects of the situation which might facilitate endeavours to turn the project to the lasting benefit of the common cause.

At the same time the memorandum went to the White House, the Russians began their great summer offensive of 1944. They soon were pressing against the boundaries of East Prussia. The war also was progressing on the Western front. On July 26 General Bradley's American forces broke through the German front at St. Lô. Four days later General Patton's newly formed Third Army captured key Avranches, and all Brittany and the Loire Valley lay open before them. On July 30 General von Kluge wired Hitler's headquarters "The whole Western front has been ripped open . . . the left flank has collapsed." It was the turning point, the climax of the Allied invasion of France. None could doubt now that the European war was in its final phases.

It was against this background that Roosevelt decided late in August that the time had come to talk to Bohr. Planning for after the war could no longer be delayed. The Bohrs returned from Los Alamos, and on August 26 Bohr entered the President's big oval office in the White House. Roosevelt welcomed him with an expansive smile, a hearty handclasp, and, with a sweeping gesture with his cigarette holder, assigned the visitor to a chair and gave him the sense that he was taking a place of honor. Bohr, like all other visitors, was seated beside the big presidential desk with its load of miniature donkeys, trick ashtrays, specially imprinted match folders, and other gadgets. Behind the president stood the American flag and his personal flag, in addition to a table crowded with photographs of members of the family and friends. In the background the south windows framed a view of the Washington Monument.

The clear-cut, easy voice of the president said that he had read Bohr's memorandum, and was generally in agreement with it. Bohr was invited to discuss his points and the president listened intently before he in his turn began to talk. He said with his famous, mischievous grin that he understood how things had gone with Churchill in London. As he well knew, this was often the way the prime minister reacted at first, though in the end he often changed. There was the implication that Franklin Delano Roosevelt often was able to persuade him. Roosevelt related several stories of his meeting with Churchill and Stalin at Teheran. He had often had to break in with a joke to relieve the heated atmosphere and to keep them from flying at one another. The stories drew the visitor into the charmed inner circle and let him share in the historic moments. Few, and certainly not Bohr with his responsiveness, could resist this intimate light joking about the great.

Seriously, Roosevelt agreed that atomic energy opened vast prospects both for good and for great perils. He said that he was confident, nevertheless, that it would make a decisive contribution to international cooperation, and he would go further—it would open a new era in history. There was enthusiasm in the president's voice and manner. Roosevelt seemed to agree too that an approach should be made to Russia. He said that he was sure Stalin was enough of a realist to understand the revolutionary character of atomic energy. He had another story of Stalin to support his point. In words that elated Bohr, he said that he would discuss the whole problem of the atom with Churchill at the meeting they had just arranged for Quebec in September, and he hoped to see Bohr soon afterward. In the meantime, he said, Bohr might write to him at any time he chose.

There was even time for a few words about Greenland and about Denmark's own position. When the Germans occupied Denmark they immediately eyed Denmark's colony, Greenland, as a possible Atlantic base of operations against the Allies. The highly regarded Danish ambassador Kauffmann had stepped in at this point, announcing that since his government was in no position to act freely, he would take upon himself the responsibility of placing Greenland under the protection of the United States. The United States promptly occupied the strategically placed island. The Faroe Islands and Iceland also were occupied by the Allies in the same manner, and Germany was barred from Atlantic bases that would have constituted a serious menace to the West and its shipping. This marked the end of the Roosevelt-Bohr talk. But an hour and a quarter had gone by. There had been no hurry, no curtailment of discussion, and no interruptions.

Bohr had heard of Roosevelt's famous charm and his tendency to let visitors believe that he was in hearty agreement with them. But this time, he felt, the president had been too explicit; Bohr had no doubts of his great interest and concern. Bohr, a man quite capable of enthusiasm, was high in spirit as he left the White House. The contrast with the Churchill interview was unbearably sharp.

Bohr had asked the president if he might report their conversation to Lord Halifax and Sir John Anderson, to which Roosevelt willingly consented. Bohr therefore hurried to repeat the momentous conversation in fullest detail to the other half of his constituency, and a long, detailed cable promptly went off to London. Bohr also wrote Roosevelt a letter of thanks, summing up his main points.

To the few others to whom he was free to speak—the scientists with whom he was working at Los Alamos—Bohr later reported his own words. He scrupulously refrained from repeating anything Roosevelt had said to him. He felt that it would not be right for him to quote the president (a scruple not shared by most visitors to the White House) and he kept his silence for more than twenty years, long after the death of Roosevelt. He broke it only when the United States and Great Britain were about to publish their hitherto secret records of the period. Mrs. Margaret Gowing, who interviewed Bohr at length in preparation for the official history of the British atomic energy effort, said that she found his memory remarkably accurate and in complete accord with the documentary evidence where such evidence existed. There was, of course, no transcript of the Bohr-Roosevelt conversation.

By September the time came when the president and the prime minister had to discuss the control of the new atomic force. Other matters, however, came first when they met for the second Quebec Conference. They had to plan the final assault on Germany, and for the ending of the war with Japan. Secretary of the Treasury Morgenthau called for a stringent limitation on the industry of postwar Germany and Roosevelt agreed to a continuation of Lend-Lease. Britain had all but exhausted her assets in fighting the war and would need continuing financial help from the United States. It was a triumphant meeting, opening and ending, as Churchill said, in a "blaze of friendship." At its close the Churchills were to visit the Roosevelts at Hyde Park, and the two leaders decided to defer the discussion of atomic energy to the greater quiet and informality of the Roosevelt home.

As Churchill, his wife, and daughter Mary rode south by train the prime minister was at one of the peaks of his career. He had swept all before him at Quebec and he was in form to continue his triumphs at Hyde Park. On Sunday, September 17, the day of the Churchill's arrival, Roosevelt drove the prime minister around Dutchess County. To give Churchill the fullest views of the lordly sweep of the Hudson far below them, the president in his hand-controlled car would drive up to the very edge of the bluffs overlooking the river. This gave Churchill some uneasy moments, though they were almost the only uneasy ones he was to have at Hyde Park.

On Monday, the two leaders shut themselves in one of the

smaller Hyde Park rooms—Roosevelt disregarding the late September warmth—and settled down to the question of the atom bomb. Churchill reviewed the military situation. Paris had been liberated and large parts of France cleared of the enemy. The armies were advancing in Italy, and the Soviet offensive, though temporarily halted, might again surge forward at any moment. Hitler's "secret weapons" were almost mastered and there was no evidence that Hitler "had learned how to make the atom bomb." Nevertheless there was no certainty the war would end in 1944.

Both leaders had talked to Bohr and they turned to his proposals. But Churchill's dislike of Bohr and his ideas had not changed at all; he was as adamant as ever. He wanted no disclosure to anyone, and Roosevelt, who only a month before had shown eager interest in attempting a bold new approach, joined him now in a flat rejection of anything other than trying to build up and maintain the American-British lead in atomic energy. They ordered a continuation of the utmost secrecy.

Churchill, by all the evidence, made a slashing attack on Bohr. There is no record of the Hyde Park conversations—though the results are clearly documented—but Churchill in ordering Cherwell to take certain follow-up action repeated what he had said to Roosevelt. "The President and I are worried about Professor Bohr," the prime minister told Cherwell. "How did he come into the business? He is a great advocate of publicity. He made an unauthorized disclosure to Justice Frankfurter who startled the president by telling him all the details. He said he is in close correspondence with a Russian professor, an old friend to whom he has written about the matter and may be writing still. The professor has urged him to go to Russia in order to discuss matters. What is this all about? It seems to me Bohr ought to be confined or at any rate made to see that he is very near the edge of mortal crimes."

The same attack must have swayed Roosevelt. Apparently the faith he had shown in Bohr was badly shaken and undermined. At the end of the day the two, the president and prime minister, drafted an aide-memoire, which reflected their complete rejection of Bohr's proposals, and the about-turn on him personally. It was headed "Aide-Memoire of the Conversation between the President and the Prime Minister at Hyde Park 19th September, 1944, and read:

1. The suggestion that the world should be informed regarding Tube Alloys with a view to an international agreement regarding its control and use, is not accepted. The matter should continue to be regarded as of the utmost secrecy, but when a "bomb" is finally available, it might perhaps, after mature consideration be used against the Japanese, who should be warned that this bombardment will continue until they surrender.

2. Full collaboration between the United States and the British Government in developing Tube Alloys for military purposes shall continue after the defeat of Japan, unless and until terminated by joint agreement.

3. Enquiries should be made regarding the activities of Professor Bohr and steps taken to ensure that he is responsible for no leakage of information, particularly to the Russians.

The English spelling throughout all parts of the Aide-Memoire, the reference to the atomic energy project as Tube Alloys, and later the instructions to Cherwell to proceed against Bohr all bespoke the determined hand of Winston Churchill. Nevertheless Roosevelt agreed to the written statement and did not again see Bohr. If the chances of turning to a new course were injured at the London meeting of Bohr and the prime minister, they were further set back at Hyde Park.

Ironically and tragically both of the bases on which Churchill rested his convictions and action were misconceptions or mistakes. How, Churchill asked, did Bohr "get into it?" Either the prime minister did not know or comprehend that the whole undertaking was developed in large part on Bohr's theoretical work, and might never have gone forward on its World War Two schedule without it. The prime minister charged that Bohr had made unauthorized disclosures to Justice Frankfurter. If Roosevelt volunteered this information, did he not know that the troubled young physicist Mrs. Roosevelt had asked him to see about a year earlier had also talked to the justice and to Bernard Baruch? The president had even phoned James B. Conant about the physicist's fears that the bomb project was floundering and might be lost. Perhaps, though, Roosevelt did not connect the two events and did not know that the justice had learned about the bomb project from the young physicist at a time when Bohr had not even left Denmark.

Churchill implied that Bohr might have been dealing improperly with the Russians. If he had asked for an intelligence report he would have learned that the Kapitza letter was answered with the full knowledge and approval of the British intelligence service and that Bohr was serving the British, not the Russians. The official British history of the period is unable to throw any light on the source of Churchill's misconceptions. They were not held by any other member of Churchill's government and could not have been derived from those upon whom Churchill relied for information.

In *Great Britain and Atomic Energy*, Margaret Gowing wrote with a historian's impartiality: "Bohr's honor and integrity were of course as great as his prowess in physics. His friends Cherwell, Anderson, Halifax and Campbell rushed in to defend them and to say that Churchill was, in effect, talking nonsense." In one letter Campbell commented that both Halifax and Cherwell "felt strongly that the great P. J. (Panjandrum) was barking up an imaginary tree."

When Cherwell was directed by the prime minister to take action against Bohr, he too told Churchill that he was wrong. "I have always found Bohr most discreet and conscious of his obligations to England to which he owes a great deal, and only the strongest evidence would induce me to believe that he had done anything improper in this matter. I do not know whether you realize that the possibilities of a super weapon on Tube Alloys lines have been publicly discussed for at least six or seven years.

"The things that matter are which processes are proving successful, what the main steps are and what stage has been reached. Most of the rest is published every silly season in most newspapers."

In addition to defending Bohr, Cherwell was striking at the other base underlying Churchill's action, his belief that the bomb was an absolute secret and could be maintained as one. Ever at the time Churchill was withholding information from his own cabinet,

Klaus Fuchs, a member of the British team at Los Alamos, was betraying much of the specific detail about the bomb to the Russians.

In addition the French scientists who had escaped after the fall of France and who had been working in the Montreal laboratories of Tube Alloys were pressing to return to France, now that much of their country had been liberated. They had all been associates of Joliot in his work on fission. If they returned to France they would take all the knowledge they had gained in working on the bomb project with them. France would not only have "the secret" but Groves feared that Joliot, whose joining of the Communist party was public knowledge, would turn it over to Russia. And yet the French scientists who had gone to great risks to transport the French heavy water to Britain and who had worked loyally on atomic energy all during the war could not be imprisoned to prevent their return to their own country.

As the sweep of the American and British armies across France increasingly portended the end of the war, scientists at the Metallurgical Laboratories at the University of Chicago, one of the sections of the Manhattan District, called for a worldwide organization to prevent the atom from "becoming the destroyer of nations." Vannevar Bush, the director of the Office of Scientific Research and Development, and Conant, the director of the National Defense Research Committee, and thus the two men ultimately responsible for the development of the bomb, had long recognized that extraordinary controls would be necessary after the war to cope with the revolution the atom would create in politics and economics. However, in the concentration required for the production of the weapon, they delayed in taking any action. Later in the summer, the progress of the war and the report of the Chicago scientists told them they could wait no longer.

On September 19, the day after Roosevelt and Churchill plumped for continuing secrecy, Bush and Conant sent a letter to Secretary of War Stimson. Emphasizing that basic scientific information would soon have to be released, and recommending a treaty with Britain to assure permanent interchange, the two scientist-

adminstrators turned to international control. It would be dangerous, they said, for the United States to assume that security lay in
holding secret its present knowledge. Conant, in notes jotted on his
own copy, indicated that they were thinking of an international
control agency, which would include Russian membership. He also
noted that in his opinion the United States and Great Britain
should avoid precipitating an arms race before an approach was
made to multilateral control.

On Friday, September 22, Bush received a call to come to the
White House. The president did not know until the letter went to
Stimson that Bush and Conant were thinking of postwar problems
and Bush did not know that Roosevelt had been working with Bohr
and Frankfurter since the beginning of 1944. When Bush went into
the president's office he met Admiral William D. Leahy, the president's personal military adviser, and Cherwell, who had come on to
Washington after finishing his work at Quebec. The admiral had
just been told of the bomb, but his pale, stony face registered no
surprise. Roosevelt then told Bush about Bohr's proposals and said
that he was very much worried that there might have been a
security leak. His confidence of a month ago had been badly shaken
at his conference with Churchill at Hyde Park. Cherwell in no
uncertain terms told Roosevelt, as he had Churchill, that Bohr was
entirely reliable and the man whose theories had made the bomb
possible. He also straightened out Churchill's distorted version of
the Kapitza letter. Bush, who knew Bohr well, joined in and completely confirmed Cherwell's opinion on all points of which he had
any knowledge.

The president then dropped the subject of Bohr, and nothing
more was heard of keeping him under suspicion or imprisoning him.
Roosevelt went on to talk generally about the bomb, whether it
should be used against Japan or be demonstrated in the United
States and held as a threat? The president also turned to the relations of Great Britain and the United States. He said the
interchange on atomic energy must be continued. Nevertheless
Roosevelt said nothing about the aide-memoire signed earlier in the
week at Hyde Park or about its provisions on the very subject he

was discussing. Perhaps the document seemed a casual one to the President; in any event it suffered an odd fate. When Roosevelt brought the aide-memoire back from Hyde Park, someone, puzzled by the reference to an unheard-of *Tube Alloys*, filed it with naval documents. *Tube Alloys* sounded as though it had something to do with ships. The American atomic energy officials, who supposedly should have been guided by the document, knew nothing about it. When much later the British brought it to the forefront, an intense search was made for the United States copy, but, until much later, it could not be found and the United States had to ask the British officials for a duplicate.

Bohr, in the meanwhile, waited anxiously to hear from Roosevelt about the Hyde Park meeting. No word came directly. When he did hear from others, Bohr was deeply disturbed by the decision to cling to the strictest secrecy and not to seek international control at a time when the United States and Great Britain would have an irresistible *quid pro quo* to offer in return for cooperation in a new kind of control—an agreement not to enter an arms race.

Bohr took gravely the attack on his own integrity and the threat to jail him, but without rancor. But he did not rest until he had done everything possible to set the facts straight. He talked at length to Cherwell, Groves, Bush, and others, giving them all the details about what had happened, including the written records. Bohr's actions and concerns could not have been more honorable, nor more scrupulous. And no one gave the slightest credence to Churchill's strange distortions of the facts. All were entirely satisfied and regretful that the aspersions had been cast.

Why had Roosevelt made such an about-face, why had he gone from his cordiality and acceptance or approval of a month earlier, to rejection of Bohr's ideas and cold suspicion, or at least acquiescence in Churchill's suspicions? There was no answer. Bohr and the few who knew could only speculate. Perhaps it was entirely a giving in to Churchill? Perhaps the president was too weary to defend his own views. Those who saw him regularly could not miss the increasing gauntness of his face and the deep circles under the eyes. Perhaps it was a growing distrust of Stalin. The Russian dictator's refusal to help the people of Warsaw when they rose against the Nazis had aroused the indignation of Roosevelt and Churchill. Could it have been Roosevelt's fear of a political reaction within America if opponents of cooperation learned that sharing of the American-British "secret" was planned. Such an uproar did come later. It was baffling. Possibly Churchill did not know about Bohr's work with the English authorities on the Kapitza letter, but Bohr had mentioned it to Roosevelt and had even mentioned it in his memorandum as part of the evidence that the Russians might be working on atomic weapons.

Oppenheimer, who knew what had happened, said many years later: "The outcome was not funny; it was terrible. For one thing it shows how very wise men in dealing with very great men can be very wrong. It worked itself out, because the English were sure this was all nonsense. . . . But the fact is that it stopped, it ended his communication with the President and it very seriously impeded his communication with our government."

Perhaps the president relented somewhat, for the word came to Bohr, "Talk to Bush." Bohr welcomed this, for in addition to defending Bohr at the White House, Bush was thinking along the same lines; the two were moving in the same direction. Bush was disturbed that the undeviating secrecy ordered by the president and prime minister might lead to extraordinary Russian efforts to develop an atomic bomb, and in a decade or two to a catastrophic conflict. Bush poured out his fears to Stimson and asked if another policy might not head off such a fearsome outcome.

Only eight days after Bush conferred with the president at the White House, Bush and Conant sent the secretary of war two memoranda, one a concise easy-to-read summary and the other a fuller development of these ideas. The scientists had first to awaken Stimson to the full magnitude of what they were talking about. The facts were set forth to speak startlingly for themselves:

The bomb (to be ready by August, 1945) will be the equivalent of a blast damage of from 1,000 to 10,000 tons of high explosive.

In the future an even more powerful bomb, a hydrogen bomb,

based on fusion rather than fission, is a possibility, and it would be a thousand times more devastating. Every city in the world would be at the mercy of any possessor of the bomb who struck first.

The present advantage of the United States and Great Britain is temporary.

Almost any nation will be able to overtake us in three or four years.

It would be folly for the United States and Great Britain to assume they will always be ahead.

American security cannot possibly be maintained by secrecy. Physicists knew all the basic facts before the work began. All facts therefore should be disclosed, except those involved in manufacturing and military development.

The best chance of forestalling a fatal atomic arms competition between nations should lie in a free interchange of all scientific information under an international office. Its staff should have unimpeded access to laboratories and plants throughout the world.

Russia might be reluctant, but the Soviets would be receiving a valuable consideration for complying. It would be to the self-interest of the Russians to go along rather than to develop their own atomic weapons.

Whatever the opposition, the danger warrants the effort.

Bush and Conant thus took up the fight where Bohr had been blocked. Bohr felt there was nothing more he could do in Washington. Without explicit directives from the American or British authorities he did not believe that he could further intervene. Bohr, a greatly saddened man, and Aage, returned to Los Alamos. There the pace was quickening.

On February 2, 1945 a small truck, escorted by two radar-equipped patrol cars, rolled onto the mesa. It carried twenty shipping cans of pure plutonium nitrate. For the first time uranium ore had been changed on greater than laboratory scale into a new man-made element. The plutonium had now to be assembled into a bomb. Another bomb was to be made out of the U-235 which also was arriving at Los Alamos.

The uranium bomb, the "Little Boy," was to be exploded by a

special type of gun. But for the plutonium bomb, the "Fat Man," another method, implosion, had to be developed to start the chain reaction. At the end of January the implosion method still presented unsolved problems, and when Bohr returned he joined in the work on this phase.

"By early in February Niels Bohr had clarified what had to be done," said Hewlett and Anderson in their official history. Thus the man who had identified the fissionable U-235 used in one bomb and had shown the possibility of creating the plutonium used in the second, gave material help in the final construction of the "Fat Boy."

Out on the mesa, as the canyons echoed with test firings and the bomb took final form, the immensity of what was happening to the world was born in upon Bohr with new intensity. Only a short time remained to achieve a new direction and head off a world-wide manufacture of atomic weapons. If an approach were made to other nations after the first bomb exploded Bohr felt certain that it would be regarded as coercion. The American and British advantage and chances of preventing an arms race would be substantially lost. The stakes were so tremendous Bohr could not give up. Despite the setbacks he had suffered, the political leaders had to be persuaded to see that the choice lay between obliteration and promise. Bohr again appealed to Lord Halifax and asked to be called to London to confer once more with Sir John Anderson.

While Bohr was waiting for an answer events were moving fast. Roosevelt, Churchill, and Stalin met at Yalta and announced that all nations would be invited to a conference in San Francisco in April, 1945, to draft plans for a United Nations. Here was a new opportunity.

Bush in Washington was also pressing steadily for appointment of a committee to work on postwar controls and policies. Other scientists, too, were becoming alarmed. Einstein, at the urging of Szilard, wrote to Bohr and President Roosevelt. Szilard had emphasized that politicians who knew nothing of nuclear power could not comprehend the threat it posed. Einstein therefore suggested in his letter to Bohr that men who did know, say Compton in the United

States, Cherwell in England, Kapitza and Joffe in Russia, and Bohr should come together to plan for the internationalization of atomic power. "Don't say impossible," said Einstein in the letter to Bohr, "but wait a few days until you have accumstomed yourself to these strange thoughts."

Bohr hastened to see Einstein to explain to him that for those entrusted with the development of the bomb to take matters into their own hands would violate all the trust placed in them and precipitate the strongest counteraction. He assured Einstein that heads of governments were indeed well aware of the problem and all its implications and were giving it their closest study. Einstein saw the point and agreed to take no independent action.

The difficulties with the French also were coming to a new crisis. The scientists were not only demanding the right to return home, but Joliot was pressing for permission to come to London, primarily to protect patents the French had taken out on some of their prewar work on fission. Even the American-British Combined Policies Committee and Churchill himself were drawn into the fray.

Churchill in one of his "minutes," on March 25, 1945, called attention to the British-American agreement at the first Quebec Conference not "to communicate any information about Tube Alloys to third parties except by mutual consent." He said that he would continue to oppose the slightest disclosure to either France or Russia. The prime minister's attitude was explicitly set forth: "You may be quite sure that any power that gets hold of the secret will try to make the article and that this touches the existence of human society. The matter is out of all relation to anything else that exists in the world, and I could not think of participating in any disclosure to third or fourth parties at the present time. I do not believe there is anyone in the world who can possibly have reached the position now occupied by us and the United States."

Bohr, by invitation, returned to Britain in March, 1945. Anderson was as much in agreement with him as ever and Foreign Secretary Anthony Eden conceded too that the issue of atomic power would have to be faced. But both Anderson and Eden were bound by Churchill's position. Anderson, who was in the middle of the

French imbroglio and thus daily reminded of the impossibility of absolute secrecy, and Bohr both feared that the great opportunity was being lost. There seemed to be only one last hope of checking the drift in the direction that could result in catastrophe—another appeal to Roosevelt. Again it was recognized that any action would have to originate from the United States.

Bohr returned to the United States on April 4, 1945, and began the preparation of a new memorandum for the president. He asked the advice of Halifax and Frankfurter on how the statement could be brought to the attention of the president. On April 12, the tall ambassador and the short, energetic justice met to consider how they could help Bohr and, as they believed, the cause of future peace. They sought privacy and security in a walk through Washington's wooded ravine Rock Creek Park. As they went down the steep embankment, it was the most superb of spring days, the sun was warm, the breezes gentle, the sky flecked with a few fleecy clouds and the green of the leaves and grass at its freshest. It was a day to glory in, despite the burden that weighed upon them. How could they get full consideration of the issues of atomic power? Should Bush or other American friends be asked to take Bohr's new memorandum to the president? Again they reviewed the basic issues. Was it more dangerous to take in a third party than to do nothing? If Russia were told, how should cooperation be managed? Both agreed that action must be taken, and they started up out of the park.

Suddenly the church bells began to toll and kept ringing until the air seemed filled with the sound. They saw hurrying people stop and speak to others walking by. Some then stood frozen, some broke into a run, some wept. Halifax and Frankfurter, in turn, stopped, transfixed. The president was dead and an era had ended. The ambassador sensed it and the justice knew it, the youngest clerk in a government department felt it, and so particularly did a guest in the United States, Niels Bohr.

After the solemn, moving ceremonies in which a stricken nation and world relinquished Franklin Delano Roosevelt to history and memory, word came through of the president's final days at Warm Springs, Georgia. Roosevelt had been working on a speech. The uncompleted text lay on his desk. He had been looking to a new era, one he did not know he would not see. "Today," he wrote, "we are faced with the preeminent fact that if civilization is to survive, we must cultivate the science of human relationships—the ability to all people of all kinds to live together and work together in the same world, at peace." The president, turning to Bohr's thesis that scientists might have a special role to take, quoted Thomas Jefferson on "the brotherly spirit of science, which invites into one family all its votaries of whatever grade, and however widely dispersed throughout the different quarters of the world."

As soon as members of the cabinet and other leading officials could reach the White House, Harry Truman was sworn in as the thirty-third president of the United States. The cabinet, grim and stricken, remained for a brief meeting with the new president. As the meeting ended the secretary of war lingered. "He asked to speak to me about a most urgent matter," said Truman. "Stimson told me that he wanted me to know about an immense project—a project looking to the development of a new explosive of almost unbelievable power.

"That was all he felt free to say at the time and his statement left me puzzled. It was the first bit of information that had come to me about the atomic bomb. But he gave me no details. It was not until the next day that I was told enough to give me some understanding of the almost incredible developments that were under way and of the awful power that might soon be placed in our hands." There could have been no more striking testimony to the stringent secrecy than these words of a former senator and vice president.

Bohr, Halifax, and Frankfurter, almost with a sense of starting anew, debated what should be done now. It was decided that Bohr should talk to Bush. Bohr took Bush the memorandum he had prepared for Roosevelt, plus an addendum prompted by the United Nation Conference soon to open at San Francisco. Bush told Bohr that he shared both his views and his sense of urgency. The bomb would soon be ready. Bohr also pointed out that the Russians might soon fall heir to some of the German work and documents on atomic energy.

Though Bohr's sentences were long and difficult to follow, his thought showed through the memorandum's tangle of words. Some excerpts demonstrate its tone:

> Above all, it should be appreciated that we are faced only with the beginning of a development, and that, probably within the near future, means will be found to simplify the methods of production of the active substances and intensify their effects to an extent which may permit any nation possessing great industrial resources to command powers of destruction surpassing all previous imagination.
>
> Humanity will, therefore, be confronted with dangers of unprecedented character unless, in due time, measures can be taken to forestall a disastrous competition in such formidable armaments and to establish an international control of the manufacture and use of the powerful materials.
>
> Any arrangement which can offer safety against secret preparations for the mastery of the new means of destruction would as stressed in the memorandum, demand extraordinary measures. In fact not only would universal access to full information about scientific discoveries be necessary, but every major technical enterprise, industrial as well as military, would have to be open to international control.
>
> . . . . .
>
> All such opportunities, however, may be forfeited if an initiative is not taken while the matter can be raised in a spirit of friendly advice. In fact, a postponement to await further developments might, especially if preparations for competitive efforts in the meantime had reached an advanced stage, give the approach the appearance of an attempt at coercion in which no great nation can be expected to acquiesce. . . .
>
> Indeed it need hardly be stressed how fortunate in every respect it would be, if at the same time the world [learns] of the formidable destructive power which has come into human hands, it could be told that the great scientific and technical advance has been helpful in creating a solid foundation for a future peaceful cooperation between nations.

How astronomically far Bohr was from advocating the handing over of the bomb to the Soviet Union is shown in the memorandum. Bohr was talking at another level entirely—one of preventing a catastrophic proliferation of atomic weapons and at the same time opening the fruits of atomic power to all who would cooperate in the peaceful use of that power. Nations would not have to impoverish themselves to obtain atomic power for peaceful uses, and simultaneously would save themselves as well as others from the unthinkable consequences of an atomic arms race.

Bush thought that a committee would be appointed to study international atomic policy, and that Bohr's memorandum should be submitted to it. Bush immediately forwarded Bohr's memorandum to McGeorge Bundy, Secretary of War Stimson's special assistant on atomic matters. Bush attached a strong endorsement, saying that he quite agreed with the general thesis and that Bohr was entirely right in emphasizing that the time for effective action was growing short. The best minds in the country should be put to work on the crucial problem, Bush declared. He urged Stimson to name an advisory committee. Even before Stimson read Bohr's memorandum, Justice Frankfurter had talked to him about the problem.

On April 25, 1945, the same day on which Bush was sending the Bohr memorandum to Stimson, the secretary of war went to the White House to give the new president a full explanation of the immense atomic energy project. Stimson had prepared carefully, marshalling his facts in consultation with his aids, Frankfurter, and Groves. He was explicit, using a one, two, three order to make the complexities easily comprehensible to a president suddenly plunged into the overwhelming problems of a nation and world. How completely Stimson had accepted Bohr's thesis can only be appreciated by examining the actual text of the memorandum he laid before the President:

1. Within four months we shall in all probability have completed the most terrible weapon ever known in human history, one bomb of which would destroy a whole city.
2. Although we have shared its development with the U.K.,

physically the U.S. is at present in the position of controlling the resources with which to construct and use it and no other nation could reach this position for some years.

Nevertheless it is practically certain that we could not remain in this position indefinitely.

a. Various segments of its discovery and production are widely known among many scientists in many countries, although few scientists are now acquainted with the whole process which we have developed.

b. Although its construction under present methods requires great scientific and industrial effort and raw materials, which are temporarily mainly within the possession and knowledge of the U.S. and the U.K., it is extremely probable that much easier and cheaper methods of production will be discovered by scientists in the future, together with the use of materials of much wider distribution. As a result, it is extremely probable that the future will make it possible to be constructed by smaller nations or even groups, or at least by a large nation, in a much shorter time.

4. As a result, it is indicated that the future may see a time when such a weapon may be constructed in secret and used suddenly and effectively with devastating power by a willful nation or group against an unsuspecting nation or group of much greater size and material power. With its aid even a very powerful unsuspecting nation might be conquered within a few days by a very much smaller one.

5. The world in its present state of moral advancement compared with its technical development would be at the mercy of such a weapon. In other words, modern civilization might be completely destroyed.

6. To approach any world peace organization of any pattern now likely to be considered, without an appreciation by the leaders of our country of the power of this new weapon would seem to be unrealistic. No system of control heretofore considered would be adequate to control this menace. Both inside any particular country and between the nations of the world the control of this weapon will undoubtedly be a matter of the greatest difficulty and would involve such thoroughgoing rights of inspection and internal controls as we have never theretofore contemplated.

7. Furthermore, in the light of our present position with

reference to this weapon, the questions of sharing it with other nations and if so, shared, upon what terms, becomes a primary question of our foreign relations. Also our leadership in the war and in the development of this weapon has placed a certain moral responsibility upon us which we cannot shirk without very serious responsibility for any disaster to civilization which it would further.

8. On the other hand, if the problem of the proper use of this weapon can be solved, we should have the opportunity to bring the world into a pattern in which the peace of the world and our civilization can be saved.

The president followed the typed copy as Stimson went over it. "I listened with absorbed interest," Truman said.

Stimson wanted Groves present to furnish some of the technical details. To prevent reporters from associating his visit with that of the secretary of war, Groves was spirited into the White House through underground entries. Admiral Leahy who was serving Truman as he had Roosevelt, as a personal chief of staff, also was present. The granitic admiral was unimpressed and skeptical: "This is the biggest fool thing we have ever done. The bomb will never go off, and I speak as an expert in explosives." Stimson and Groves acknowledged that they could not say positively that the whole gigantic effort would be a success. No single bomb had yet been assembled or exploded. But both assured the president that they, unlike the admiral, were confident of results. The secretary recommended that the president authorize the immediate appointment of a committee to advise the government on the new force, and Truman did so.

A few days after the White House "briefing session," George L. Harrison, president of the New York Life Insurance Company, who was serving as special assistant to Stimson, and Bundy saw the secretary. They had Bohr's memorandum in hand. "If properly controlled by the peace loving nations of the world this energy should insure the peace of the world for generations," they said, emphasizing Bohr's point. "If misused it may lead to the complete destruction of civilization."

The committee, to be known as the Interim Committee, was

appointed on May 4 with Stimson as Chairman and Harrison as his alternate. Byrnes was named as a special White House representative, and Compton, Ernest Lawrence, Oppenheimer, and Fermi were appointed to a special scientific panel.

At last, a year and a quarter after Bohr had begun pressing for action and six months after Bush and Conant had urged the necessity, some formal machinery was established to consider the most momentous danger and opportunity ever to confront man. So far did political capability lag behind technical advance.

Roosevelt and Churchill had shown a keen awareness of the potentialities; they knew even as Bohr talked to them that the world and its prospects were being changed. But Churchill had assumed that the new day could be met, at the outset at least, with an Anglo-American monopoly, perhaps with a new Pax Britannica-Americana based on overweening power and a benevolent mastery of the determinative weapon and power.

Roosevelt had permitted himself to be swayed toward the Churchill viewpoint, or at least had hesitated to act on Bohr's counter proposal. Stimson was fearful of a partnership that would include a nation as intransigent as Russia. Adding to their caution (the stakes were enormous) was the weariness of both the Americans: they had been through years of the most crushing strain; Roosevelt's health was failing; Stimson was nearing 80.

The time of great climaxes had come. At 2:41 A.M. in the early dark of May 7, the German High Command surrendered unconditionally. The war in Europe, the greatest war in destruction and extent, ended. People broke into a delirium of joy. In Washington, tens of thousands converged on the White House; they filled the streets until not a car could move; they climbed the White House fence; they shouted for the president. Bohr watched the pandemonium. He was a part of it, with an elation matched only by those who had seen their own country overrun by the now conquered enemy.

At the Danish embassy the Bohrs sought the first news from Copenhagen. Soon they learned that on May 4 the German High Command in Denmark had surrendered and that on May 5 an

Allied "holding force" had come into Copenhagen by air. The rest of Denmark was soon freed by fast-moving American and British armored columns. Before the Allied troops arrived, reports had come in that the Russians were dropping parachutists on Danish soil. It turned out that there were only two men who had descended on Danish land. However, the report caused the British and Americans to move fast. In a cable to Anthony Eden who was in San Francisco attending the organization meeting of the United Nations, Churchill reported: "I think, having regard to the joyous feelings of the Danes, and the abject submission and would-be partisanship of the surrendered Huns, we shall head our Soviet friends off at this point too." They did.

Denmark was again free. Bohr's thoughts immediately turned to home. He could not, however, leave at once. Transportation was impossible, and there were obligations to be fulfilled. Beneath the joy of victory Bohr was somber. He knew that troubles lay ahead and that the future might be dark. The atom loomed there, and Japan was not defeated.

Two days after the final European victory was won, the Interim Committee held its first meeting. Bush supplied the group with a copy of his September 30, 1944, memorandum on international control, and copies of Bohr's memorandum to Bush were made available. Bush strongly recommended to Harrison, the co-chairman and working head of the committee, that he talk to Niels Bohr.

In the closely guarded sessions, part of the membership argued that the other great powers would never permit the United States and Britain to enjoy an atomic monopoly, and would certainly undertake to develop atomic arms of their own. This they feared would lead the world into a "flaming inferno." Groves, however, advised the committee that the Russians could not "catch up" in twenty years. Most of the scientists estimated that the Russians could do it in three or four. As Oppenheimer put it: "Our monopoly is like a cake of ice melting in the sun." Byrnes, who once had been a New Dealer, but whose conservatism and suspicion of change were constantly deepening, intervened decisively. Stalin might insist on coming into a partnership with the United States and Great Britain, he said. The Russians should be told nothing and the United States should push ahead with research to make certain that it stayed ahead.

The committee at this point gave in. Though other meetings were held, and warnings of the futility of such a course were sounded, the day was lost again. On June 6, 1945, Stimson reported to President Truman that the committee recommended that no information should be given to Russia or anyone else until the first bomb had been used successfully against Japan. The Committee's only suggestion on international control was mild: each country should be asked to make public all work being done on atomic energy. To assure fulfillment, an international committee should be empowered to inspect the work in each country. But, Stimson said, there should be no disclosures until the control was established.

Many of the scientists in the laboratory were becoming frantic at the imminence of the bomb and the apparent absence of any methods of controlling it. A committee at the University of Chicago headed by Nobel Prize-winner, James Franck, warned in the most solemn of words: "Nuclear bombs cannot possibly remain a 'secret weapon' at the disposal of this country for more than a few years; the United States could maintain its position of moral leadership if nuclear bombs were not used unannounced against Japan but were first revealed by a demonstration in an uninhabited area; and international control is essential to the safety of the world."

Bohr did not stand alone; a large part of the laboratory scientists were with him. Bush and Compton saw that the future rested on international control. Stimson believed it. On the British side Halifax and Anderson worked wholeheartedly for world cooperation rather than a futile secrecy. "Many were clear about this," said Oppenheimer. "But there was a difference. Bohr was for action and timely action. He realized that it had to be taken with those who had the power to commit and to act. He wanted to change the framework in which the problem would appear early enough so that the problem itself would be altered. He was for statesmen; he used the word over and over again; he was not for committees and the Interim Committee was a committee."

But against Bohr were tradition, fear, and caution, and a very few men who could sway others. When it became clear that the fight was lost, Oppenheimer went over to the British mission offices where Bohr was working and tried to comfort him. "He was much too wise and he would not be comforted," said Oppenheimer.

One man had tried to change the course of the world. Whether his new framework might have avoided the arms race that he so clearly foresaw and the proliferation of nuclear weapons that he feared, no one will know. The attempt to pull the twentieth century political world up to twentieth century science had failed. It was one of the most valiant and bravest of battles.

A few days later Bohr sailed for England, on the first leg of his long journey home.

18 /

Copenhagen Again and the Bomb

MARGRETHE WAS WAITING in England. Twenty-one tumultuous months had passed since the Bohrs had parted in Stockholm. War and the uncertainty of not knowing had added to the strain of separation. "Even at the end we knew much less than everyone supposed," said Mrs. Bohr.

But the Bohrs were together again. Margrethe was as slender as ever, her softly curling hair was a little grayer, but the years only added a remarkable loveliness to her nearly classical features and to a gentleness and character that could cope with upheavals, public demands, and the rearing of children. During the war years in Sweden, Mrs. Bohr had seen to the continuation of their youngest son's education. She had also kept in touch with family and friends left in Denmark, and had worked with refugees in Sweden.

As soon as the war ended she hurried back to Denmark. In the short time she was in Copenhagen, before she left for England to meet her husband, she found that the institute was undamaged. A few personal possessions of the staff, cameras, principally, had disappeared during the months the Gestapo was in possession. But the buildings and equipment had come through unscathed. This was also true at Carlsberg. The Bohrs had a second cause for rejoicing.

Bohr was eager to return to Denmark, but he had to wait until after the bomb was tested to make certain that his services were no longer required by England and the United States. The prospects for international cooperation then looked dim, but he did not want to neglect any steps that might influence a change in policy. Although the first battle had been lost, Bohr would continue to work to avert the disaster that he felt misuse of the atom would bring.

Bohr did not have long to wait. On July 16, 1945, the Big Three, now Truman, Churchill, and Stalin, assembled for their first postwar meeting at Potsdam. Boundaries and occupation zones and all the urgent problems created in Europe by its greatest war in history had to be faced. And then there was the bomb, of which Stalin, the third ally, supposedly knew nothing.

In order that President Truman might know with certainty whether the world's most powerful weapon or the world's most expensive dud was in his hands, a test of the bomb was planned to coincide with the opening of the Potsdam Conference. Just before dawn on July 16 the first atomic weapon ever made by man was exploded successfully in the desert near Alamogordo, New Mexico. At the instant of explosion, the dry surrounding hills were bathed in a luridly brilliant light. Frisch, who was watching from the scientists' post some twenty miles away, said that "It was as though somebody had turned the sun on with a switch." When the scientists, whose backs were turned to protect their eyes, dared to turn and look they saw a perfect round ball as red and glowing as the sun. A gray stem connected it with the ground. The incandescent ball rose slowly, then a knob formed atop it, and the protuberance burst forth as a second mushroom. Only as the mushroom turned into a purplish blue did the scientists hear the sound—not a deafening clap of thunder, but a long rumbling "as of huge wagons running through the hills."

A message was dispatched to Truman at Potsdam: "Babies satisfactorily born," and just a little later; "Operated on this morning. Diagnosis not yet complete but results seem satisfactory and already exceed expectations." In the afternoon Secretary of War Stimson called on Churchill and laid the messages before him too. "It means," said Stimson, "that the experiment in the New Mexico desert has come off. The atomic bomb is a reality."

As the details were brought in by special plane, Truman and Churchill learned that the devastation within a one mile circle was complete. The steel tower that held the plutonium bomb had turned into a gas. Windows had been shattered 125 miles away, and the energy was estimated as the equivalent of 15,000 to 20,000 tons of TNT. "Here is a speedy end to the Second World War, and perhaps much else besides," said Churchill, as he heard the first news.

Truman invited Churchill, General Marshall, and Admiral Leahy to a conference to discuss what would now have to be done. Plans had been made for an assault on the Japanese homeland by terrific air bombings and invasion, both scheduled to begin in the fall of 1945. "We had contemplated the desperate resistance of the Japanese fighting to the death with Samurai devotion, not only in pitched battle but in every cave and dug out," said Churchill, with his characteristic drama.

I had in mind the spectacle of Okinawa where many thousands of Japanese rather than surrender had drawn up in line and destroyed themselves by hand grenades after their leaders had committed hara-kiri.

To quell Japanese resistance man by man and to conquer the country yard by yard might well require the loss of a million American lives and half that number of British, if we could get them there; for we were resolved to share the agony.

Now all this nightmare picture had vanished. In its place was the vision—fair and bright indeed it seemed—of the end of the whole war in one or two violent shocks.

I thought immediately myself of how the Japanese people whose courage I had always admired, might find in the apparition of this almost supernatural weapon an excuse which would save their honour and release them from the obligation of being killed to the last fighting man.

"'Moreover, we would not need the Russians.'"

So it looked to Churchill with his sublime assurance and channeled vision that excluded all but the one truth he saw. The thought was inevitable, but no one gave voice to it at the moment, and the conference ordered a continuation with the plans for the invasion.

Another problem confronted Truman and Churchill: what to tell Stalin. The Interim Committee had recommended even before the bomb was tested that Truman should inform the Russians that the United States and Britain had an extraordinary weapon. Now the question had to be reviewed in the light of the nearly incredible reality. "We both felt," said Churchill, speaking of himself and the president "that Stalin must be informed of the great New Fact which now dominated the scene, but not of any particulars." But how then should he be told? In writing or verbally? At a special meeting, or during one of the scheduled conferences, or after one of the meetings had adjourned? And what if Stalin should demand "the particulars" the two leaders were determined not to give? "I think" Churchill quotes Truman as saying, "I had best just tell him after one of our meetings that we have an entirely novel form of bomb, something quite out of the ordinary which we think will have decisive effects on the Japanese will to continue the war."

Truman waited for the touchy critical moment until the conference was approaching its end. On the 24th of July, as the three heads of state got up from the round table at which they worked and everyone stood about talking informally, Truman went up to Stalin. "I casually mentioned to Stalin that we had a new weapon of unusual destructive force," Truman reported in his memoirs. "The Russian premier showed no special interest. All he said was that he was glad to hear it and hoped that we would 'make good use of it' against the Japanese." When Truman was asked in an interview on his eightieth birthday about Stalin's reaction, he added: "There wasn't any. He didn't know what I was talking about."

Churchill has provided a more colorful account of how the Russian leader finally was told about the great New Fact:

I was perhaps five yards away and I watched with the closest attention the momentous talk. I knew what the President was going to do. What was vital to measure was its effect on Stalin. I can see it all as if it were yesterday. He seemed to be delighted.

A new bomb! Of extraordinary power! Probably decisive on the whole Japanese war! What a bit of luck! That was my impression at the moment, and I was sure he had no idea of the significance of what he was being told.

Evidently in his intense toils and stresses the atomic bomb had played no part. If he had had the slightest idea of the revolution in world affairs which was in progress his reactions would have been obvious. Nothing would have been easier than for him to say "Thank you so much for telling me about your new bomb. . . . May I send my expert in these nuclear sciences to meet with your expert tomorrow morning?" But his face remained gay and genial and the talk between the two potentates soon came to an end.

As we were waiting for our cars I found myself near Truman. "How did it go," I asked? "He never asked a question," he replied. I was certain therefore that at that date Stalin had no knowledge of the vast process of research which the United States and Britain had been engaged in for so long and of the production for which the United States had spent over 400 million pounds [$2 billion] in an heroic gamble.

Thus no attempt was made, as Bohr had urged, to negotiate with the Russians and to draw them out of the secrecy to which they tended, or to divert them from the atomic course they were virtually certain to follow, if they were not already launched upon it. The opportunity to avoid the dangers Churchill so dramatically described dwindled to the few bland words and Stalin's seemingly naïve acceptance of them.

Both Churchill and Truman wrote their accounts of the anticlimactic Potsdam drama many years later, at a time when both knew that Fuchs and probably others had betrayed essential information to the Russians. According to Groves and the trial record, Fuchs had been feeding important information to the Russians from 1943 on. Both Truman and Churchill still thought that Stalin was a surprised listener rather than a consummate actor who did not want to reveal how much he knew, or possibly that Russia itself was engaged in the production of an atomic bomb. Certainly at Potsdam both of the allied leaders persisted in their belief that the bomb was an absolute secret up to that time. Even after the test they kept the New Fact, as Churchill called it, from other members of their staffs.

On July 26 Truman, Churchill, and the president of China called upon Japan for unconditional surrender. The alternative, they warned, with a meaning not made explicit, would be "prompt and utter destruction." Leaflets were dropped warning the Japanese to evacuate twelve of their largest cities. On August 6, 1945, a single plane dropped a single bomb through the thin clouds drifting over the city of Hiroshima. A swirling cloud of smoke mushroomed 40,000 feet into the sky. Beneath its rolling, tumbling, boiling billows lay ruin; four square miles of the city had been completely devastated, eight square miles severely damaged. Within four months 64,000 died.

As terrible as was the destruction, it was far from the worst ever wreaked upon a modern city. One incendiary attack on Tokyo in March of that same year had killed 83,000 and devastated sixteen square miles. But Hiroshima was, as Churchill said, almost a supernatural thing; a "second coming in wrath."

When the bomb was dropped Truman was aboard the *Augusta*, returning from Potsdam, but the news was released from the White House. The bells on the wire service machines in newspaper offices jangled wildly—news of the utmost importance was coming through. The great secret was suddenly the world's most surprising and shocking knowledge.

"The fact that we can release atomic energy ushers in a new era in man's understanding," said the president in his official statement. Even in this first announcement that a new force had been born, Truman turned to the core problem of secrecy—thus far had Bohr persuaded the leaders to recognize that it was a core problem: "It has never been the habit of the scientists of this country or the policy of this government to withhold from the world scientific knowledge," the President continued. "But under present circumstances it is not intended to divulge the technical processes of production or all the military applications, pending further examination of possible methods of protecting us and the rest of the world from the danger of sudden destruction."

The president said he would recommend the establishment of a commission to control atomic energy in the United States and that he would take further steps to determine "how atomic power can become a powerful and forceful influence toward the maintenance of world peace."

In these days of incredible developments, Winston Churchill's government had been defeated in the English general election; to Churchill as a war leader the people of Britain had given their fullest measure of devotion, but they did not want him to lead them into the peace. Thus when the bomb shattered Hiroshima and shook the conscience of the world, it was Churchill's successor, Clement Attlee, who cabled President Truman: "There is widespread anxiety as to whether the new power will be used to save or destroy civilization. I consider therefore that you and I as head of the governments which have control of this great force, should without delay make a joint declaration of our intentions to utilize the existence of this great power, not for our own ends, but as trustees for humanity in the interests of all peoples in order to promote peace and justice in the world."

Confronted with a force beyond all past concepts of power, the conscience of the world was indeed aroused. Bohr suddenly felt new hope. Up to the time of Hiroshima he had based his hopes for a constructive solution solely on the chosen leaders of the free world. It was in fact all that could be done—no one else except the scientists was cognizant that a new force was altering all past patterns of society. It was one of the few times in the modern history of the democracies when one man could exert such influence on an issue of national life and death.

Even though the leaders now were showing a will to use atomic power constructively, Bohr thought that the time had come to speak out, to give the citizen an understanding of the issues involved and some perspective for judging them. Soon after his arrival in England Bohr started working on a letter to the *Times*. Before he could complete it and get it to the newspaper, the plutonium bomb was dropped on Nagasaki. It brought the death of 39,000 people, and abruptly ended the war in Japan. Bohr's letter appeared on August 11 under a two column head: "Science and Civilization." His words could not have been graver:

> The possibility of releasing vast amounts of energy through atomic disintegration, which means a veritable revolution of human resources, cannot but raise in the mind of everyone the question where the advance of physical science is leading civilization.
> . . . it is evident that the formidable power of destruction which has come within the reach of man may become a mortal menace unless human society can adjust itself to the exigencies of the situation. Civilization is presented with a challenge more serious perhaps than ever before, and the fate of humanity will depend on its ability to unite in averting common dangers and jointly to reap the benefit from the immense opportunity which the progress of science offers.
>
> . . . . .
>
> Against the new destructive powers no defense may be possible. The issue centers on world wide cooperation to prevent any use of the new sources of energy which does not serve mankind as a whole. The possibility of international regulation for this purpose should be ensured by the very magnitude and the peculiar character of the efforts which will be indispensable for the production of the formidable new weapon.
> It is obvious, however, that no control can be effective without free access to full scientific information and the granting of the opportunity of international supervision of all undertakings which unless regulated might become a source of disaster.
> Such measures will, of course, demand the abolition of barriers hitherto considered necessary to safeguard national interests but now standing in the way of the common security against unprecedented danger. Certainly the handling of the precarious situation will demand the good will of all nations, but it must be recognized that we are dealing with what is potentially a deadly challenge to civilization itself.

There was little more that Bohr could do. The man whose understandings had made the bomb possible had exerted every effort that a human could to turn the force unleashed into a blessing rather than a potential disaster for humanity. So far he had failed and yet the ideas that he planted—of international control and the abolition of secrecy that would make the new power beneficial—were alive. Both of the new leaders who were taking over, Truman in the United States and Attlee in Britain, had cited his ideas in their first statements.

Late in August the Bohrs returned to Denmark. On August 25 the Dannebrog, Denmark's hearty red flag with its large cross of white, flew high over the Institute for Theoretical Physics. The lawn was green and raked; the vines that had grown over the temporary building in the forecourt, well trimmed; the casement windows swung wide; and a small crowd had gathered on the graveled driveway at the front gate. It was a gala day; Bohr was coming back.

Down the broad bicycle path that ran between the sidewalk and the street, came Bohr, riding his bicycle exactly as he had on hundreds of mornings before there had been a world war and an atom bomb. His trousers were rolled up and his hair was blown. A cheer went up—fervent, excited, heartfelt. And tears mingled with the cheers.

"Bohr was deeply moved and he was received by a similarly moved staff at a little gathering in the institute's drawing room," said Rozental. "It was not easy to express one's feelings in words; we could hardly believe that what had been hoped for during all this time had now taken place. Then as a symbol that the institute had at last got its chief back Bohr was handed the new keys to the house."

The keys to the house had more than the usual significance. At the institute all doors of entry as well as many others were locked, and keys were a necessity, always were one of the first acquisitions of a new student. For Bohr they were indispensable as well as a symbolic and a meaningful welcome home.

Bohr walked into one of the laboratories. On a shelf with all the other bottles sat one filled with a drab-looking fluid. It was dissolved Nobel medals, which were in due time recast. In his office the plaque of Rutherford set into the fireplace hood and the photographs of the fellows of each year—stiff group photographs with every one lined up and Bohr in the middle of the front row—all were in place. Bohr looked at them carefully with the appreciation of a man who had not been sure that he would ever see them again. Even to sit at his desk was a meaningful act. Bohr smiled and laughed and embraced old friends. It was one of the happiest days of his life and yet there were long thoughts too, of some who were gone, of the changes that had come, or what they had all been through. Too much had happened for joy to be unalloyed.

Bohr was beseiged with questions about the astounding atom bomb and his part in it. He turned aside all questions about his own work. He did not want to talk about it, but in addition the disclosure of classified information was forbidden under the American Espionage Act. Whether or not the restrictions applied to a Danish citizen, Bohr intended to observe them scrupulously. Bohr would shift the subject to what atomic power could mean to the world, or to the new opportunities the work had opened to physics.

During the long debates over secrecy, Bush and Conant had strongly urged, when the bomb was dropped and thus announced to the world, that a report describing its development and the scientific efforts should be published. H. D. Smyth, chairman of the department of physics of Princeton University, and one of the scientists of the Manhattan Project, was assigned to write it. Mimeographed copies of the hundred-page report were handed to the press on August 11, 1945. Bohr had brought a copy of the Smyth Report with him and gave it to the scientists at the institute to read. It did not answer all of a scientist's questions, but it answered most of them.

General interest in the subject of the atom kept increasing. Bohr finally was persuaded to give a lecture before the Engineers' Association. He presented some of the material from the Smyth Report, for it had been designed, as Smyth said, for professionals who in their turn could explain the principles of atomic energy to the citizen who needed information if he were properly to discharge his responsibilities.

But this was Denmark's first public discussion of the atom, and it was a discussion by the country's foremost citizen, a man who had been spirited out of occupied Europe to lend his aid to the esoteric project. The press turned out in force for the speech. One of Copenhagen's newspapers used its biggest headlines to proclaim: "Bohr Divulges Secret of the Atom Bomb." The wire services picked up the story and it was featured in newspapers all around the world. Bohr divulged nothing that had not already been released in the Smyth Report. Nevertheless, inquiries were made from America. "The speech caused much difficulty and unpleasantness," said Rozental. "The American authorities had to be reassured about what had been said."

On October 7 Bohr was to observe his sixtieth birthday. There was a tradition of celebrating each of Bohr's "round number" birthdays, but in 1945 the tradition took on special meaning. The birthday would be a welcome home. Invitations were sent to the institute alumni. Europe, however, was still in the midst of postwar resettlement. Ships and railroads were jammed with returning troops and others engaged in official business. Only a few were able to come to Copenhagen.

Bohr, as he caught a few glimpses of the hush-hush arrangements going on around him, had considerable misgivings about the occasion. It was to begin with a morning reception at the institute. As Bohr rode up on his bicycle he saw that other friends had joined the institute staff. The honoree greeted everyone cordially and steeled himself for something of an ordeal. Laudatory speeches can be difficult to take.

With a show of solemnity, Rosenfeld stepped up to deliver the first address.[1] He announced his title "My Initiation," and Bohr relaxed a little. He suspected that what was coming would be far from solemn. "The first message I got from Bohr was a telegram announcing that the Easter Conference was postponed two days," Rosenfeld began (he was in Goettingen at the time of which he was speaking). "When we arrived in Copenhagen, Bohr informed us of the reason for the postponement. He had to complete (with

[1] See the chapter by Leon Rosenfeld in *Niels Bohr His Life and Work*.

Klein's kind help) a Danish translation of some of his recent papers. Bohr explained that it would have been a catastrophe if the work had not been ready in time.

"That struck me as a hyperbolical way of stating the matter. How little I imagined the tragedy hidden in this seemingly innocuous procedure of putting the finishing touches to a paper. How little I knew that it was to be my destiny to play a part in a lot of such tragedies."

By this time Bohr had brightened. He could see that he was going to enjoy himself after all.

Rosenfeld went on: "Take the case of the Faraday lecture. Bohr arrived in London with a manuscript practically finished. Only a few pages were lacking. The plan was to seek isolation in the romantic atmosphere of an English inn and in a week's time [with Rosenfeld's kind help] the speech would be ready. After a week's hard labour in a crowded unromantic hotel . . . the ten odd lacking pages had actually been written.

"We had furthermore gained the insight that a great improvement could be obtained by the mere addition of some twenty more pages."

The audience fully appreciated the irony. There was not a one of them who did not know Bohr's methods of writing, or rather of dictating and revising a paper. The audience was laughing with pleasure and so was Bohr. He was entirely reconciled to the birthday celebration.

Rosenfeld continued the story of his initiation. At his first Copenhagen conference—the one postponed for the finishing touches—one of the speakers presented some views that seemed quite erroneous to Rosenfeld. "Bohr however, opposed them rather feebly (as I thought)," Rosenfeld went on. "In his [Bohr's] answer the phrase 'very interesting' recurred insistently, and [Bohr] concluded by expressing the conviction "we agree much more than you think.

"I was worried, the more so as the high brow bench seemed to find it all right. I ventured later to explain my doubts to Bohr. I began cautiously to state that the speaker's argument did not seem to me quite justified. 'Oh,' said Bohr 'it is pure nonsense.' So I knew that I had been led astray by a mere matter of terminology."

The audience and the object of the story now were roaring with laughter. The words were so undeniably Bohr's and so typical that everyone understood. From this point on scarcely a foible of the professor escaped the shafts of his fellow workers. In nonsensically scientific and scientifically nonsensical papers and verse the institute members "took off" their leader.

The papers were preserved for posterity in Volume II of

> Journal of Jocular Physics
> October 7, 1945
> Nicht um zu kritisieren
> Nur um zu lernen
>> It's not to criticize,
>> It's only to learn.

The words quoted on the cover were Bohr's favorites in rising to comment on some paper just presented at one of the institute colloquia or another scientific meeting.

One of the speakers and contributors to the journal made his contribution entirely in "b's"

> Bohr, Blegdamsvej,
> Blandtandet beromt Bolgemekaniker
> Bandsatte brunskjorters brutalitet
>> bringer Pallade
> Bohr's bortrejse besluttes behjaepsomhed
>> behoves.

The B's went properly with Bohr on his sixtieth birthday. He was benevolent, benignant, balanced, beaming, beneficent, brave, a builder. The anxieties of the war and the struggle for peace had saddened, but mellowed, him. His gray thinning hair, the jowls that draped but did not conceal the massive jaw, the heavy beetling eyebrows, and the swags of shadow emphasizing the intelligence and kindliness of the eyes, all made Bohr entirely the figure of

"Uncle Nick," though that was a title used only by the Los Alamos contingent.

That night, as the light faded from the sky, the university students of Copenhagen came marching to Carlsberg. Each carried a torch, until there was a river of light and flame winding through the tree-lined streets of Copenhagen. As the Bohrs came out on the steps to welcome them, the students broke into song. It was another welcome home to suffuse the heart with gratitude and to obliterate many of the trials and griefs that had been a part of the sixty years Bohr had lived.

Bohr returned from America with no doubts whatsoever that the age lying ahead would be an age of the atom. Denmark had no uranium or thorium ore at all, but if ore ever became available, Bohr knew that atomic power might be an answer for a country without fossil fuel resources or water power. Bohr started thinking at once about preparing Denmark and Scandinavia generally for the future. At the moment there was only a question of preparation. It would be some years before atomic power could become a practical reality. Bohr was convinced, however, that there should be no delay in training young men and women to work with it. This meant the institute had to be expanded and its equipment thoroughly modernized.

Bohr marshalled his data and went to see his friends in the government. He requested permission to expand the institute, and asked for the financial help to do it. Bohr had little difficulty in convincing his fellow Danes to start without delay. The government agreed to defray about half of the cost of expanding the institute. The Carlsberg Fund—the fund that the choice Danish beer built— and the Thrige Fund agreed to contribute substantial sums. Planning could begin at once.

Bohr began by examining exactly what he wanted to accomplish and then turned to how the institute could be adapted to that goal. "Nothing was left to chance," said Rozental, a participant in the process. "Every detail was discussed. Bohr insisted 'No dividing line can be drawn between the grand scale and the details. Both are equally important for the whole.'"

Finally, though, plans were put down on paper and the architects called in. One key to the plan was expansion underground. A low workshop at the back of the three front buildings, was rebuilt into a five-story laboratory. The laboratory, however, was set so low into the ground that it did not dominate the buildings facing the street. Its presence was largely invisible from the front.

To make additional room the excavations were extended under the entire site, and into a small strip of park land donated by the City of Copenhagen. This both created new space and made possible the underground connection of the entire institute. The expansion provided new room for the cyclotron. It had developed such strong radiation that it had to be moved and placed behind thick concrete walls. Space also was created for the high tension equipment and the isotope separator. The Van de Graaf generator, which had been built during the war, was moved and a new pressure chamber with a more powerful generator was built in the space it formerly occupied. When the work began Bohr was on hand every day. He watched almost every scoop of earth taken out and was constantly up on the scaffolding to examine the placement of every beam. "The architects and engineers turned gray as their plans were changed again and again," said Rozental.

In the end, though, the institute came out the way Bohr wanted. From the street the institute remained a gracious, well-balanced group of three buildings well scaled to Copenhagen's modest three- and four-story heights. They formed a group fitting well into the big trees of the park. Good planting and bright flower boxes in the best Copenhagen style made the institute even more a complementary part of its setting and its city.

Physics laboratories with their giant equipment were becoming more and more like factories, but not at the institute. It did not take on the air of a huge mechanical workshop. Nevertheless all the required massive equipment was there, neatly and safely hidden away underground. The institute was like an iceberg in the sense that most of its bulk was out of sight. Only a tour could disclose how much there was of it.

Bohr was pleased. "See how nice it turned out," he said to

some of the shaken victims of the change-orders and some of the skeptics who thought that it would never get built at all. "That's the symmetry of it." Bohr was always certain that the symmetry was there, if it could only be discovered and brought out. When it was found, all was right and harmonious.

In 1946 graduate students from abroad once more began flocking to the institute and life at Carlsberg flowed back into its gracious prewar patterns. The new students were invited, a few at a time, to Sunday dinner. After a stroll through the grounds dinner was served—a hearty Danish meal ending frequently with a piece of cake heaped high with whipped cream.

During the evening Bohr would lead a new fellow—in 1946, A. Pais who had just emerged from occupied Holland—into his little office just off the big living room. Pais told Bohr about the studies he had carried on during his years in hiding.

Bohr smoked one of his dozen or so pipes, and generally kept his eyes on the floor while Pais put some of his formulae on the blackboard. When Pais finished, Bohr said very little and the young physicist felt somewhat disheartened. Only later did he learn that Bohr's interest had been aroused. Bohr had not dismissed what Pais was saying with his usual equivalent of disapproval, "very interesting, very interesting."

After a little private session that gave Bohr and a new student a chance to become better acquainted, they would wander back to the living room to join Mrs. Bohr—who often was working on needlepoint—and other guests. Soon everyone was grouped around the Bohrs, some sitting on the floor.

Bohr talked about the unique opportunities atomic power could offer to overcome the hostilities and suspicions of the world. "It would be wrong to suppose, however, that an evening at the Bohrs was entirely filled with the discussion of such weighty matters," said Pais. "Sooner or later Bohr would tell one or two stories. "I believe that at any one given time, Bohr had about half a dozen favorite jokes. He would tell them; we would get to know them. Yet he would never cease to hold his audience. For me, to hear again the beginning of a favorite tale, would lead me to anticipate not so much the denouement as Bohr's own happy laughter at the conclusion. At that time he had one joke of which for the life of me I can remember only the punch line 'The question is not whether the Irish are humans, but whether humans are Irish.'"

It was such evenings at Carlsberg that the Institute fellows never forgot. There they got to know Bohr in a way that was difficult in the more formal atmosphere of the institute. There was a warmth, a closeness, a sense that this was the way civilized human beings should enjoy one another's company. Good talk, fun, inclusion in a circle—all of it was combined at Carlsberg—and there was Margrethe to lead the shy along, and to make all feel not only welcome, but at home, in this gracious house so far from their own homes.

Frequently some of the Bohr sons and their families were at Carlsberg for the evening. Hans had become an orthopedist; Erik was a chemical engineer; Aage, continued to work closely with his father at the institute; and Ernest had become a lawyer. The Bohrs had encouraged each to follow his own interests, and were tremendously proud of "the boys" and the young grandchildren who came to romp happily through the Pompeian court at Carlsberg or in the little playhouse built for them at Tisvilde.

These were days when celebrations tended to be frequent— they were an outlet for profound feelings of thanksgiving. On March 2 the institute was 25 years old, and this called indisputably for a special observance. It was decided that it would be a celebration of their own, an intimate rather than a large public gathering. There were a few short speeches and then Bohr reminisced of heroic and tragic days, of friends and enemies.

That night Parentesen, the club of the graduate students, gave a dinner. It ended with the singing of *Videnskabens Faedre* (The Fathers of Science). As always the last verse was rendered with all present standing on their chairs and lifting a stein high to "Nobelmanden Niels Bohr, ved vej blandt alle vildspor" (Nobelman Niels Bohr who knows the way among all false tracks.)

What with new building and celebrations, welcoming new students and watching the not reassuring news, Bohr was hard pressed to get on with his scientific work. He brought back with him the article on the passage of charged particles, the original of which Rozental had lost on his flight to Sweden. New experiments had provided data to round it off. Bohr was plunging into this task when an invitation came to speak at a conference at Cambridge on particle physics. The subject acknowledged the birth of the next era in physics.

Bohr had largely created the atomic age and had set the direction of nuclear physics. He thus had dominated physics for more than thirty years. Now the post-Bohr era was beginning, and Bohr was asked to help in giving it a fair start. The visit to England also gave him an opportunity to talk again with Sir John Anderson, who then sat on the opposite bench. It was clear by this time that the seeds Bohr had planted in the United States and Great Britain were showing signs of growth.

Stimson, who had become fearful at Potsdam that the democracies might be unable to deal with the Russian police state, had again returned to his original positions: that sharing action could prevent the Russians' development of a bomb of their own. On September 11, 1945, a month after the two atomic bombs were dropped on Japan, the secretary of war recommended to President Truman that a direct approach should be made to Russia on atomic energy. Unless the Soviet nation was "voluntarily invited into partnership on a basis of cooperation and trust," Stimson wrote, Russia undoubtedly would undertake the development of a bomb of her own, and the result would be a "secret armaments race of a rather desperate character."

Whether Russia gets control of the necessary secret of production in a minimum of say, four years or a maximum of twenty years, is not nearly as important to the world and civilization as to make sure that when they do get it they are willing and cooperative partners among the peace loving nations of the world," said Stimson. "It is true if we approach them now, as I would propose, we may be gambling on their good faith and risk their getting into production of bombs a little sooner than they would otherwise.

I emphasize, perhaps beyond all other considerations the

importance of taking this action with Russia as a proposal of the United States—backed by Great Britain, but peculiarly the proposal of the United States.

Stimson added that after the nations which won the war had agreed to it, there would be ample time to introduce France and China into the covenants and finally to incorporate the agreement into the scheme of the United Nations.

Stimson, who had been in close touch with Frankfurter, had come to the ideas that Bohr advocated. Byrnes, then secretary of state, went to Moscow in December for a meeting of the Big Three foreign ministers to make the fateful approach, in part at least. For once the Russians showed little of their usual obduracy. An agreement was reached that the three countries would support the creation of an atomic energy commission under the United Nations. The step was a small one, not the large one called for by Stimson. "What really mattered" though was that the United States had made a special approach to Russia and Russia had received it favorably. On January 3, 1946, the United Nations General Assembly unanimously set up the commission. Bernard Baruch, the 76-year-old financier and adviser of presidents, was named the American representative on the commission. The other eleven members all were members of the U.N. Security Council.

In the meanwhile the United States was considering an atomic energy bill, vesting all control over fission in a commission appointed by the president. The State Department named a special committee made up of the Under Secretary of State Dean Acheson, Bush, Conant, Groves, and John J. McCloy to draw up proposals for the United States to submit to the United Nations commission. The committee in turn named a panel of consultants, under the chairmanship of David E. Lilienthal, former director of the Tennessee Valley Authority. In about two months of work the consultants produced the "Acheson-Lilienthal Report" proposing international control, based not on policing, but on a common effort to develop atomic energy for peaceful uses. An international agency was recommended to take control of all raw materials.

It was in some senses a revolutionary report, moving far in the direction of international control. Baruch, however, insisted upon writing in a statement of controls, and above all, punishments and sanctions. "The people want a program not composed of pious thoughts but enforceable sanctions—an international law with teeth in it," declared Baruch.

On January 14, the tall, 6 foot 4 figure, looking the very embodiment of the elder statesman, laid the drastically amended proposal before the U.N. Bohr was far from happy about it.

"It was not centered enough in the absolutely central theme of openness," said Oppenheimer, a member of the Acheson-Lilienthal panel. "I may say that when this became part of the United States position at the United Nations, our military staff officers said that if this were to be open, there could be no secret military installations, or it would not work. So in a way Bohr would have had his openness if it had come into being. But Bohr said further, and a very true reproach, 'It called for action. It was an action to make the bomb.'"

A little more than two weeks later, on July 1, 1946, an explosion shook even this tenuous hope for action. With a whole shipload of correspondents looking on, and a television correspondent describing the explosion from Dave's Dream, a plane flying overhead, a test atomic bomb was set off in the lagoon of Bikini Atoll. A fireball shot upward, a cosmic mushroom formed in the sky.

"Testing the bomb with one hand and seeking to control it with the other was bound to lay the United States open to charges of conducting atom diplomacy," said Hewlett and Anderson, in the official history of the U.S. Atomic Energy Commission. Pravda immediately charged that the United States wanted to perfect and not to restrict the bomb. At about this time the disclosure of a Soviet spy ring operating in Canada aroused suspicion and antagonism on the American and British sides. In the United Nations, Baruch was insisting too that there must be no veto on the punishments provided for any violation of international controls. This proved another stumbling block of the largest proportions.

In October Bohr came to the United States to attend a

symposium of the National Academy of Sciences on "Present Trends and International Implications of Science." It was held in Philadelphia and on October 21, Bohr gave a paper on "Atomic Physics and International Cooperation." He again emphasized that the fundamental discoveries of atomic physics "originated in lines of inquiry pursued in various countries by scientific schools with different traditions and outlooks, the confluence of which gave rise to international cooperation of an intensity and enthusiasm which has indeed only a few counterparts in the history of science."

When the meeting ended Bohr went on to Washington to talk to his friends there. Following one interview, a high official of the State Department undertook to drive Bohr to his hotel, but as they talked they kept driving. They drove around for hours while Bohr "talked in that familiar talk in which silences seemed as essential and active as words." Finally when both were worn out, Bohr was dropped at his hotel. He apologized for having talked so much, saying "You see the problem is never to speak more clearly than I think."

The technical experts of the U.N. commission, including two Russians, had just produced a unanimous "feasibility" report, setting forth the technical requirements for an effective control system. The scientific group was headed by Kramers, a member of the Dutch delegation to the U.N. Atomic Energy Commission, and Bohr's longtime friend. It was Kramers who had wandered into Copenhagen during the First World War and who had stayed for ten years as Bohr's chief assistant. Through the years they had kept in the closest touch. They happily met again in New York in 1946 and talked at length about "the necessity of common control of the formidable power which through the development of atomic physics had come into the hands of man." And thus once again through a man trained in the Copenhagen school and spirit, progress was made in coping with the dominant new force.

But this was the last atomic document to receive unanimous support. When Baruch submitted the technical report, along with the "American plan," to the U.N. General Assembly, the Soviet Union and Poland abstained when it came to a vote in committee.

Once more there had been a close approach to the kind of agreement that Bohr and Stimson sought. Again it had been lost, this time largely by provisions inserted by Baruch. But the Russians changed in one year from an apparent willingness to cooperate to opposition to a plan that commanded worldwide respect. Hewlett and Anderson in their study of the record, ask if the Politburo had not "decided to trust to Russian scientific and technical resources." Was Russia now rejecting a system of control because she soon expected to have a bomb of her own? Bohr did not know, but the opportunity of bringing the Russians into the world community before they undertook the development of atomic weapons had slipped away. Again Bohr made a sad, worried crossing of the Atlantic.

19 /

Another Course

I n 1948 BOHR RETURNED to the United States to spend the spring
semester at the Institute for Advanced Studies in Princeton. He had
been a permanent nonresident member since 1939 and thus was
free to come and go as he might wish. His arrival in the late winter
of 1948 brought under the one roof—and colonial clocktower—the
two men who had revolutionized twentieth century physics and the
twentieth century. Bohr and Einstein, the two perpetual though
respectful opponents, were within immediate arguing distance.

Their discussions soon began. Bohr would logically destroy
Einstein's position, only to have Einstein insist that he did not
"like" the kind of universe Bohr was creating, and that its duality
and indeterminacy were only temporary, only expedients until the
final answer could be found. Bohr would seethe with frustration.

After one of these sessions Bohr came bursting into the office
of Pais, who was then a temporary member of the institute. "I am
sick of myself," Bohr exclaimed. "I am sick of myself."

Bohr still thought that he should somehow be able to convince
Einstein that it would be impossible to go beyond the two often
contradictory aspects of the atom. Einstein had himself been the
first to point to the dual nature of light—to light's being at once a
wave and particle. Why had Einstein lost faith in his own method?

Bohr was then working on a written account of his running
dispute of more than a quarter of a century with Einstein. When he
had previously been in Princeton for the bicentennial of Princeton

University, Paul A. Schilpp, the philosopher, asked him to contribute to the book *Albert Einstein—Philosopher-Scientist* with which he was planning to celebrate Einstein's seventieth birthday in 1949. Bohr gladly assented, then undertook an historical account of their clashes that began, face to face, in 1920 and were now continuing down the hall.

One morning Bohr came into Pais's office and with one of his ingratiating smiles began "You are so wise." Pais laughed (No solemnity ever was called for with Bohr, he noted). Bohr wanted Pais to come to his office to help him with a part of the answer to Einstein. Soon Bohr was furiously pacing around the oblong table in the center of the room. He would dictate a few sentences and Pais would take them down. "It should be explained that at such sessions Bohr never had a full sentence ready," said Pais. "He would often dwell on one word, coax it, implore it, to find the continuation. This would go on for many minutes."

This time Bohr halted at the word "Einstein." He was nearly running around the table, repeating "Einstein. . . . Einstein. . . . Einstein. . . ."

Suddenly he stopped to look out the window, still saying "Einstein. . . . Einstein. . . ."

At that instant the door opened softly and Einstein stuck his head in. As soon as he grasped the situation, he put his finger to his lips in a signal to Pais to say nothing, and to Pais's mystification tiptoed over to the central table. He was heading for Bohr's tobacco pot. Einstein's physician had forbidden him to buy tobacco, but had said nothing about borrowing some. At that instant Bohr, with another firm "Einstein," turned around.

"They were face to face as if Bohr had summoned him forth," said Pais. For an instant Bohr stood there in frozen shock. He was speechless. Pais, who had realized what was coming but was helpless to stop it, had a sense of the uncanny. Only after a few minutes was the tableau broken. Then they all burst into laughter.

With the continuing argument to sharpen his thoughts and with the remindful presence of Einstein on the floor below or even materializing in his office, Bohr completed most of the work on the

Einstein piece for the honorary volume. It was a review of the modern physics the two had largely created and a searching analysis of its very bases as the two giants had differently seen them.

Bohr began: "The many occasions through the years on which I had the privilege to discuss with Einstein the epistomological problems raised by the modern development of atomic physics have come back vividly to my mind and I have felt that I could hardly attempt anything better than to give an account of these discussions, which even if no complete accord has so far been obtained, have been of the greatest value and stimulus to me."

Without the controversy, it is doubtful that Bohr might have pursued his thoughts to the same depths. And only an opponent of Einstein's caliber could so completely have engaged him and demanded so great and continuing an effort. Bohr knew this himself. In concluding his article he acknowledged it: "Whether our actual meetings have been of long or short duration, they have always left a deep and lasting impression on my mind and when writing the report I have, so to say, been arguing with Einstein all the time, even when entering on topics apparently far removed from the special problems under debate in our meetings."

When Einstein was asked to reply to the comments made by the twenty-six contributors to the seventieth birthday volume (there were six Nobel Prize winners among them) he went straight to the arguments of Bohr and the Copenhagen school.

"They [Bohr, Pauli, and others] deprecate the fact that I reject the basic idea of contemporary statistical quantum theory insofar as I do not believe that the fundamental concept will provide a useful basis for the whole of physics," said Einstein.

"I am in fact firmly convinced that the essentially statistical character of contemporary quantum theory is solely ascribed to the fact that this [theory] operates with an incomplete description of physical systems."

Einstein insisted that he would be satisfied only with a complete description of a "real situation as it supposedly exists irrespective of any act of observation or substantiation." Einstein charged that the quantum physicist's [Bohr's] resort to "ensembles of systems" for a complete description of nature is "egg-walking."

There was no agreement, no complete meeting of the two greatest minds of the century. Nevertheless their disagreement produced a dialogue that illuminated the depths and subtleties of modern quantum physics, and the modern version of the world, to a degree otherwise impossible. It was a perfect illustration of complementarity.

Oppenheimer called Bohr's report of the continuing twenty-eight year long discussion "the richest and deepest" dialogue since Plato's *Parmenides*. Only Socrates debating his theories with the great Eleatic philosophers, Parmenides and Zeno, had explored reality as profoundly and subtly.

Five years earlier, when Bohr had learned after his escape from Nazi-occupied Denmark that an atom bomb was under development, he foresaw that it would cause trouble with Russia, unless the Russians were made partners rather than rivals. Now the former allies—the United States, Britain and Russia—were clashing headlong. The iron curtain was ringing down. Bohr's fears were being realized. During the spring of 1948 in Princeton, Bohr, with grave misgivings, watched the growing quarrel between East and West.

The suspicions and enmity were highly in evidence in the United Nations Atomic Energy Commission. After the Soviet rejection of the first United Nations plan for control of atomic energy, the U.N. Atomic Energy Commission under Kramers's chairmanship brought in a more detailed plan. It emphasized the dangers of large accumulations of fissionable materials in any place where they might be seized by national governments, and proposed controls on atomic power plants and on fissionable materials. Elaborate safeguards were recommended for the inspection of atomic plants and the detection of clandestine activities.

On June 11, 1947, the Soviet Union had come in with a plan of its own. It included inspection by a U.N. commission, with the right of free entry into all countries and access to all atomic installations. It also provided that sanctions against violators should be reserved to the Security Council where the U.S.S.R. had the right to veto. This proposal was rejected by the West, and the work for control came to a dead stop. So complete was the stalemate that on May 17, 1948, the U.N. Atomic Energy Commission recommended the suspension of its own activity.

At Princeton, Bohr talked to his colleagues about what changes in the world could break the stalemate and make possible some security. The answer to which he came with increasing emphasis was the abolition of secrecy. Only if there was exchange and openness could there be any basis for confidence, Bohr argued.

From his convenient base in Princeton, Bohr went to New York a number of times to talk to Baruch and others. He also took the other direction, to Washington. Bohr and McCloy had continued in close touch and had become friends. On one of Bohr's visits to Washington, the two drove around Washington until the early hours of the morning talking, and talking more, about how the impasse could be broken and the fearful race halted. McCloy then arranged for Bohr to see Secretary of State Marshall. The thoughtful, gracious general-turned-statesman was at once interested in what Bohr had to say. The man whose name was given to the plan that was restoring the economic stability of much of the world appreciated the need for atomic security and asked Bohr to prepare a memorandum setting forth his main proposals.

Bohr immediately went to work, and completed his memorandum on May 17, the very day on which the U.N. commission conceded its defeat. It was, however, June before Bohr could resume his conversations with Marshall and present the memorandum.

As a beginning Bohr recommended an offer, on a well-timed occasion, of immediate measures toward mutual openness. "Such measures should in some suitable manner grant access to information of any kind desired, about conditions and developments in the various countries and would thereby allow the partners to form proper judgment of the actual situation that is confronting them."

Bohr freely admitted that such an offer might seem beyond the scope of conventional diplomatic caution, but he argued that if it were accepted a radical improvement in world affairs would result "with entirely new opportunities for cooperation in confidence and far reaching agreement on effective measures to eliminate common dangers."

Bohr, facing the civilian-suited former general, urged that the United States with its lead in atomic armament was in a particularly favorable position to take the initiative, but that such a course would also be in the deepest interest of all nations. There was no mistaking the interest of Marshall. The secretary of state could uniquely appreciate large and soundly altruistic concepts. The interview was long, thoughtful, and serious.

"Bohr hoped that Marshall would say 'We are for doing away with secrets and we are prepared to do it.'" said Oppenheimer. Marshall, however, did not make such a statement. If the secretary of state had been inclined to seek a new basis for international agreement along the lines proposed by Bohr, the time could not have been less propitious. Tension was mounting then in Berlin, and later that month, June, the Russians closed off all Western rail access to the city.

To keep the former German capital alive (it was an island in the Russian-controlled zone of East Germany) the United States and Great Britain had to institute their famed air lift. Even coal was flown in by a shuttle of planes. Then the Russians threatened to operate fighter planes in the U.S.—British flight corridor, and almost any incident could have touched off a fight and war. Twelve menacing months passed before an uneasy settlement was reached in June, 1949, and Berlin was reopened to ordinary traffic.

No sooner did the Berlin crisis abate than another shock changed the world outlook and altered political balances. On September 3, 1949, a plane operating in the United States Long Range Detection System collected a sample of radioactive air. The suspicious cloud from which it was taken was tracked from the north Pacific to the British Isles where the Royal Air Force took up the trace. The evidence was conclusive. Sometime between August 26 and 29 an atomic explosion had occurred somewhere on the Asiatic mainland.

On September 23 President Truman announced to a startled world:

> I believe the American people to the fullest extent consistent with national security, are entitled to be informed of all developments in the field of atomic energy. That is my reason for making public the following information: We have evidence that within recent weeks an atomic explosion occurred within the U.S.S.R. Ever since atomic energy was first released by man, the eventual development of the new force by other nations was to be expected. This probability has always been taken into account by us.
>
> Nearly four years ago I pointed out that "scientific opinion appears to be practically unanimous that the essential theoretical knowledge upon which the discovery is based already is widely known." There is also substantial agreement that foreign research can come abreast of our theoretical knowledge in time. And in the Three Nations Declaration of the President of the United States, and the prime ministers of the United Kingdom and Canada dated November 15, 1943, it was emphasized that no single nation could in fact, have a monopoly of atomic weapons.
>
> This recent development emphasizes once again, if indeed such emphasis is needed, the necessity for that truly effective enforceable international control of atomic energy which the government and the large majority of the United Nations support.

The atomic arms race was on. This was the race that Bohr and the other scientists had foreseen and feared, and the outcome of which still is unknown. As the first radioactive cloud floated out of Asia and on around the world the American-British monopoly of atomic weapons had ended, only a little more than four years after the first atomic bomb exploded at Hiroshima.

Bohr heard the announcement without surprise. He had always acted on the assumption that the Russians had the scientific skill and the economic power to develop a bomb of their own and that they would do so if they were not otherwise persuaded by being given access to the economic fruits of atomic power. Nor was he surprised about the timing. The British and many of the American scientists had estimated that the Russians could produce a bomb in four or five years. Bohr characteristically wasted no time on "I told you so's." Despite the set-backs he had suffered, Bohr's only question was what could be done. The discussions, grim discussions, ran late into the night at Carlsberg.

In Washington the reaction was immediate. An expansion of American bomb production had been under consideration and Lewis L. Strauss, a member of the Atomic Energy Commission, and some of the scientists, particularly Teller, were calling for the production of a "super," a super-hydrogen, fusion bomb, in which the explosion would come from the fusion rather than the splitting of atoms. It would be a thousand times more destructive in its blast than a fission weapon.

"Where will this lead?" asked Lilienthal in his diary. "Whether it will lead to something good is difficult to see. We keep saying 'We have no other course.' What we should say is 'We are not bright enough to see any other course.' "

The top American political atomic and military advisers recommended the expansion and the production of the super-bomb. Truman told Lilienthal that "the way the Russians are acting" he felt he had no other course. On January 31, 1949, the president announced publicly that he was directing the hydrogen bomb program to proceed "until a satisfactory plan for international control of atomic energy is achieved."

The president said later that he believed that anything that would assure the United States of a continuing lead in the development of atomic energy for defense "had to be tried out." The alternative, Truman noted with evident regret, "was a long way from clear cut." Thoughts were long and the regrets were poignant as the United States entered upon the next lap in the atomic arms race.

During this month when the United States was deciding to step up its production of atomic weapons, John A. Wheeler of Princeton University, who had worked with Bohr on the identification of U-235 as the fissionable part of uranium, was in Copenhagen to work with Bohr on a paper on the collective model of the nucleus.

Several times as the debate raged in the United States atomic circles about the "super," Wheeler breakfasted with Bohr at Carlsberg. The table was always set up in the dining room near Thorvaldsen's statue of Hebe. Before they settled down to discuss the mightiest of explosions Bohr would glance up at the goddess of youth and the cup bearer of the gods. When Wheeler noticed his glance, he explained: "The wise old philosopher who was my predecessor in this house once said that he looked up at Hebe each morning to find out from her inscrutable expression if she was satisfied with him or not. Not the worst preparation for these days."

The subject was enough to discomfort Hebe. Wheeler, like Bohr, had taken an important part in the production of the atom bomb and he was soon to be called back to the United States to work on the thermonuclear weapon.

"Bohr said that whatever immediate decision might be required, the long term aim had to be 'the open world.'" said Wheeler. "At the same time—aware as always of political realities— he emphasized again what he had often said before: 'How could Western Europe have remained free and at peace after World War II if America had not had the atomic bomb?'"

Bohr pondered the words he had so often heard from heads of states and other public officials, and that were then being re-uttered: "There is no other course." It was true that the head of a great state with his weighty, crushing responsibility would find it difficult to see that there might be another course. His vision often was narrowed by the very responsibilities resting upon him.

At the moment, with Russia exploding an atomic bomb, with the iron curtain, and with an iron dictatorship, it was possible to understand that a president might believe that the only answer was to expand and speed the arms race. Perhaps the hope for another course that might transform the deadly arms race into cooperative progress would have to lie not with the leaders but with the people.

On another trip to the United States and in more conferences with officials, Bohr's conviction was strengthened. He began to work on an open letter to the United Nations. In this way, he thought, he could reach the largest number of people in the largest number of nations.

Upon his return to Copenhagen, Bohr and several assistants went to work in a special room where they could shut themselves away from all distractions. The letter was delivered to the Secretary General of the U.N. on June 12, 1950. At the same time Bohr took the almost unheard-of step for him of calling a press conference in Copenhagen. Copies of the letter were distributed. Bohr did not invite questions. On the contrary, he explained that everything he could possibly want to say was contained in the 3,000-word letter. To clarify the need for action and to put the formidable problem into perspective, Bohr for the first time revealed publicly his own role in the struggle to find a way to live with "this development which holds out such great promise for the improvement of human welfare" and yet menaces the whole of civilization.

"My association with the American-British atomic energy project during the war gave me the opportunity of submitting to the governments concerned views regarding the hopes and dangers which the accomplishment of the project might imply as to the mutual relations between nations," Bohr explained.

"While possibilities still existed of immediate results of the negotiations within the United Nations . . . I have been reluctant to take part in the public debate on this question. In the present critical situation, however, I feel that an account of my views and experience may perhaps contribute to renewed discussion about these matters so deeply influencing international relationship."

Bohr also explained that he was acting entirely on his own responsibility without consultation with the government of any country. His aim, he emphasized, was to point to the unique opportunities created by "the revolution in human resources brought about by the advance of science," and "to stress that despite previous disappointments these opportunities still remain and that all hopes and all efforts must be centered on their realization." Bohr then made public the key sections of the memorandum he submitted to President Roosevelt on July 3, 1944, as a basis for discussion. These were the ideas that so favorably impressed the president that he authorized Bohr to tell Churchill that he would welcome discussion of control measures.

"Looking back on those days," he continued, "I find it difficult to convey with sufficient vividness the fervent hopes that the progress of science might initiate a new era of harmonious cooperation between nations, and the anxieties lest any opportunity to promote such a development be forfeited."

What could one man do? Why did one man fight indomitably for a new day of peace? Bohr knew what the consequences of an atomic explosion could be and felt an awful dread—a dread shared by many of those who developed the bomb. "Until the end of the war I endeavored by every way open to a scientist to stress the importance of appreciating the full political implications of the project and to advocate that before there could be any question of use of atomic weapons, international cooperation be initiated on the elimination of the new menaces to world security."

Bohr's letter to the U.N. dispassionately stated the facts. The bomb was dropped without any agreements having been reached, and the end of the war, though it brought immense relief, aggravated the new controversies he had feared. New barriers were created; distrust and anxiety increased.

Bohr conceded that his call for openness might appear Utopian, but he emphasized "it would be in the deepest interest of all nations." But he argued that all was not yet lost—the coincidence of a great upheaval in world affairs and a veritable revolution in technical resources still could offer the world "unique opportunities." And on the other hand, Bohr emphasized, the increasing tensions had increased the danger that "great countries may compete about the possession of means of annihilating populations or large areas and even making parts of the earth temporarily uninhabitable.

"The situation calls for the most unprejudiced attitude towards all questions of international relations," Bohr declared as he moved toward his conclusion.

Indeed proper appreciation of the duties and responsibilities implied in world citizenship is in our time more necessary than ever before. The progress of science and technology has tied the fate of all nations inseparably together.

. . . . .

The arguments presented suggest that every initiative from any side towards the removal of obstacles for free mutual information and intercourse would be of the greatest importance in breaking the present deadlock and encouraging others to take steps in the same direction. The efforts of all supporters of international cooperation, individuals as well as nations, will be needed to create in all countries an opinion to voice with ever increasing clarity and strength the demand for an open world.

There was a flurry of attention and comment. But twelve days later, on June 24, the Communist-supported North Koreans invaded South Korea, where American troops had been stationed since the end of the war. At the appeal of the United States, the Security Council of the United Nations unsuccessfully called upon the invading troops to cease hostilities and withdraw to the 38th parallel.

"On the contrary," announced the President of the United States, "they have pressed the attack. The Security Council called upon all members of the United Nations to render every assistance to the United Nations in the execution of this resolution. I have ordered United States air and sea forces to give the Korean troops cover and support." The Korean War had begun.

Whatever chance Bohr's new appeal might have had was again smothered. Animosities increased and the atomic arms race accelerated.

On November 1, 1952, less than four years after the "super" go-ahead was ordered, the United States exploded the first "thermonuclear device" on the island of Elugelab in the Eniwetok area of the Pacific. A flaming dome more than three miles in diameter spread over what had been a tranquil coral islet. When the mushroom cloud drifted away an incredible sight greeted the shipboard watchers. The island had disappeared. A crater a mile long and 175-feet deep had been dug into the coral. The man-made star had exploded with a force equal to that of several million tons of TNT, and the United States was again in the atomic arms lead. And again the lead was short-lived. Less than a year later, on August 12, 1953, Malenkov, Stalin's successor, announced that the United States no longer had a monopoly of the thermonuclear bomb. Four days later the skies carried radioactive samples of the explosion of another hydrogen bomb.

The race was moving with a vengeance and a speed that boded ill for humanity. No prophet of doom had ever been more accurate than Bohr when he warned that a race would develop unless the world adopted a new policy of cooperation. That Bohr might be as right about the possibilities of averting the menacing competition as he was about its fateful occurrence did not seem to impress the otherwise occupied world.

"Physics," said Einstein, "is an attempt conceptually to grasp reality." Bohr's definition was different, almost an opposite: "It is wrong," he said, "to think that the task of physics is to find out how nature is. Physics concerns what we can say about nature." Traditional philosophy, and Einstein, had always regarded reality as primary, and what humans can say or discover as secondary. Bohr thought the distinction inappropriate.

"When one said to him," Aage Petersen related, "that it cannot be language which is fundamental, but that it must be reality which, so to say, lies beneath language and of which language is a picture, he would reply, 'We are suspended in language in such a way that we cannot say what is up and what is down. The word "reality" is also a word, a word which we must learn to use correctly.'"

Bohr did not in the least surrender in the struggle to prevent the suicide of the world. Through all of the struggle he continued to study the effect of physics and the insights it yielded on other fields of knowledge.

As a young student, fired with the ideas Höffding was opening to him, Bohr had dreamed of "great inter-relationships" between all areas of knowledge. He had even considered writing a book on the theory of knowledge or epistemology. But physics has drawn him irresistibly, only for him to discover himself back in philosophy as he and the physicists of the world plunged deeper and still deeper from surfaces to the solar-like atom, to the nucleus-sun, to the interior of the nucleus itself down to the ultimates, the essences, the structures, and forces that accounted for the stability of the earth and its form. Quantum physics brought him again to his starting point and gave him a whole new approach to the great interrelationships.

The tremendous fruitfulness of international cooperation in physics suggested that the same approach would be equally productive in politics. The finding that absolutely contradictory states could both be true, and together could constitute the full truth disclosed analogies to the conflicting elements in other fields. Dualities that could be reconciled if the framework were widened as it had been in physics existed in biology, psychology, language, the cultures of the world, politics, economics, the arts, and religion. In politics the lesson of quantum physics suggested that even such conflicting theories of government as communism and democracy might each represent a usable concept of government and that each could contribute to the world. In biology the same lesson showed that neither vitalism nor mechanism had to be proved wrong; that both could be combined into a greater whole. In language and psychology, both subject and object could exist together; in psychology, man could be both actor and spectator.

When he visited Japan, Bohr had been greatly struck by the various aspects of Mount Fuji. In the sunset, on the evening when he first saw the sacred mountain, its top had disappeared behind a curtain of gold-fringed clouds. The mass of the mountain, surmounted by the shining crown, conveyed an impression of awe and majesty. The next morning the great mountain offered another spectacle entirely. The pointed summit, covered with white and glistening snow, emerged from the mists cloaking the lower slopes and valleys. The aspect was one of brightness and joy. The "two mountains," Bohr liked to say, did not simply equal one; each, the great mass with the glowing crown and the unembodied white peak was an individual expression, and the two were complementary. Neither one nor the other constituted the mountain; together they pictured it as it really was.

Ultimately Bohr discussed "the unity of knowledge" in more than twenty lectures and papers. In each of these presentations his approach always was the same, historical, logical, allusive.

When one friend reminded Bohr that he had glimpsed the possibility of fundamental relationships in all fields of knowledge even before he came to quantum physics and found the approaches there, Bohr answered with a mea culpa smile and a story about Socrates. A greek Sophist who had been away from Athens for a long period on his return discovered Socrates in the same place engaged in the same discussions. "Are you still standing there, Socrates, saying the same things about the same things?" he inquired with an edge of sarcasm. "Don't you ever say the same things about the same things?" Socrates asked. The question was both answered and incontestable.

In all of his discussions of the unity of knowledge, Bohr always began with a review of the steps in physics that had led to the discovery of the new approach. First, he explained, Newton clarified the problem of cause and effect by showing that if the state of a system and the forces acting upon it were defined at a given instant, its state at any subsequent time could be predicted. A simple consistent description was provided of the behavior of the bodies surrounding us, and even of "the regularities exhibited by the grand display of the sky." A new sense of certainty pervaded the world.

Einstein's demonstration that observers moving relative to one another, say on passing trains, will coordinate events differently did not shake or confuse this certainty. "Einstein succeeded in lending to our world picture a unity surpassing all previous expectations," said Bohr.

The upset came only when science penetrated into the world of the atom. Here effect no longer followed inevitably from cause.

The answer came, Bohr reminded his audience, when these discoveries were made: That all processes occur in quanta or packets—that the loss of a quanta or the gain of a quanta, or the change in position of a unit, all make changes possible without loss of stability.

That in the infinitely small world of the atom the very act of observation—the impact of the photon of light used for measurement—disturbs the phenomenon being observed. Further certainty is thus forever barred by the very nature of the circumstances.

Though no one final, ultimate answer can be found—the place or velocity can be determined. The two aspects may then be considered as complementary in the sense that they represent equally important aspects, or equally essential knowledge, of atomic systems, and only together do they provide complete knowledge.

"The history of physical science thus demonstrates how the exploration of ever wider fields of experience, in revealing unsuspected limitations of accustomed ideas, indicates new ways of restoring logical order," said Bohr.

Bohr thus established his base. "As we shall now proceed to show," he said, "the epistemological lesson contained in the development of atomic physics reminds us of similar situations with respect to the description and comprehension of experience far beyond the borders of physical science, and allows us to trace common features promoting the search for unity of knowledge."

Bohr often liked to consider first the "problem of our own position in existence." In the early days of the Greeks and even before there was no sharp distinction between animate and inanimate matter. To Hesiod the wind was Boreas wrapping himself around a shivering shepherd. By the renaissance, however, a mechanistic concept of nature prevailed; if enough were known, it was believed, all organic functions could be accounted for by the physical and chemical properties of matter. "If, however, sufficient account is taken of the circumstance that the free energy necessary to maintain and develop organic systems is continually supplied from their surroundings by nutrition and respiration, it becomes clear that there is in such respect no question of the violation of general physical laws," Bohr said.

On the other hand, the notion of life is elementary in biology, and is, like the quantum, not to be explained in the ordinary terms, he reminded. "The main point," he emphasized, "is that only by renouncing an explanation of life in the ordinary sense do we gain a possibility of taking into account its characteristics." Two irreconcilables were reconciled in a larger frame.

Bohr had followed his father's interest in biology and he kept abreast of the new developments. The discovery that DNA (deoxyribonucleic acid) with its remarkable coil structure is the material of heredity, paralleled at a later time the discovery of the structure of the atom. Research on mutations also disclosed a special and striking application of the statistical laws of quantum physics. The exact time and place of a mutation proved unpredictable. Only the probability of a mutation's occurring could be calculated.

"We must realize," said Bohr, "that the stability of the atoms composing the tissues, and the fine molecular structures in the cell with which the hereditary properties of the organism are associated, bring us right into the domain of nuclear physics. "Also the fineness of our senses has been found to go down to the atomic level."

Bohr's arguments were subtle and can best be appreciated in his own words:

> Returning to the general epistemological lesson which atomic physics has given us, we must in the first place realize that the closed processes studied in quantum physics are not directly analogous to biological functions for the maintenance of which a continual exchange of matter and energy between the organism and the environment is required. Moreover, any experimental arrangement which would permit control of such functions to the extent demanded for their well-defined description in physical terms would be prohibitive to the free display of life.
> This very circumstance, however, suggests an attitude to the problem of organic life providing a more appropriate balance between a mechanistic and a finalistic approach. In fact, just as the quantum of action appears in the account of atomic phenomena as an element for which an explanation is neither possible nor required, the notion of life is elementary in biological science.

Thus Bohr argued again there is no one answer; it is not a case of either man or machine. Rather, Bohr maintained, the vital and machine aspects together make man, and offer a complete description of him.

Bohr found the same contradictions and the same possibilities of accepting them in psychology. As a student Bohr was led into the conflicts of subject and object and mind by reading Poul Martin Moller's *Tale of a Danish Student.* Thinking about thinking the Danish student became dizzy: "and then I come to think of my thinking about it; again I think that I think of my thinking about it, and divide myself into an infinitely retreating succession of egos observing each other. I don't know which ego is the real one to stop at, for as soon as I stop at any one of them it is another ego again that stops at it. My head gets in a whirl with dizziness as if I were peering down into a bottomless chasm, and the end of my thinking is a horrible headache."

The student's no-nonsense cousin to whom all of this was said, answered practically: "I cannot help you in sorting your many 'I's'. It is quite outside my sphere of action and I should either be or become as mad as you if I let myself in for your superhuman reveries. My line is to stick to palpable things and walk along the broad highway of common sense; therefore my 'I's' never get tangled up."

Aside from delighting in the humor and the light tone of the tale, Bohr was convinced that it would have been impossible to find a more pertinent account of the essential aspects of "the situation with which we all are faced." Bohr emphasized that we are detached observers, but that our mental observations alter what we observe—much as the light of the microscope knocks out of the atom the very electron the light was supposed to illuminate.

"An especially striking example is offered by the relationships between situations in which we ponder on the motives for our action and in which we experience a feeling of volition," Bohr said. "In normal life such shifting of the separation is more or less intuitively recognized, but symptoms characterized as 'confusion of the egos' which may lead to dissolution of the personality, are well known in psychiatry."

Biologically, Bohr pointed out, every conscious experience may be an irreversible recording in the nervous system. It is at the same time the outcome of processes "which are hardly adapted to exhaustive definition by mechanistic approach." Bohr again summed up the duality and the conflicts with the observation that we are both actors and spectators in the great drama of human existence.

Bohr liked to tell the story of the young man who was sent by his own village to a neighboring town to hear a great rabbi. He was to bring back a report in which all could share. When he returned he told his eagerly waiting fellow citizens: "The rabbi spoke three times. The first talk was brilliant; clear and simple. I understood every word. The second was even better, deep and subtle. I didn't understand much, but the rabbi understood all of it. The third was by far the finest; a great and unforgettable experience. I understood nothing and the rabbi himself didn't understand much either."

When Bohr turned to the problem of language few followed him with ease, or understood him at all.

What, Bohr would ask, is it that human beings ultimately depend on? His answer was "words." "Our task is to communicate experience," he said. "We must strive continuously to extend the scope of our description, but in such a way that our messages do not thereby lose their objective or unambiguous character."

Words, Bohr explained, cannot be understood and communicated without being fixed in a logical frame—a frame made up of the way we characterize and combine experience and that we call "ourselves." The frame at times must be expanded, he argued, as was the frame of physics by the discovery of the quantum. "More and more deeply explored presuppositions may reveal relationships of greater and greater scope," he said.

Bohr also pointed out that language sets man aside from the animals who have not yet put together a simple declarative sentence. Bohr even asserted that a child does not become a full human until it learns to talk. At the same time he pointed out that words are far from precise (unambiguous was the word he preferred). But precise or imprecise, Bohr emphasized again, "we are suspended in them" and that even the most exacting experiments, if they are to be conveyed to others, ultimately must be described in words as well as mathematics.

The pattern extended particularly to cultures. In each there were contradictory aspects, each true in its own way and each inherently limited. "In comparing different national cultures we meet with the special difficulty of appreciating the culture of one nation in terms of the traditions of another." Bohr said.

But in culture and almost every other human endeavor Bohr maintained that by recognizing the contradictions as varied aspects and by bringing them together the full picture might be obtained. In any other course lay frustration and stalemate, and in politics, wars of destruction.

All depended upon an interplay and upon how one looked at it. Bohr often told the story of the three Chinese sages who were brought together to taste vinegar, the Chinese symbol for the spirit of life. The first to taste the vinegar, Confucius said "It is sour." The second, a philosopher of the Schopenhauer school tasted and pronounced the vinegar "bitter." Then Laotze sampled the vinegar. "It is fresh," he declared.

Bohr, unlike the recognized philosophers did not set up a system for seeking the truth. He did not even formally define his philosophy; no one sentence or paragraph encompasses it. Nor did he attempt to reduce the laws of nature to a few basic principles. Bohr's philosophy was, rather than a system, an approach or an attitude.

Bohr always spoke of his "argument," or his "point of view," and insisted "Every sentence I utter is to be understood as a question and not as an affirmation." The French were shocked by this general "breach of the Cartesian rules," and complained that Bohr shrouded himself in "les brumes du Nord," the mists of the North. Bohr nevertheless presented a cohesive philosophy and theory of knowledge. He joined the small company of thinkers who, from Descartes on, have "pondered on the meaning of human knowledge."

20 /

Into the Unknown

BOHR HAD APPEALED to the people of the world to make atomic power a benefit rather than a fatal scourge. Little had been done, but in the absence of political action, Bohr proposed that science itself make a beginning. By strengthening and extending the bonds forged in the prewar years, Bohr argued that physics could again show the promise of international cooperation and even lay a foundation for a joining of nations.

In 1951 Bohr called all of the institute alumni back for another of the gatherings that were held annually before the war. The meetings had been suspended during the war and had since been limited because of continuing disturbances and travel difficulties, but by 1951 the physicists could and did come from all corners of the earth. At the meetings the new developments and the new paradoxes of physics were discussed and debated with all of the old Copenhagen spirit. But at lunch and over a glass of beer at the party Bohr gave at Carlsberg, the physicists talked too about an international physics center where pure research could go forward and young physicists from many countries could be trained.

One year later, in 1952, representatives of fourteen European countries met at Copenhagen to plan such a center and to set up the Conseil Européenne Recherche Nucléaire, thereafter to be known as CERN. The problem was to decide what research equipment would be needed and its size. A program then was laid out for building two large accelerators, a synchro-cyclotron and a proton synchrotron at Geneva, Switzerland. The cost was estimated at about \$28,000,000. A schedule of about seven years was agreed to for the construction of what would be Europe's most advanced and largest establishment devoted exclusively to the exploration of nature without any commercial or military uses.

Bohr became the chairman. To obtain the cooperation and support of fourteen nations was as difficult as gaining agreement in the United Nations. Until the Cern establishment could be completed the section on theoretical physics was situated at Copenhagen, to avoid delay in training men for the center and related work. About fifteen students a year from all over Europe were fitted into the regular work and activities of the institute on Blegdamsvej. Once again, in a somewhat different context, Bohr began to shape the next generation of European physicists, and once again they were formed in a school that superseded all national boundaries and drew the physicists of Europe into a scientific community.

The low flat land of Denmark, ending in its fringe of islands, had no waterfalls for power and no coal deposits to supply fossil fuels, nor did Denmark have uranium supplies. However, if uranium could be obtained, atomic power potentially could be a great boon to a nation with a growing need for power. The progress being made suggested that atomic power might be a possibility in the not distant future. The Danish Academy for Technical Sciences on February 23, 1954, appointed a special committee for atomic energy. Bohr, of course, was made the chairman. At about this time Bohr heard from his good friend Sir John Cockcroft that England soon might be able to make concentrated uranium ore available to countries prepared to use it for research and possible power production. This would, however, require a formal government agreement. Thus the time had come to approach the government of Denmark on this special question.

In the early autumn of 1954 Viggo Kampmann, minister of finance and later prime minister received a call that Prime Minister Hans Hedtoft wanted to see him. He hurried through the long dark maze of corridors in Denmark's Christianborg Palace. As Kampmann entered the prime minister's office he saw Bohr and Hedtoft sitting at a small table on which the light fell. Hedtoft asked him to join them, and Kampmann took his place in what he mentally called the "petitioner's seat." The prime minister explained that he and Bohr were discussing atomic energy and that he wanted Kampmann to familiarize himself with the problem and work with Bohr on it. Kampmann agreed, though with some private misgivings. Hedtoft had once told him that it required much patience to talk to Bohr. As he recalled this remark, Kampmann also thought to himself, he related, that "patience is a commodity in short supply in the case of prime ministers." (Kampmann was not thinking of Bohr and Churchill.)

Soon afterward, in December, 1954, the United Nations established the International Atomic Energy Agency (IAEA). After the earlier stalemate the international organization was again bestirring itself about the atom. It had been spurred by a speech by Dwight D. Eisenhower, who had succeeded Harry S Truman as president of the United States. Eisenhower had suggested that some sharing of the knowledge about the atom would promote the peaceful use of atomic energy.

The new commission, with this backing, called the first Atoms for Peace Conference to meet at Geneva in August, 1955. When the call went out Bohr saw Kampmann to discuss Denmark's participation and the organization of a Danish delegation. As he talked Bohr also began to lay before Kampmann the European atomic energy program, and world need for control.

Kampmann's doubts about working with Bohr or his fears that it would be a time-consuming process soon began to change into the near reverence Bohr inspired in most of those who knew him well and worked with him closely. He explained:

> As I gradually became familiar with Niels Bohr's methods, I understood a little of how he obtained his great results. He simply left nothing to chance, but let his thoughts try out everything they touched upon, and particularly dealt in detail with things which others took for granted—a phrase often synonymous with neglect.

In the government offices as well as at home or at the institute Bohr had to pace as he talked. He could not sit still and explain. It seemed to Kampmann that Bohr's pacing and his logic followed something of the same pattern. Bohr would start on the periphery of a subject and work his way, in gradually decreasing circles, to his point. Once you accustomed yourself to it, Kampmann thought, you realized that it was a good way of "grasping things."

But Kampmann was a minister, and ministers of finance, as well as prime ministers, had their problems of time. He found that it helped if he knew beforehand what Bohr was going to discuss. As a matter of routine, Kampmann had his secretary obtain a list of topics in advance, a list that broadened as the years went along and it became important to change the public attitude toward science and research.

While preparations were under way for the Geneva meeting Bohr was shocked and grieved by the death of Einstein. Opponents are sometimes more closely linked than even close associates, and all through the years Einstein was seldom absent for long from Bohr's thoughts. Their profound disagreement went only to quantum physics and the view it imposed of the universe. Each appreciated the stature and the challenge of the other, and thus a disagreement extending over thirty-five years created none of the acrimony that has marked some of the famous disputes of science.

"With the death of Einstein," Bohr wrote, "a life of service to science and humanity which was as rich and fruitful as any in the whole history of our culture has come to an end.

"Mankind will always be indebted to Einstein for the removal of the obstacles to our outlook which were involved in the primitive notions of absolute space and time. He gave us a world picture with a unity and harmony surpassing the boldest dreams of the past."

A scientist's sincere tribute to another scientist is a revelation of his own standards of values and priorities. Bohr cited Einstein's "logical clarity and creative imagination," the verifiability of his work on relativity, the breadth of his views, and the openness of his mind. Bohr also emphasized Einstein's scientific intuition in identifying the photon as the carrier of energy in radiative processes and his "forcing of a reconsideration of our most elementary concepts."

"The same spirit that characterized Einstein's unique scientific achievements also marked his attitude in all human relations," Bohr continued. "Notwithstanding the increasing reverence which people everywhere felt for his attainments and character he behaved with unchanging natural modesty and expressed himself with a subtle and charming humor." Bohr in his own modesty had no idea that he was also describing himself. He saw his own ideals and codes in Einstein, and thus could not fail to comment on Einstein's sense of responsibility in public affairs.

"He was always prepared to help people in difficulties of any kind, and to him, who had experienced the evils of racial prejudice, the promotion of understanding among nations was a foremost endeavor. His earnest admonitions on the responsibility involved in our rapidly growing mastery of the forces of nature will surely help to meet the challenge to civilization in the proper spirit.

"To the whole of mankind Albert Einstein's death is a great loss and to those of us who had the good fortune to enjoy his warm friendship it is a grief that we shall nevermore be able to see his gentle smile and listen to him."

Einstein had valued the same qualities in Bohr. In speaking of Bohr he had said: "What is so marvelously attractive about Bohr as a scientific thinker is his rare blend of boldness and caution; seldom has anyone possessed such an intuitive grasp of hidden things combined with such a strong critical sense. With his knowledge of the details, his eye is immovably fixed on the underlying principle. He is unquestionably one of the greatest discoverers of our age in the scientific field."

At another time Einstein had written that Bohr's work on the structure of the atom was "the highest form of musicality in the sphere of thought."

One of the two giants of physics was gone. For many years afterward physicists would argue how two men of such minds and characters happened to be associated and whether the heroic development of physics could have occurred if they had not lived when they did. Did they happen to live at a time when the gold mine lay open for development or was the mine found because they were there? Whatever the answer the supra-brilliant advances that they led have not since been duplicated in any succeeding period, and in physics, were not equaled previously except in the time of Newton.

Though Einstein was gone, Bohr did not cease to argue with him. Bohr paid him the compliment of continuing to seek the words and the formulations that would settle the view of the universe about which they had always disagreed. Nowhere was this more in evidence than at the Atoms for Peace Conference. In August, 1955, 1,200 delegates from 72 nations met at Geneva. Sixteen years after the atom was split for the first time and Bohr informed the world of the literally incredible feat, some of the benefits were at last to be opened to all. The Atoms for Peace Conference was a long reach from the openness Bohr advocated, but Bohr regarded it as a promising step in the right direction.

To open the unprecedented gathering of scientists and others concerned with the atom, Bohr was invited to speak, not about the technical achievement, but about the subject the scientists felt much more deeply: "Physical Science and Man's Position." Ultimately it was man's position in the universe that had been endangered and the threat weighed heavily on the minds and consciences of the world's physicists. If anyone could put the upheaval into perspective it was Bohr.

"It is a great privilege to be given the opportunity of addressing this assembly convened by the United Nations Organization in order to promote international cooperation on the use, to the benefit of humanity, of the vast new energy sources made accessible by the exploration of the world of atoms," Bohr began with a meaning that went far beyond the conventional "great privilege" introduction. He defined his role as that of speaking "about the general lesson to be drawn about our positions as observers of that nature of which we ourselves are part."

"We are all of course deeply aware of the responsibility associated with any advance of our knowledge resulting in an increased mastery of the forces of nature," said Bohr.

"Indeed in the present situation our whole civilization is confronted with a most serious challenge demanding an adjustment of the relationship between nations to ensure that unprecedented menaces can be eliminated and all people in common can strive for the fulfillment of the promises offered by the progress of science for the promotion of human welfare all over our globe."

During the first three days the conference considered the atomic power that might potentially solve the power needs of men. The need was estimated, the economics considered, and the capital requirements discussed. The United States exhibited a model of its "swimming pool" reactor with the nuclear source at the bottom of a pool; and the U.S.S.R., its atomic power station reactor. The conference then divided into three sections, on reactors, on chemistry and technology, and on the life sciences and the use of radio-isotopes. In as far as it went, the openness was all that Bohr had fought for. A total of 474 papers were presented. Bohr hoped, though he was not so sanguine as to expect, that the lesson of Geneva would carry over into other fields.

Bohr returned from Geneva more convinced than ever that Denmark must proceed with the development of atomic power as part of the great international effort. Hedtoft had died early in 1955 but H. C. Hansen, his successor, had not only been well informed when he served as foreign minister, but also was an old friend of Bohr's.

A temporary commission was appointed to prepare for Danish participation. The commission included representatives of industry and science, but as Kampmann said, Bohr "as a matter of course" became chairman.

A grant of 2,000,000 kroners was made to initiate the commission's work. Bohr sounded out the United States and Great Britain about possible information and uranium for a reactor in Denmark. He then applied to the prime minister for permission to start formal negotiations. The construction and maintenance of a reactor would involve a large expense for Denmark. Hansen did not want to commit the country without widespread support and therefore solicited the views of all political parties.

Would an atomic plant be too expensive for Denmark? Would

it not be better for all Scandinavia to join in such an undertaking? Was the time right? Many of the questions that arose in the public and party discussions came back to Bohr. He pointed out that the necessary enriched uranium could now be acquired, as well as information on the construction of reactors. Bohr also supported the point of view that Denmark should contribute to atomic technology. In the end even the most strident critics hesitated to say that Denmark should not move ahead in this field that might be the future. Formal negotiations were started with the United Kingdom Atomic Energy Authority and with the United States Atomic Energy Commission. Explorations also were made into the possibility of obtaining uranium or thorium ore from Denmark's colony Greenland. A little later Bohr himself went to Greenland and hiked through the possible ore areas to study the search being made there.

Bohr and the commission were then ready to start looking for a site for the Danish reactor. Bohr went at it exactly as he would have approached the drafting of a paper or the solution of a scientific problem. He began with a map, studying the whole country for an island, peninsula, or headland where the water supply would be ample and ships could easily come or go. It did not have to be in a remote spot, for the reactor would be no more dangerous than a gas works or many other industrial plants. Nevertheless, if possible, Bohr thought that it would be desirable to have space for a security zone. He also recommended that the reactor should not be too far from Copenhagen, where most of the personnel who would man it lived. With the criteria set, Bohr began driving around to examine the areas he had marked as possibly meeting the requirements. The driving was done by another member of the commission or staff, for Bohr was quite likely to become more preoccupied by the countryside than by the road.

One day Bohr rounded up H. H. Koch, the permanent undersecretary of finance, and others to view a site on Roskilde Fjord. It met the general criteria for a site and belonged to the National Trust, and thus was already nationally owned. Choice of such a site would reduce the cost. It was largely an island, though at low tide connected with the shore by a marshy strip of land.

Off with the shoes and socks, Bohr ordered. Roll up trouser

legs. Bohr was following his own instructions and in a few moments was leading his little barefooted delegation over the water-covered neck of land and up onto the sharp stones of the island.

It was easy to see that the site was unsuitable. Even though it was nationally owned and no money would be required to buy it, they could in good conscience cross it off the list. At the same time the little group learned something about Bohr's methods of work.

Not long after Bohr came striding happily into the Ministry of Finance. He was excited and delighted. He had found another site that might be the right one. It was near the town of Roskilde. Another expedition set forth. Roskilde was less than an hour's drive from the center of Copenhagen. It was an old town with neat houses clustered around an ancient high towered cathedral where the kings of Denmark are buried. Along the roads leading down to the fjord, summer cottages were just beginning to be built. The site —Riso—lay less than a mile beyond the town; the land sloped in low, pillow-shaped rolls down to the curving edge of the fjord. Here too at high tide the little isthmus was under water, but with some raising, a road could be constructed and power lines run across. The headland was large enough for reactor buildings. On the shore lay ample space for many buildings and for expansion in the future. Furthermore Bohr had ascertained that the land was available for purchase. All the requirements were met.

Bohr did not fail to point out in addition that the site was a beautiful one, a factor that he rated very high. In the distance lay the forest; in the foreground the gray-green fields met the lavender waters of the fjord in an unbroken harmony. The commanding towers of the cathedral rose in the distance as the only vertical note. As Bohr pointed up all the advantages of this tranquil site for the most advanced of modern construction he did not fail to note that the Vikings might have come ashore at this very point. A Viking boat with its high carved prow had recently been retrieved from the nearby waters of the fjord.

Plans were made for a research establishment with three atomic reactors, an ambitious undertaking for a country of Denmark's size. During the public discussions of the building of Riso and the establishment of the Danish Atomic Energy Commission, some parts of the public obtained the idea that the new plant would immediately solve all of Denmark's power problems, and perhaps cheaply. Bohr insisted that the commission must make clear that atomic power would not soon compete in cost with traditional power. As this idea in its turn permeated the body politic, opposition parties began to attack the expenditures being made at Riso.

As commission chairman, Bohr made use of every opportunity to emphasize that power was only one phase of the development of an atomic energy program. It was important, he explained, to increase the knowledge of nuclear physics and to train men and women for the kind of future that undoubtedly lay ahead. But the money was coming out of current budgets. Bohr ran head-on into the political penchant for considering the moment and its costs rather than the future and its yield.

As Bohr listened to one debate in the parliament he remarked to Kampmann: "Scientists argue to reach an agreement or at least to agree to disagree. The aim of political discussion seems to be to reach the greatest possible disagreement." Nevertheless the budgets were approved and the construction of Riso began.

During the summer of 1956, while Bohr worked hard to demonstrate the peaceful use of atomic energy—to begin where a beginning was possible and to prepare his own country for the future—the world situation went from bad to worse. In June more than 50,000 workers in the Polish city of Poznan called a general strike. On July 26, Egypt announced the seizure of the Suez Canal. In Hungary in October students rose against the Communist government and as the revolt spread the government was overthrown. Each action, though, brought its counteraction. In Poland the "bloodless revolution" succeeded to the extent that a somewhat less restrictive Communist government was established. At Suez, Israel moved and Britain and France landed troops in the Canal Zone, before the United Nations succeeded in halting hostilities. In Hungary in October the full force of Soviet military power descended to crush the revolt. More than 100,000 Hungarians fled their country.

Bohr again watched with acute anxiety the resort to arms and brutal suppression. In a world where there were nuclear arms the danger had multiplied. Bohr once more felt that he must make a public appeal in the hope of establishing a climate in which disputes could be settled peaceably and the way opened to a future livable world.

In another letter to the United Nations, addressed to his friend Dag Hammarskjold, the UN Secretary General, Bohr said:

November 9, 1956

In view of the widespread anxiety called forth by serious divergencies among member states of the United Nations Organization, as regards the interpretation of their rights as sovereign nations and of their obligations towards the organization, I take the liberty of addressing myself to you with some considerations in continuation of my letter to the United Nations of June 9, 1950.

As stressed in the letter, full access to information regarding conditions in every country for life in all its apects and free intercourse and exchange of opinion across all boundaries must form the foundation for that cooperation in confidence between nations, which in our time is so vital for the future of mankind. Indeed, it is only by such cooperation that the promises for improving the welfare of people all over the globe, held out by the developments of science, can be fulfilled and the menace to civilization from the new powerful means of destruction can be eliminated.

. . . . .

All nations might be called upon to declare their readiness to assist the Organization in obtaining the information necessary for impartial judgment on any issue with which the Organization may be confronted. Such assurances might not only serve to relieve present tensions but might also by helping to harmonize the interests and obligations of individual nations, be a step towards the realization of world-wide cooperation.

In the hope that these considerations may in some modest way be of aid to your untiring endeavours in the service of the United Nations Organization,

Respectfully yours,
Niels Bohr

Bohr now was suggesting a relatively small beginning. If an agreement were made to open information about one critical issue, he urged, the results would be so beneficial that the system could be expanded. Hammarskjold was interested and sympathetic, but the UN was so occupied in trying to extinguish fires that it could spare no time to prevent them. Once again Bohr's effort was lost, though once again the rare kind of sense he proposed impressed a man in a position to influence the course. It also touched the minds of others. The Bohr proposals continued to grow slowly, if not to triumph.

The Ford Motor Company Fund in 1957 established the Atoms for Peace Awards to make annual awards "to the individual or individuals judged to have contributed most to the peaceful uses of atomic energy." Each year for ten years an individual selected by a distinguished panel was to receive a gold medal and a cash award of $75,000. The fund stipulated that the awards should be truly international, that the recipient should be chosen by a jury "without regard to the recipient's political inclinations or nationality." The jury, headed by James R. Killian, Jr., president of the Massachusetts Institute of Technology, had a very easy task in giving the first award, unanimously and without question, to Niels Bohr. Bohr virtually personified the struggle to use the atom for the benefit rather than the destruction of mankind; the wording of the terms of the award might have been his.

For the presentation on October 24, 1957, the President of the United States and a notable audience of scientists, educators, and government officials gathered at the National Academy of Sciences in Washington. In opening the meeting Killian placed the action in its larger context: "The presentation of the first Atoms for Peace Award to Niels Henrik David Bohr symbolizes the transition of modern culture into the atomic age." Though the influence of the new science on man's personal and political decisions still might be only dimly seen, he added, "It is certain that all men will be influenced by the knowledge of the nature of the universe and its control given us by the modern scientist. Niels Bohr personifies the modern advances in science and the concern of the man of science for the broad human implications of scientific knowledge."

The applause rolled through the room as Bohr rose to accept the medal inscribed "For the Benefit of Mankind." The volume of sound mounted, it continued, shouts could be heard above the uproar, for Bohr was not only known and loved by that audience, he stood for all that they wanted physics to represent. Bohr bowed gravely in acknowledgement and began in response: "It is a great privilege to me in this distinguished assembly, honored by the presence of the President of the United States, to express my deep gratitude for having been selected as recipient of the Atoms for Peace Award, which is dedicated to a cause embraced by the hopes of humanity. . . .

"It has been my good fortune at close hand to follow how new vast fields of knowledge, hitherto beyond the reach of man, have been opened through a most intensive cooperation of scientists from all over the globe."

John Wheeler of Princeton, one of the speakers, had noted that Bohr's far-reaching ideas could be seen leavening foreign policy. They could also be seen exerting their influence on the words of Dwight D. Eisenhower:

> The world now has a choice between the technology of abundance and the technology of destruction—between the use of power for constructive purposes or for war and desolation.
> There is no question in the minds of the people of the world as to which choice is to be desired: the constructive use. Our country has sought to encourage the application of atomic energy in the arts of peace—toward the end of happiness and well-being for all men and women.
> So in saluting and honoring Dr. Bohr . . . we give recognition to a scientist and a great human being who exemplifies principles the world sorely needs—the spirit of friendly scientific inquiry and the peaceful use of the atom for the satisfaction of human needs.

The award of the Atoms for Peace prize to the famous Dane aroused the interest of the American and world press. The magazines followed with full accounts and special biographies of Bohr.

"U.S. Honors Bohr," said *Time* magazine; "Bohr Accepts Peace Prize," headlined *Science News Letter*; "Pioneer of the Atom," said the line in the *New York Times Magazine* over a long, carefully written biography; "Mr. Atoms for Peace," read another headline; "Award for Bohr," said *Scientific American*; and "Bohr Wins $75,000 Atoms for Peace," said *Science*. *Life Magazine* titled a picture essay "Erudite Human."

Without exception the award of the prize to Bohr was praised. Only *The Nation*, a liberal journal of opinion, added another note. After telling what Bohr had accomplished, the journal went on: "Having done so much, we wish he had refused the award. The real purpose of the gift is to make us think that significant progress is being made toward disarming the atom and turning it into a complement of peace. Dr. Bohr had the opportunity to warn humanity that neither he nor anyone else is making much progress, and to label the award for what it is—wishful thinking."

With the Atoms for Peace Award, Bohr, in his seventy-second year had received probably more awards, prizes, decorations, honorary degrees and memberships than any other living scientist. The honors came throughout Bohr's career, and not at one time for one accomplishment. The Hughes Medal of the Royal Society was presented to Bohr even before he received the Nobel Prize in 1922. The H. C. Oerstad Medal was given to him in 1924, the Norwegian Gold Medal from the University of Oslo in the same year; the Metucci Medal of the Societa Italiana della Scienze in Rome in 1925, the Franklin Medal of the Franklin Institute in Baltimore in 1926, the Faraday Medal from the Chemical Society of London in 1930, the Planck Medal of the Deutsch Physikalische Gesellschaft in 1930, and the Copley Medal of the Royal Society in 1938. These were the highest honors the scientists of Great Britain, Norway, Italy, the United States, Germany, and Denmark could confer. There were also many others not so widely known.

Universities joined in honoring Bohr. He was given honorary degrees by Cambridge, Oxford, Manchester, Oslo, Edinburgh, and innumerable others. He was elected to membership in more than twenty scientific societies. The medals, degrees, and memberships from almost every country observing such scientific traditions were testimony to the lofty character of Bohr's scientific achievement, for they represented the appraisal and judgment of colleagues who knew and could judge.

Others honors came to Bohr the man and patriot. Foremost among them was Denmark's Order of the Elephant, his own country's highest honor, in 1949. At Frederiksborg Castle, a fairy tale castle of tall spires and cobbled courtyards reflected in a mirror lake on which float swans and white water lilies, is an alcove dedicated to the order. A window is inset with stained glass plaques bearing the coats of arms of each member. Not far from Churchill's plaque is Bohr's, and thus the two men who contended over atomic policy here are represented side by side. Bohr's coat of arms shows the ancient Chinese symbols of Ying and Yang, the two sides of truth, or complementarity.

Following the brilliant and moving Washington ceremony, Bohr went to Boston to spend the month of November at the Massachusetts Institute of Technology and to deliver a series of six lectures, the Carl T. Compton Lectures, on "Quantum Physics and Complementarity." Bohr had prepared carefully as he always did for his lectures. He worked steadily on the M.I.T. series during most of the preceding summer at Tisvilde. At M.I.T. Bohr spoke to a learned special audience, but even for such a selected group a Bohr lecture was not easy to follow.

Bohr often departed from his prepared paper. "But," he would say, and pause deep in thought. "And," he would resume. Bohr had reasoned out a whole sequence in his own mind, but sometimes forgot to tell his audience about it. They were left dangling. When he had finished Bohr usually had only one question or comment: "I hope it was tolerable." He knew that he did not have the gift of swaying large audiences.

Bohr had to return to Denmark as soon as possible. An important change was occurring. After nearly seven years of planning and construction the powerful new accelerators of Cern, the largest in Europe, were completed. The theoretical and training phases of the work could then be transferred from Copenhagen to Geneva.

No gap was to be left at Copenhagen. Two years before, Sweden had invited the physicists of Scandinavia to a conference at Göteborg to discuss the possibility of a Scandinavian nuclear institute. It would permit the countries to cooperate in research and the training of young physicists. The conference decisively held one idea: the Scandinavian effort should be centered in Copenhagen in close conjunction with the Institute for Theoretical Physics. In this way the new institute would profit by the international activity going on there. Bohr accepted the proposal with the greastest of pleasure and plans moved steadily forward.

As soon as the Cern activities could be transferred to Geneva, the new Scandinavian institute would move in (there would not be sufficient space any earlier). The Scandinavian institute was called Nordita, Nordic Institute for Theoretical Atomic Physics. Bohr became chairman of Nordita's administrative committee and his close associate Rozental, the administrative director. Nordita took over the former Cern offices at the institute in the fall of 1957. Scarcely a day was lost.

In addition to training about fifteen physicists a year, Nordita was dedicated to furthering cooperation among the physicists of the five participating countries, Norway, Sweden, Denmark, Finland, and Iceland. Each country contributed financially in proportion to its population. In carrying out its assigned task Nordita began to send lecturers to each of the countries and to arrange special conferences. It also became an active publisher of scientific material.

Bohr had dreamed of such a Scandinavian institute when the institute in Blegdamvej was established. It had taken many years for the dream to come to fruition. When it did, it not only linked all of Scandinavia but showed on a small scale what might be done in the world at large. Nordita was a proof of what might be if cooperation were substituted for rivalry.

Bohr did not in the least relinquish his interest in Cern after all operations were moved to Geneva. As Weisskopf the director once said, Bohr worried about every detail. Bohr could be called in to assist with any serious problem that arose, whether it was scientific or budgetary. "He came and did a great deal," said Weisskopf. "If Bohr had not supported Cern and participated actively in its founding and development, and if he had not sat together with others and worried about every detail Cern would never have come into existence."

But Cern was a reality, a full reality. Nordita had brought all Scandinavia together to further science and to strengthen the already strong bonds among the northern countries. Bohr's son Aage was taking over much of the administrative direction of the institute, and Riso was nearing completion. Bohr had much to rejoice about as 1957 neared its end.

June 6, 1958, was set as the day for the dedication of Riso. The buildings were fitted into the rolling landscape and surrounded by grass and flowers. The reactors stood out large and bold, almost as an extension of the rocky headland itself. Above the new establishment flew the flag of Denmark, one of the few notes of red in that land of soft colors. The king and queen were there and members of the government and representatives of all the political parties. There were foreign notables. At a meeting in the comfortable auditorium Bohr bid them all welcome. All were invited to see the buildings and grounds; here there were no secrets. Other buildings were still to be added, and soon some of the heavier equipment of the institute was moved to another new group of buildings just down the road from Riso.

June 6 was a memorable day for Bohr and for the Danish nation. Nevertheless a debate was scheduled in the Lower House. The large expenditures for Riso had to be approved. Kampmann anticipated some opposition and criticism. To the surprise of everyone, but in the way frequent to legislatures, almost the only questions dealt not with the large sums spent, but with the cost of the flagpole from which the Dannebrog flew so proudly, and with a kennel provided for a watchdog. The unexpected turn of the debate again made Riso a news story, and the kennel became famous throughout Denmark.

Bohr believed that Riso should be known to the Danish people. Visitors were invited and thousands came to tour the astounding new research station. "Thousands have had an opportunity to see this symbol of Danish ability," said Kampmann, "and have realized that a center of research has been created here which can bear comparison with any abroad, and which represents a little of our contribution to that universal effort to gain better insight into the world of atoms."

Many of the distinguished visitors who came to Denmark for the dedication of Riso were entertained at Carlsberg. For such gala occasions and on the days when the Bohrs entertained the King and Queen of Denmark, the King and Queen of Sweden, and Queen Elizabeth and Prince Philip of Great Britain, the house was decked with flowers. Eigil Kiar, who was in charge of the grounds, almost covered the columns and friezes of the peristyle with green branches and garlands of flowers. More flowers and plants were banked around the central fountain.

When the season permitted, as it did for the visit of Elizabeth and Philip, lilies were used on the tables and throughout the house. The blue-leather framed portraits which the British sovereigns presented to their hosts and the silver framed photograph of the Danish King and Queen thereafter sat on top of the piano, handsome reminders of memorable days and a point of interest to other visitors.

The House of Honor reflected the life of the Bohrs. On their trip to the Orient Niels and Margrethe had been enthralled with rubbings of a group of horses carved in low relief on the walls of a Chinese temple in Peking. On the walls of the peristyle, they made an unforgettable montage along the arcade. Bohr took the greatest of pleasure in the rubbings and delighted in showing them to visitors. A number of Chinese figurines and vases, some of them presented to the Bohrs, occupied places of honor in the drawing room, and among the many paintings in the house were a number of the misty, mysterious forest landscapes painted by their neighbor at Tisvilde, William Scharff. But the Bohrs also delighted in Scharff's painting of a group of white chickens, and always had it in a place where everyone could see it. Another work of art stood out in the generally conventional collection of the Bohr's. It was an abstract work "Lady with a Horse" by the French artist Metzinger. Visitors invariably would notice and comment on it, often aciduously, if they disliked contemporary art. Bohr was not in the least fazed by such art criticism. He would explain that he liked it very much because it defied all the laws of mathematics. The critics of non-representational art usually were stopped completely by that remark. To a few who, like Rosenfeld, knew Bohr very well, Bohr's liking for bold abstract painting was no mystery. The quantum theory was often compared to Picasso's painting. Both recognized nature outside of its ordinary and familiar contexts. Bohr did not at all object to the comparison; in fact he thought it discerning.

One wall in the House of Honor had special significance. Here were the portraits of those nearest to Bohr, grouped reverentially together. There was a portrait of Bohr's father, staunch, mustached, hand thrust into one pocket, and looking much like a less belligerent Theodore Roosevelt. Just beneath it hung a portrait of Bohr's mother as a young woman, her perfect oval face accented by the dark hair parted in the middle and drawn simply back. She wore a black silk dress, a lace collar, and a brooch at the throat. In this same special grouping were portraits of Harald Bohr, of Bohr's grandfather Adler, a bearded gentlemen of affairs, and of Bohr's teacher, Höffding. If any doubts existed of Höffding's influence on Bohr's life it was settled by the placement of his portrait.

At Carlsberg there were many books, the Goethe, Schiller, and Shakespeare bound in richly tooled leather. An Oriental rug was draped across one table. On another sat a replica of the Gunnestrup Karret, one of the treasures of the Danish National Museum. The large silver bowl with its hammered frieze of early Norse gods or kings and its inner bands of spear-wielding hunters and the fierce animals they pursued was one of the great works of art of the iron age. Bohr admired it so greatly that the institute had presented the reproduction to him on his seventieth birthday.

By the late 1950s Carlsberg and Tisvilde again rang with the laughter and shouts of children. All four sons were married and their children often were at their grandparents' houses. They raced around the Carlsberg peristyle and along the vine-covered pergola just outside. When a photographer appeared to photograph Bohr seated at a table in his small home office, the blond head of grandson Thomas unexpectedly peered out from under the table. Without anyone's noticing he had crept into the scene.

Never was Bohr more the grandfather than at Tisvilde. In this fairy tale setting on the edge of the forest Bohr read the children the old fairy tales and particularly the Icelandic sagas, as he had a few years earlier to their fathers. He took them on walks through the forest where, as everyone knew, the trolls dwelt, and, putting the little ones on the handlebars of his bicycle, rode them to the beach. Should a youngster refuse to eat his cereal Bohr would take a hand. A photographer caught him going through all the motions of delighted tasting to convince an unconvinced granddaughter that the cereal was the most delectable of foods.

There was a playhouse for the children at Tisvilde and Bohr entered into the games that went on there "I love playing with children," Bohr explained to anyone who caught him in the middle of a romp.

Then, turning philosophical, as he was prone to do, he went on "In their joyful eagerness one witnesses the potentialities of human life at its best and is constantly reminded of what one's self has experienced and what luck and shortcomings one's own expectations may have met with. Children have expectations of everything." Then philosophy was forgotten in shouts of "Grandfather, grandfather, let's go swimming, please, please."

The years were adding up, incredibly it seemed. In September, 1961, a group of English scientific societies were planning the "Rutherford Jubilee International Conference" to celebrate the fiftieth anniversary of Rutherford's discovery of the atomic nucleus, 1911–61. Bohr was invited to be one of the principal speakers along with Sir Charles Darwin, Sir Ernest Marsden, Chadwick, E. N. daC Andrade, and others. Bohr was asked to speak first on the general significance of the discovery of the nucleus, and at length on his reminiscences of "the founder of nuclear science and of some developments based on his work."

Much of the work of preparing the addresses was done at Tisvilde that summer. Soon after breakfast Bohr and whoever was helping him would retreat to the "pavilion." The pavilion was a tiny round cottage not much more than twelve feet in diameter on the end of the same old dune on which the house stood. It was then hidden from the house by the lilacs; a one-time small clump of the bushes that flourish with a special felicity in Denmark here had grown into a thicket larger than the pavilion. The pavilion was equipped with a table, a few wooden chairs, a blackboard, and a small black stove. It fitted Bohr's definition of a work room—a "room where no one can stop you from working."

The dictation went on—Bohr was reliving some of the most cherished days of his life. But how to make it exactly right, exactly a reflection of a man who was so alive, so hearty, so perceptive, so brilliant that no words were apt enough to recreate him for those who had never had the good fortune of knowing him. Occasionally, in despair, Bohr would announce that he wanted to go outside for a while. He would plunge furiously into pulling weeds or into chopping down one of the trees in a small area he was trying to clear.

As Aage and some friends watched one of these sessions from the grass on which they had stretched out, they saw the bowl fall from Bohr's pipe. They waited in anticipation to see what would happen. In a few minutes Bohr reached for a match and held it up to the bowl only to discover that there was no bowl there. At his look of stupefaction the watchers shouted with laughter.

The work would go on again at night until finally all were weary. To unwind after the strenuous day there might be another walk or they might even play at an English crossword puzzle. As two of Bohr's aides were falling asleep one night they heard a soft knock on the door. Bohr thrust in his head to say "I have the name of the English city ending in 'ich.' It's Ipswich." Or Bohr might come in to say that he had just thought how one sentence could be improved.

To be back in Cambridge again was always a heart-stirring experience for Bohr. The memories crowded back, and never more than this time when he had come to talk about Rutherford. In his reminiscences Bohr traced the development of Rutherford's work and its determinative effect on all the atomic discoveries that it released with nearly explosive speed. But with this thorough scientific review Bohr wove in many of the little unforgettable acts and words and quirks that were so characteristically Rutherford.

At a Royal Society dinner shortly after Rutherford had been raised to the peerage, Bohr had referred to him in the third person as Lord Rutherford. Rutherford overheard it. "Rutherford furiously turned on me," said Bohr. " 'Do you lord me?' " he yelled. Bohr never did again.

And the evenings came back to Bohr. If long discussions took a turn that did not especially interest Rutherford, he was apt to fall asleep. "One then just had to wait until he woke up and resumed the conservation with his usual vigor and as if nothing had happened," Bohr fondly recalled.

There was the time too when Bohr and Rutherford were walking home after a dinner with a group of the Cambridge humanists. "Rutherford had great esteem for his learned colleagues, but I remember how he once remarked on our way back from Trinity that to his mind the so-called humanists went a bit too far in expressing pride in their complete ignorance of what happened in between the pressing of a button at their front door and the sounding of a bell in the kitchen," said Bohr. It was a masterful account and a fond one. In a way it was Bohr's own life.

A month later in Brussels at the twelfth Solvay Conference Bohr reviewed the history of physics written at each of the meetings. He called his review a "cursory" one. It was in fact an account of science and its problems over fifty years and of the discussions that helped to create that science.

The invitations to the Bohrs to visit other countries for lectures and dedications of various kinds poured in irresistibly, and they went to the U.S.S.R. and India, whose art, traditions, and problems deeply interested Bohr, and to Israel. Back in the Manchester days of 1911 Bohr had known a young chemist named Chaim Weizmann. It was the same Weizmann who became the first President of Israel. Bohr took especial pleasure therefore in working with the Danish committee to raise funds for a science institute honoring Weizmann. When Bohr visited Israel he was asked to assist in setting up the physics department and became a member of the board of governors. After this he and Mrs. Bohr toured through the country to see the old and the new.

In June, 1962, Bohr went to attend a conference at Lindau, Germany. While he was there he suffered a slight cerebral hemorrhage. At his doctor's recommendation he and Mrs. Bohr went on to Amalfi for three weeks of recuperation. In that setting of sun, cobalt blue sea, and mountain terraces vivid with their crop of lemons, Bohr made a rapid recovery. His doctor said that he might resume his normal activities. Bohr was then 77.

On Friday, November 16, 1962, Bohr was quite well enough to preside as he had for many years, at a meeting of the Danish Royal Academy of Sciences and Literature. Two days later on November 18 after a lunch with Mrs. Bohr and a few guests, Bohr said that he had a slight headache and would lie down for a while. Death came shortly afterward.

A man of the century had gone; the last of the giants, as many said. Bohr even more than Einstein had transformed the century. By his profound discoveries he had guided the century into the scientific era that is its character, its genius, and its despair. But Bohr did not stop there. He fought without surcease to direct the world into a course that he believed would bring peace and a new abundance to all. He did not succeed, but whether the openness and cooperation Bohr urged would have changed the course of history none will ever know. The alternatives that he feared and of which he warned—proliferation of nuclear weapons, the increase in nuclear armed nations, the distrust and enmity—came to pass.

To have changed the world's course once and to have nearly changed it again is a score rarely equaled. Niels Bohr twice found the path into the unknown.

Index